Ambient Networks

Ambient Networks

CO-OPERATIVE MOBILE NETWORKING FOR THE WIRELESS WORLD

Editors

Norbert Niebert
Andreas Schieder
Ericsson GmbH, Germany

Jens Zander
Royal Institute of Technology, Sweden

Robert Hancock
Roke Manor, UK

John Wiley & Sons, Ltd

Telephone (+44) 1243 779777

Email (for orders and customer service enquiries): cs-books@wiley.co.uk
Visit our Home Page on www.wiley.com

Other Wiley Editorial Offices

John Wiley & Sons Inc., 111 River Street, Hoboken, NJ 07030, USA

Jossey-Bass, 989 Market Street, San Francisco, CA 94103-1741, USA

Wiley-VCH Verlag GmbH, Boschstr. 12, D-69469 Weinheim, Germany

John Wiley & Sons Australia Ltd, 42 McDougall Street, Milton, Queensland 4064, Australia

John Wiley & Sons (Asia) Pte Ltd, 2 Clementi Loop #02-01, Jin Xing Distripark, Singapore 129809

John Wiley & Sons Canada Ltd, 6045 Freemont Blvd, Mississauga, ONT, L5R 4J3, Canada

Wiley also publishes its books in a variety of electronic formats. Some content that appears in print may not be available in electronic books.

Anniversary Logo Design: Richard J. Pacifico

British Library Cataloguing in Publication Data

A catalogue record for this book is available from the British Library

ISBN 978-0-470-51092-6

Typeset in 10/12 pt Times by Thomson Digital
Printed and bound in Great Britain by Antony Rowe Ltd, Chippenham, Wiltshire
This book is printed on acid-free paper responsibly manufactured from sustainable forestry in which at least two trees are planted for each one used for paper production.

Contents

Acknowledgements

This book was written as a result of the research done in the Ambient Networks project [7] that is running in the 6th Framework Programme of research of the European Commission. The editors would first like to thank for the encouragement given by the people in the Commission which were supportive to both the Ambient Networks concept and the idea to write this book, in particular Andrew Houghton our project officer.

The Ambient Networks idea needs a broad industry consensus and has to incorporate innovative concepts born in the academic world. Leading edge companies and research institutions have partnered to make the Ambient Networks project happen. We thank for their support, namely Ericsson as project coordinator, the telecommunication's sector industrial partners (in alphabetic order): Alcatel SEL, Critical Software, Lucent Technologies Network Systems, NEC Europe, Nokia Corporation, Panasonic European Laboratories, Siemens, DaimlerChrysler as a partner from the technology user side, the operators or research institutes associated with them: British Telecommunications, DoCoMo Communications Laboratories Europe, ELISA, France Telecom, Netherlands Organisation for Applied Scientific Research – TNO, Telefonica Investigación y Desarrollo, Telenor Communication, TeliaSonera, Telecom Italia, Vodafone Group Services and the research institutes: Swedish Institute of Computer Science, RWTH Aachen, Budapest University of Technology and Economics, Fraunhofer Gesellschaft zur Förderung der angewandten Forschung, Instituto de Engenharia de Sistemas e Computadores do Porto, Kungliga Tekniska Högskolan Stockholm, TU Berlin, University College London, Universidad de Cantabria, Consorzio Ferrara Ricercha, University of Surrey, Technical Research Centre of Finland, and the partners from outside Europe: Motorola Japan, National ICT Australia (University of New South Wales), University of Ottawa, Concordia University.

Of course, in the end the results are attributed to the people who have contributed their ideas and proposals, have discussed and agreed on concepts and architectures as well as written code to proof the viability of the ideas. More than 120 people have been involved with the project – too many to name them here although we want to extend our thanks to all of them. We would like to mention here Henrik Abramowicz and Lars Lundgren from Ericsson as the project manager and assistant manager who have held the project together.

As editors we have to thank in particular our authors of the various chapter contributions, namely (without affiliations)

- Irena Grgic Gjerde and Bryan Busropan for Chapter 3,
- Alf Zugenmaier; Michael Georgiades and Peter Schoo for Chapter 5,
- Martin Johnsson for Chapter 6,
- Cornelia Kappler, Nadeem Akhtar and Paulo Mendes for Chapter 7,

- Johan Lundsjö and Peter Karlsson for Chapter 8,
- Jochen Eisl, Jukka Mäkelä, Ramon Aguero Calvo and Shintaro Uno for Chapter 9,
- Frank Hartung, Jose Rey, Stefan Schmid and Thomas Petersen for Chapter 10,
- Alex Galis, Raffaele Giaffreda and Theo Kanter for Chapter 11 and
- Alex Galis, Róbert Szabó and Marcus Brunner for Chapter 12.

A special thank you to Aneliya Hoelper for her day and night support with the formatting and integration of the book. She has done all the nitty-gritty details which make up the consistency which you can expect from this book.

Finally, we would like to thank our families for their support and understanding during the production of this book.

Norbert Niebert
Andreas Schieder
Robert Hancock
Jens Zander

1

Introduction

One traditional view of how wireless networks evolve is of a continuous, inevitable progression to higher link speeds, combined with greater mobility over wider areas. This standpoint certainly captures the development from first and second generation cellular systems focused on voice support, and the early short-range wireless data networks, through to today's 3G cellular and mobile broadband systems; there is every confidence that the trend will continue some way into the future. Such a picture neatly summarizes a massive body of research and development of radio technologies, from antenna design to link coding to radio resource optimization. Pictures such as Figure 1.1 are well known from discussions of future wireless systems.

However, this book takes quite a different perspective. Instead of starting from the physical layer problems of wireless systems, the focus is on the networking issues that arise as the communications world moves towards offering ever more sophisticated services in more complex commercial environments. Furthermore, although these questions arise most prominently for the mobile and wireless domain – partly because the very nature of wireless communication encourages diversity in the business relationships and partly because the technical challenges require diversity in the physical layer solutions – the same issues will arise in any networking context. The resulting trend towards increasing technological and administrative heterogeneity is the one which has to be addressed primarily at the network level, and the pressures that it causes may lead communications systems to look radically different in the future from how they look today. This book presents a snapshot of current research into a set of new networking concepts that will enable such a vision.

1.1 The Current Communications Environment

There are already successful standards for mobile and wireless networking which fully address today's markets and existing air interfaces. The standardization bodies responsible for cellular systems, primarily the 3rd Generation Partnership Project (3GPP) [1], and for data networks, primarily the IEEE 802 LAN/MAN Standards Committee [2], both maintain their current standard systems and have a continuous programme of enhancements and

Ambient Networks: Co-operative Mobile Networking for the Wireless World Norbert Niebert (Ericsson GmbH), Andreas Schieder (Ericsson GmbH), Jens Zander and Robert Hancock

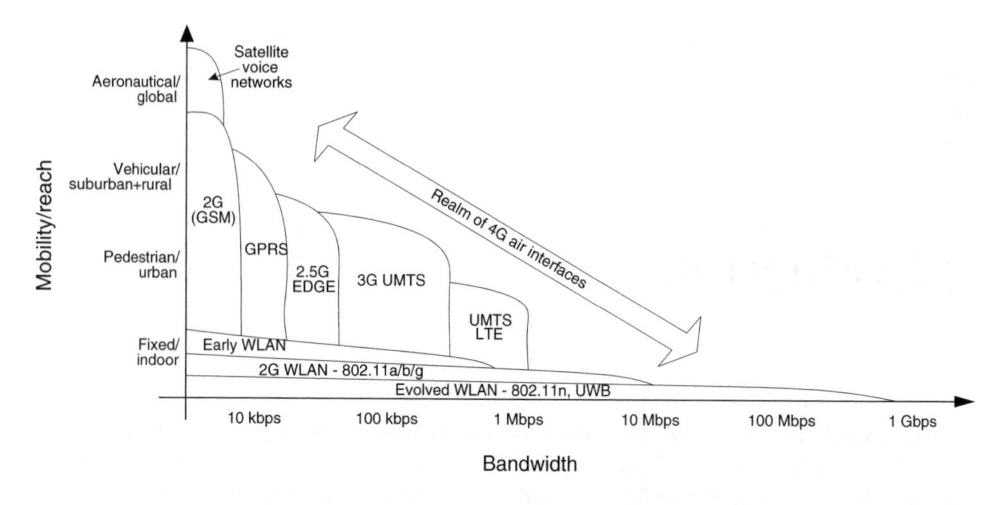

Figure 1.1 Trends in the wireless physical layer

system evolution. For example, 3GPP networks are already rolling out high-speed uplink and downlink packet access (HSDPA/HSUPA) as an extension of the current third-generation air interface. At the same time, they are working on a long-term evolution for the radio access network, and the evolution of the system architecture in general, activities referred to as LTE/SAE respectively [3].

This book is not primarily about these new developments *per se*; they can be seen as relatively low-risk incremental changes to current architectures and deployed networks, as is quite appropriate work for standards bodies to undertake. Rather, they are examples of the increasing complexity that will eventually require a new way of thinking about the way the mobile communications networks are put together. This growth in complexity is actually the result of two more fundamental, underlying trends.

Foremost among these is a change in the business environment. The starting point has been a vertically integrated model, where a complete end-to-end service, including access provision and infrastructure management, is provided by a small number of operators, supplemented by international roaming. The trend is towards much more complex models. The first aspect of this is a lengthening and fragmentation of the traditional value chain, allowing entities to focus on and specialize in particular activities such as service creation and marketing, or infrastructure operation. This already happens in the cellular market, where a set of new interoperator interfaces has had to evolve to support it. The same trend is visible in the integration of new access technologies such as WLAN, where service and access provision are almost invariably split in the 3GPP interworking case, bringing the additional complexity of the need to offer the same services over radically different bearer types. The rise of the hotspot market also presents new scaling problems, as the number of individual operators is larger by some orders of magnitude compared to the cellular world. Finally, to maintain growth, there is a need for the mobile world to extend to embrace new communications markets – not just the enterprise, but also home and personal networks. Along with the issues of scaling and heterogeneity already mentioned, any such development will create further difficulties for internetworking: the relatively open and unmanaged nature of the environment and the wide variation in business models will mean that traditional forms of interoperator agreement will no longer be sufficient.

The second major motivation is the rate of change in the technological environment, a rate which shows no sign of decreasing. Along with the introduction of new air interfaces (mentioned above) there is also evolution, driven by basic economic and engineering requirements, in the configurations in which access infrastructure needs to be deployed. Examples here are vehicular networks, which can insulate user terminals from the special problems of high physical mobility, and meshed wireless networks, which reduce the cost of achieving area coverage with very short range air interfaces. Along with encouraging the business evolution referred to above, these developments present challenges for existing systems as they cannot be reconciled with the assumptions about air interface behaviour or functionality distribution that are implicit in the network architectures. At the other end of the protocol stack, there is similar if not more rapid change in the range of services that networks are expected to accommodate. The changes encompass both the type of service (from voice to data and multimedia) and the users (extending to peer–peer operation). These developments make additional demands on the flexibility of the network in the efficient mapping of services to very heterogeneous physical resources, and the routing and control of traffic within the networks. One common feature is the trend towards the Internet Protocol (IP) as a universal network layer, which is visible in both its use as the basis of these advanced services in 3GPP networks and its adoption even for wholesale replacement of fixed-line voice networks [4]. However, the core Internet standards do not offer the level of control that is necessary for the advanced scenarios that are being considered. Thus, when we consider either of these aspects of technological change, current system architectures are not able to adapt at the speed which the marketplace demands.

The combination of these trends – increasing heterogeneity at all levels and increasing demands for service complexity and control – can be seen as the networking counterpart to the physical layer trends that are more commonly used to mark out the evolution through the mobile network generations. The goals for the Ambient Networking concept can best be shown by an analogous picture, Figure 1.2.

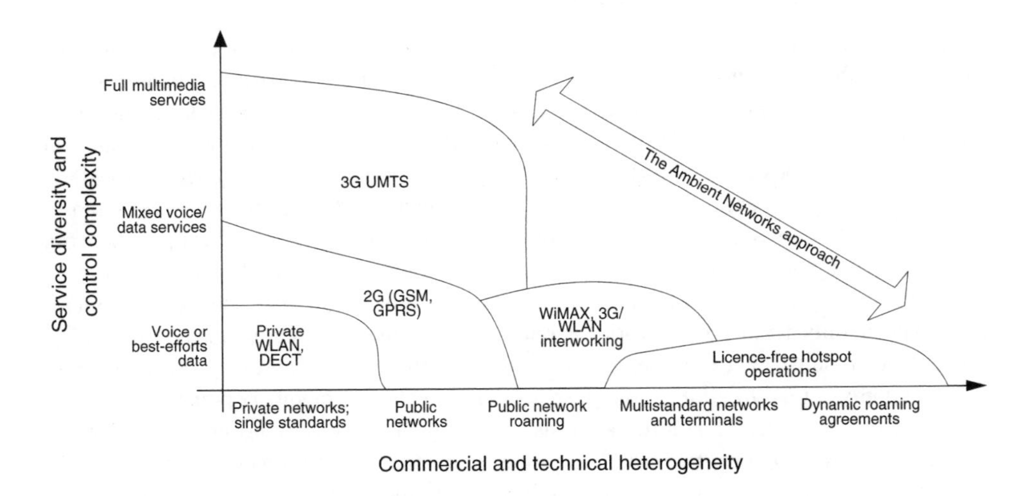

Figure 1.2 Networking aspects of system evolution

1.2 The Ambient Networking Concept

The Ambient Networking concept is a unification of a wide variety of new ideas from across the communications research community; however, it is not driven solely by scientific or technological goals. It also places a premium on developing solutions which are commercially exploitable, by taking into account the necessity to develop from current systems and building a consensus between the multiple different communities which make up the mobile world. In this sense, it certainly does not start from a clean sheet, but it does take the license to consider a more radical set of possibilities than would be encompassed by pure incremental evolution from current standards.

The core of the approach is the development of a set of control functions, which operate primarily at the network level. These functions can be implemented as an overlay on top of existing or new connectivity technologies, provided those technologies expose certain basic data transfer capabilities. This concentration on control functionality as an overlay is the key to addressing the twin problems of convergence between different technology types and migration from existing systems. Such an approach partly addresses the trend towards heterogeneity and service richness as described above, but introduces the risk of an explosion of complexity in network operation, especially as the commercial environment becomes more complex. The aim is to handle this issue by considering the set of control functions for a given network as a coordinated set. This goal is that this integrated Ambient Control Space will present a simpler interface to the outside world for external interworking. More ambitiously, given such a universal internetworking interface in the control plane, it should be possible to develop techniques whereby the control spaces of different networks can be recursively combined to support more complex scenarios in the same basic way. In the Ambient Networking context, this concept is referred to as *network composition*; it can be seen as the generalization to the control plane of the best efforts internetworking in the user plane (i.e. for forwarding and routing) that has been the foundation of the success of the Internet.

1.3 The Ambient Networks Project

It has been widely recognized for some time in the research community (see for example the references gathered by the NewArch project [5] and the final report [6]) that there is a need for new thinking to handle heterogeneity and control problems in the network layer and to enable the demands of different business actors to be arbitrated within a common technological framework. This book does consider trends in the research landscape in general, especially in particular technical areas such as resource management, context, overlay networks and so on. However, the core of the work, and the overall architectural concept, is based on the work done in the first phase of the collaborative project 'Ambient Networks' [7].

Ambient Networks is a joint industrial and academic research project. The industrial members come mainly from the network operator and equipment manufacturer communities, although other industries are also represented, following the view that networking functionality will eventually spread to all technology sectors. The academic members come from a wide variety of both universities and research institutes. The activity is sponsored under the 6th Framework Programme of the European Union as an 'Integrated Project' – in other words, a project which takes a set of related research activities and both develops them individually and integrates them into a unified conceptual vision. Indeed, this book has the same two-part structure, describing an overall architecture and a set of specific technology solutions. As befits its origins, the consortium members are mainly based in Europe; however, the nature of the subject matter and the industry is that the perspective is entirely global, and the consortium also includes members from North America and the Asia/Pacific region. The first phase of the project involved over 150 different researches over two years of work, carrying out mainly conceptual investigations with initial simulations and demonstrations.

The Ambient Networks project should not be viewed in isolation. Historically, projects under previous EU Framework Programmes have been instrumental in setting the stage for major commercial developments, especially in the communications world. For example, in the 4th Framework, the FRAMES and RAINBOW projects (see e.g. [8,9]) provided the foundation for what rapidly became the air interface and network architecture for UMTS, and there are also links to current complementary work. In particular, although Ambient Networks focuses on network layer issues, it is recognized that there will be significant interactions with future air interface evolution and future terminal and service concepts. These are the subjects of sister projects within the 6th Framework Programme, under the general umbrella of the Wireless World Initiative (WWI). Further information on the WWI projects, and its associated open research community, the Wireless World Research Forum, can be found in [10,11].

1.4 How to Read This Book

The structure of this book follows the structure of the thinking behind the Ambient Networking approach itself: a conceptual framework and unifying architecture, within which a set of specific research topics is developed in more detail.

The first part of the book treats Ambient Networking at the overall level. Chapter 2 provides a technical perspective on the origins of the concept, in current thinking about convergence between different types of networks (fixed, mobile, wireless), treating the case of the introduction of the Internet Multimedia Subsystem in 3G networks as an example. From this starting point, the need for a new architectural approach and the requirements on that new architecture are derived. This is complemented in Chapter 3 by a discussion of the business perspective and economic drivers, including a description of a generic business model within which the Ambient Networks results can be analysed, with a particular focus on the issues that influence migration and deployment of new networking technologies. The architecture itself is presented in Chapter 4. The Ambient Networks architecture is in many respects deliberately minimalist, and the chapter begins with five basic principles from which most of the more specific architectural decisions have been derived. It then describes the two basic features of the architecture around which the details are arranged: the Ambient Control Space which provides an environment within which the various control functions are organized, and the Ambient Layer Model which captures how interactions with connectivity infrastructure and services and applications are codified. Most readers will find all of these three chapters relevant, albeit with a different level of importance depending on whether their focus is business or technical; the material of Chapter 4 is a prerequisite for following the remainder of the book.

The first part of the book continues with three chapters on specific technical aspects of the overall Ambient Networks concept. Chapter 5 describes the approach taken to security, starting with a survey of the problem space and assumptions about feasible security mechanisms, leading to a definition of the fundamental building blocks of the Ambient Networks security architecture – in particular, secure identification and authorization, and their application to some specific security problems. Chapter 6 provides a detailed discussion of the network composition concept, first from a procedural perspective (how the composition process might actually take place) and then considering which types of composed networks might be produced as a result. Network composition is one of the key Ambient Networking concepts and implies a new set of requirements for creation and management of control relationships between networks; these control relationships will require support from a new family of signalling protocols. Chapter 7 presents the ambient signalling solution in the context of current IETF signalling protocols and concludes with a detailed example of the application of the signalling to the specific problem of internetwork service level agreement negotiation.

The second part of the book consists of five chapters which cover specific technical research which has been carried out within the Ambient Networks framework. These chapters can be read largely independently of each other and in any order, although in the book they are presented roughly in the sequence of the protocol stack. Chapter 8 presents work on multi-radio access, specifically the problems of integrating multiple different radio access technologies into a single system. There are two key concepts: an architecture for coordinating the resource management functions, and protocol components to unify a set of diverse link layers. A specific aspect is an analysis of the commercial benefits from multi-radio integration. Chapter 9 continues with a detailed study of the mobility management functions required in the network layer, again with a particular emphasis on methods for combining different mobility mechanisms in different network types. Chapter 10 considers the use of overlay network techniques to provide value-added functionality with the network infrastructure, to meet resource optimization requirements for media delivery, which are particularly critical in the wireless domain. The work includes detailed consideration of the scalability issues in management of large-scale overlay networks. Finally, Chapters 11 and 12 consider architecture for the integration of context information into network operation, including the definition of a common framework for context awareness across all functions in the Ambient Control Space, and the application of new ideas in network management to the Ambient Network environment.

1.5 Outlook

As we write these words, the material which this book describes is already being developed from its original conceptual form. The project itself has entered its second phase; here, the major emphasis is on formalizing the system architecture and its interface definitions and also on building a set of simulators and demonstration systems that can be used to show the concepts in action and quantify their benefits. At the same time, the first steps have been made in taking the work to the major standardization bodies, both for specific protocols and at the overall conceptual level. The Ambient Networking concept is itself evolving to meet the real challenges of implementation and deployment.

2

Ambient Networks – The Consequence of Convergence

Acknowledgements

This chapter is based on the joint experiences and efforts of the researchers in the first phase of the AN project and particularly the following people listed as contributors and authors (i.e. in alphabetical order): Bengt Ahlgren (Swedish Institute of Computer Science), Antonio Alves (Critical Software SA), Ulrich Barth (Alcatel), Hendrik Berndt (DoCoMo), Marcus Brunner (NEC), Bryan Busropan (TNO Telecom), Lars Eggert (NEC), Svante Ekelin (Ericsson EAB), Anders Eriksson (Ericsson EAB), Hannu Flinck (Nokia), Robert Hancock (Siemens (RMR)), Frank Hartung (Ericsson EED), Eiko Heuer (Ericsson EED), Geert Kleinhuis (TNO Telecom), Takashi Koshimizu (DoCoMo), Lars Lundgren (Ericsson EAB), David Moro (Telefonica Investigatiŏn y Desarrollo SA Unipersonal), Luis Munoz (University of Cantabria), Norbert Niebert (Ericsson EED), Gunnar Nilsson (Ericsson EAB), Toon Norp (TNO Telecom), Borje Ohlman (Ericsson EAB), Manuel Quadros (Critical Software SA), Juergen Quittek (NEC), Jarno Rajahalme (Nokia), Simone Ruffino (Telecom Italia Lab), Andreas Schieder (Ericsson EED), Mikhail Smirnov (Fraunhofer FOKUS), Michael Soellner (Lucent), Heiner Stuettgen (NEC) and Olle Viktorsson (Ericsson EAB).

2.1 Convergence Leading Towards Ambient Networks

In the rapidly evolving communications market, fixed and mobile operators are facing a new challenge termed fixed–mobile convergence(FMC),[1] which visualizes the trend to achieve converged services and networks. FMC is used by the telecommunications industry to describe the integration of wireline and wireless access technologies in a common services world.

[1]Fixed–mobile convergence has been defined largely from a fixed operator's point of view. The opposite, seen from mobile operator's side, is termed fixed–mobile substitution, where the wireline access is replaced by a wireless access. We mostly consider FMC from a mobile user's point of view in this book.

Ambient Networks: Co-operative Mobile Networking for the Wireless World Norbert Niebert (Ericsson GmbH), Andreas Schieder (Ericsson GmbH), Jens Zander and Robert Hancock

Furthermore, convergence can be considered from four different points of view: user services convergence, device convergence, network convergence and business convergence.

User services convergence takes a strong user-centric perspective on the communications services package. Its main paradigm is that users should be able to remain technology agnostic and get their communication needs fulfilled at any place using the best possible means. This is often referred to as 'always best connected' [12] – the users can reach and be reached by both mobile and fixed access via the same user interface.

Device convergence is a trend fuelled by Moore's law and the advancements of microelectronics, which coincides with a broader services portfolio and higher mobility in an all-digital world. Although it was sufficient in the 1990s to carry a phone and a laptop when travelling, digital cameras, music players and soon mobile TV sets will add to the devices in pocket format. In order to enable the first wave of new services to be offered to the customers, specialized devices appear in the communications market. The main goal attending the second wave of mobile and fixed user services is to have these devices integrated into other devices and objects to be carried anywhere and anytime. Often the mobile phone is the prime target for devices integration as it features the main mobile communication channel and is being replaced at a good pace, often with subsidies from a network operator.

Network convergence has been triggered by the success of the Internet and its proven suitability for all kinds of services. The Internet Protocol (IP) and the Internet paradigm are being introduced in all areas of communications allowing the evolution to so-called all-IP networks, networks that fully utilize the capabilities of the IP protocols in both transport and control. Advancements in the provision of access and core bandwidth for both mobile and fixed technologies (3G followed by HSPA and LTE, WiMAX, DSL followed by ADSL2+ and VDSL as well as fibre to the home) have recently accelerated the convergence towards the IP networking world.

Business convergence is less obvious at a first glance and rather a consequence of the other convergence trends mentioned above. Vertically integrated businesses offering a single service via a single access channel will face a difficult case in a more dynamic converged marketplace. Nowadays business convergence means to offer a tailored set of services using any access channel to selected customer sets. Such business convergence goes in hand with mobility and personalization. It also means to face a more open marketplace where opportunities have to be seized more quickly and dynamically. Business decisions will need to be implemented quickly and often the cooperation of several business partners is required to realize the full potential of converged services. This will be possible only when supported by technology in a cost-efficient manner.

Such realization of the full potential of convergence will make it necessary to deploy not only a patchwork of IP-based solution but also a dynamic, cooperative and business-aware consistent network control layer. This was the basic idea behind the Ambient Networks approach which is outlined later in this chapter.

2.2 Realization of Convergence

Nowadays the term convergence stays for convergence between media, data communication and telecommunication industries, as shown in Figure 2.1.

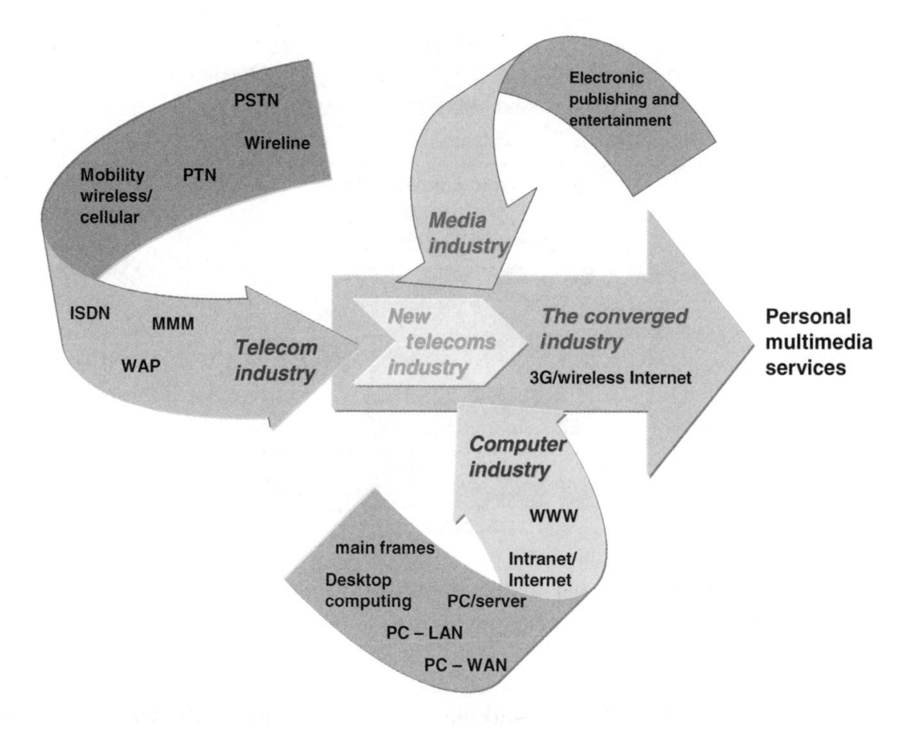

Figure 2.1 Industry transformation and convergence

Network operators already offer convergent services, which can be realized relatively independent of the evolution of networks and devices. Services such as 'one number' and 'follow-me' are available as, for example, Telia/Sonera Connect and Telia mobile@home (TeliaSonera in Nordic countries), O2 Genion (O2 in Germany), BT Fusion (British Telecom) and Cerillion solution for convergent telecommunications providers (Cerillion in the United States). These services, together with the bundling of subscriptions (fixed, mobile, broadband), can be an opportunity to gain new revenue streams and decrease churn.

Fixed and mobile users are offered a single numbering plan with access to the same PBX services, video telephony from both PCs and mobiles, push-to-talk and other combinational services (such as combining telephony and multimedia sessions).

To allow interoperability and provide opportunities for upgrading, standards should be followed. When the underlying structure is systematically standardized, other areas have more room for variation. This means that customized services, which of course still can be convergent according to the definition, can be provided more efficiently.

2.3 Converged All-IP Networks

Nowadays fixed and mobile operators are facing a double challenge: to create and deliver attractive (IP-based) multimedia services and to evolve their current networks to an architecture that can deliver such services in a highly adaptable and cost-effective way.

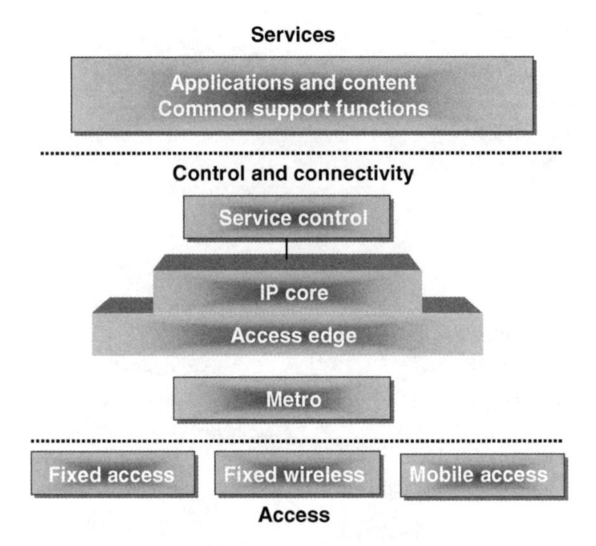

Figure 2.2 Industry transformation and convergence

The key principles of the all-IP network architecture integrating different types of networks and applications are:

- all services and applications are based on IP technology;
- service control is mainly handled by Session Initiation Protocol (SIP) signalling;
- packets form the whole network traffic;
- IP connectivity is supported by many different access types.

The target architecture of all-IP implying convergence of networks and efficient provision of new services – including new voice and data, high-speed Internet, TV and video, and gaming services – whenever and wherever users demand is illustrated in Figure 2.2.

2.3.1 Operator Challenges for Converged All-IP Networks

A converged all-IP network should enable operators to plan for the unexpected, to create an adaptive high-availability network, to support different service characteristics and to leverage the cost efficiency of IP.

2.3.1.1 Flexibility – The Key to Plan for the Unexpected

Maintaining flexibility in connectivity solutions allows fixed and mobile operators to plan for the unexpected. Technical advances such as optical fibre communications provide a bandwidth explosion that gives the opportunity to converge on IP as a general networking technology. The usage of transport resources leads to increased flexibility as more and more interfaces and services are sharing the same IP connectivity.

The new network should aim at taking changing customer demands into account and at creating spontaneous, adaptive services that can be delivered anytime, anywhere, to any device the user prefers. This behaviour is possible only if changing and reconfiguration of the network resources are internally supported by the network resources.

Besides the generic mechanisms and over-provisioning in the core of the network, the unexpected demands in certain access locations (e.g. due to sports or cultural events) for temporal capacity increases make it necessary to foresee technologies for exploiting all available access methods. This is currently possible in the GSM networks for emergency calls, but a new method is needed to exploit peaks of demand for business. Our approach to this problem is called network composition and is explained later in Chapter 6.

2.3.1.2 Towards Adaptive and Highly-Available All-IP Networks

IP-based network convergence will make the full set of fixed and mobile services available across all accesses. In the future, a continuous stream of innovation and progress in the access methods and protocols in use is expected. It will include existing technologies like GSM/WCDMA with their evolutions, SIP/IMS-based multimedia, broadband access and wireless LAN, many emerging technologies, such as WiMAX, DVB, and others yet not invented.

The main challenge is to use cost-efficient IP technology in a way that maintains the required characteristics for features like availability and scalability, but can also request demands for delivering high-volume, high-performance, real-time services in a way that maintains the high reliability and quality of traditional telecom services. A recognized architectural approach to achieve this is applying IP/MPLS nodes in the more complex, dynamic core network. A mixture of IP routing, MPLS/VPLS tunnelling and Ethernet switching technology is often used to build the metro and access parts of the transport network. MPLS provides an autonomous control plane for efficient handling of resilience and changes in the resource allocation to the architecture. Connectivity is no longer solely best-effort Internet access served at a 'single edge' of the aggregation network, but rather a set of Triple Play services,[2] offered at different aggregation levels. The resulting network is optimized for cost, flexibility and scalability for future growth.

The IP infrastructure should be scalable, i.e. designed to address the needs of small networks, comprising only one or two sites, to medium and large networks, comprising tens or thousands of sites.

From this emerging structure we can conclude that an all-IP network will still show heterogeneity in the control mechanisms deployed in various parts of the network. The need for end-to-end service control across domain boundaries remains an issue to be solved.

2.3.1.3 Support of Service Characteristics in the All-IP Network

IP and its companion technologies make it possible for the all-IP networks to support flexible and generic QoS, traffic handling and bandwidth management and thus to enable user services, conserve network resources where needed and reduce costs. Regardless of the access method and network being used, services will also be adaptable and aware of the type of access in order to meet user expectations in different scenarios. Maintaining the service quality users are

[2]This term is used for provisioning of the three services: high-speed Internet, television and telephone services over a single broadband connection.

accustomed to, e.g. for voice service, presents a key challenge all-IP networks have to deal with.

2.4 Network Convergence with the IP Multimedia Subsystem

The next-generation solutions for converged networks are expected to use a common multi-service layered architecture. The networks will have a layered structure with a service layer, a control layer, a backbone layer and access networks. Later in this book, we will focus more on the network side of the control layer. Here, we put more attention to the service layer interface. In the following paragraphs, a more detailed description of IMS and its functionality is given. See for instance [15] for a comprehensive discussion about IMS.

2.4.1 The IP Multimedia Subsystem (IMS)

The IMS standard defines a generic architecture for offering Voice over IP (VoIP) and multimedia services. It is an internationally recognized standard, first specified by the 3rd Generation Partnership Project (3GPP/3GPP2) and now being embraced by other standards bodies including ETSI/TISPAN. The standard supports multiple access types – including GSM, WCDMA, CDMA2000, wireline broadband access and WLAN. IMS enables convergence in all its forms by supporting services independent of the access.

2.4.1.1 IMS Architecture Overview

As can be seen from Figure 2.3, IMS provides an open, standardized way of using horizontal, layered network architecture.

The *application layer* comprises application and content servers to execute value-added services for the user. Generic service enablers as defined in the IMS standard (such as presence and group list management) are implemented as services in a SIP application server (AS).

The *control layer* comprises network control servers for managing call or session set-up, modification and release. The most important of these is the CSCF (call session control function), also known as a SIP server. This layer also contains a full suite of support functions, such as provisioning, charging and operation and management (O&M). Interworking with other operators' networks and/or other types of networks is handled by border gateways.

The *connectivity layer* comprises routers and switches, both for the backbone and for the access network.

IMS takes the concept of layered architecture one step further by defining a horizontal architecture where service enablers and common functions can be reused for multiple applications. The horizontal architecture in IMS also specifies interoperability and roaming, and provides bearer control, charging and security.

2.4.1.2 Service Creation and Delivery

With the introduction of the IMS architecture, many functions can be reused for fast service creation and delivery. IMS services are hosted by application servers, which means that they are implicitly placed in the IMS application layer and that various aspects of service control are defined. For example, IMS defines how service requests are routed, which protocols are supported, how charging is performed and how service composition is enabled.

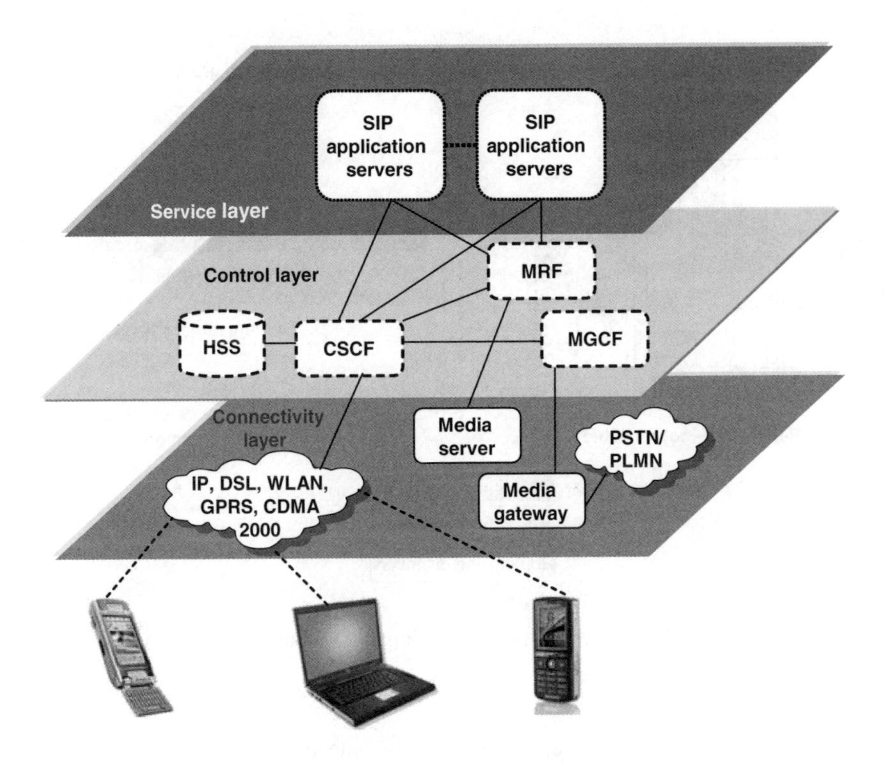

Figure 2.3 Simplified view of the layered architecture in IMS

The horizontal architecture of IMS enables operators to move away from traditional vertical 'stovepipe' implementations of new services, as shown in Figure 2.4.

This traditional network structure – with its service-unique functionality for charging, presence, group and list management, routing and provisioning – is very costly and complex to build and maintain. Separate implementations of each layer must be built for every service in such a network, and the structure is replicated across the network, from the terminal via the core network to the other user's terminal.

IMS provides for a number of common functions that are generic in their structure and implementation and can be reused by virtually all services in the network. Examples of these common functions are group/list management, presence, provisioning, operation and management, directory, charging and deployment.

Another advantage is that the operational competence required across services is more generic – and can be overlaid with service-specific knowledge – rather than demanding specialist operational competence for each service.

2.4.2 IMS Features

2.4.2.1 Service Enablers

IMS facilitates the creation and delivery of multimedia services based on common enablers in a 'write once, use many' way. These key elements in the IMS architecture are so-called

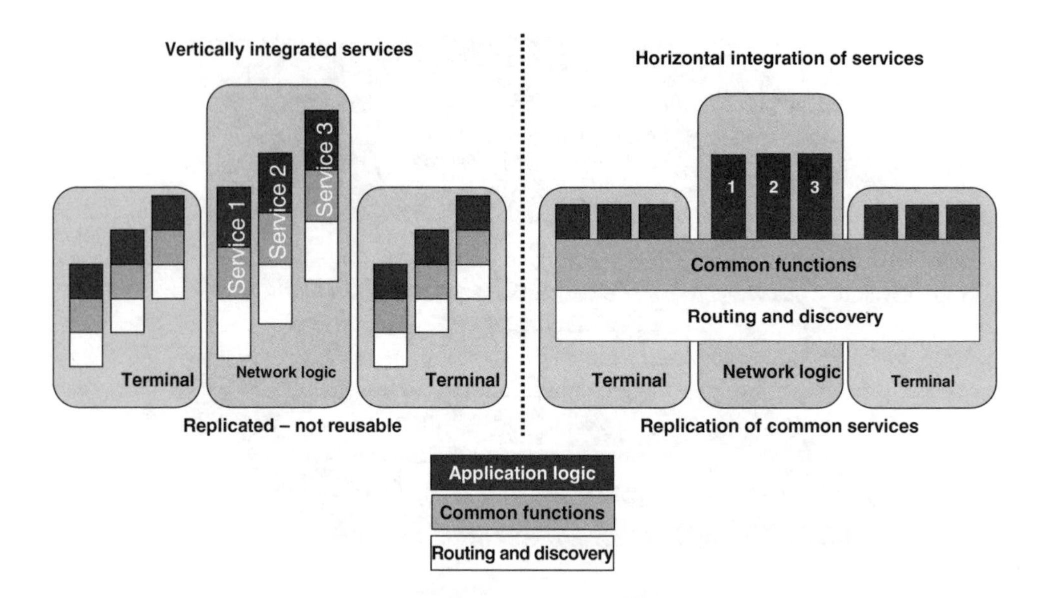

Figure 2.4 Vertical 'stovepipe' service implementations versus horizontally layered architecture

service enablers. The service enablers developed for successful applications can become 'global enablers' that are automatically included in new applications and services. The two most important service enablers are *presence* and *group list management*.

The *presence* service enabler allows a set of users to be informed about the availability and means of communication of the other users in the group. It enables a paradigm shift in person-to-person and other communications – for example, by enabling users to 'see' each other before connecting (active address book) or to receive alerts when other users become available.

In IMS, presence is sensitive to different media types, users (requestors) and user preferences. The IMS presence function is also aware of what terminals the user can be reached on across the various wireline and wireless networks. Different rules can be set by the user to define who can view what information.

The *group list management* service enabler allows users to create and manage network-based group definitions for use by any service deployed in the network. Application examples for group management include personal buddy lists, 'block' lists, public/private groups (for example, the easy definition of VPN-oriented service packages), access control lists, public or private chat groups and any application where a list of public identities is required.

2.4.2.2 Service Delivery

IMS enables a much more user-focused approach to deliver personal services than traditional networks. In the pre-IMS world, users access personal services from one or more service-specific, user-independent access point(s).

With IMS, users access personal services via a dynamically associated, user-centric, service-independent and standardized access point, the CSCF. The CSCF is dynamically allocated to the user at log-on or when a request addressed to the user is received. Routing to the server is service independent and standardized. The service architecture is user-centric and is highly scalable.

2.4.2.3 Simple Access to Services

IMS greatly simplifies the sign-on and authentication process, for both operators and users.

Once authenticated through an IMS service, the user is able to access all the other IMS services that he is authorized to use. Authentication is handled by the CSCF as the user signs on. When it receives a service request, the SIP authentication server can verify that the user has been authenticated.

When an end user logs on to his mobile phone or PC software client, the system is automatically updated on the user's new presence state.

IMS enables the reuse of interoperator relations. Rather than developing different interconnect relations and agreements for each service, IMS enables a single interoperator relationship to be established and built upon for each service. Once IMS is in place, access to other users' services is an IMS network issue, common to all IMS personal services, as shown in Figure 2.5. The requesting user's operator service does not need to be involved in routing the request.

The interoperator network-to-network interface is established in IMS, and the general IMS interoperator service agreement, routing, service network access point and security are all reused.

2.4.2.4 Service Creation in Terminals

IMS services require an IMS/SIP client (including GUI, service logic, routing and discovery functionality) in the user equipment to communicate with the network servers – in a sense,

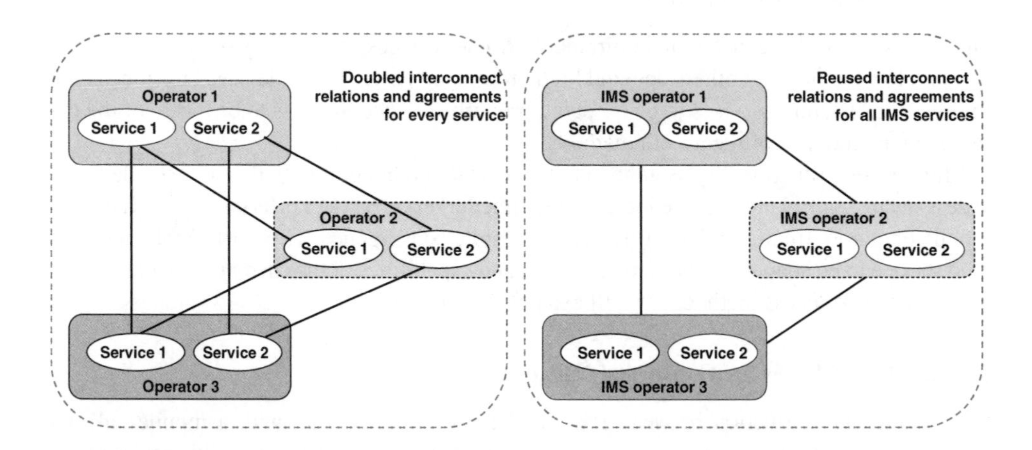

Figure 2.5 Service interoperability in a common network and in an IMS-enabled network

mirroring the service logic in the network. The IMS/SIP client is structured in such a way that the core functions are reused for many applications and that many applications can be co-located on the same user equipment.

2.4.2.5 Interworking with Legacy Networks

The possibilities for interworking between legacy services and IMS-based services will vary, according to the actual services supported in each domain and in the user terminals. Any interworking must have end-user experience as its key focus. As an example, presence in IMS must support interworking between different presence server domains, enabling different users to subscribe to the relevant parts of each others' presence services – and have them in their contact lists.

Another important interworking case is between IMS and existing Intelligent Network (IN) services like VPN. This would enable, for example, IMS services to use existing VPN short numbering: the SIP AS would interrogate the IN VPN for the full number to complete the application.

2.4.2.6 Convergence Enabling Features

Designed for mobile operators and adapted for wireline requirements, IMS presents a standard enabler for the fixed–mobile convergence.

2.4.2.7 Common Control and Application Layer

The application and control layers are ready to handle both fixed and mobile communications from the start. Common functions and service enablers are equally equipped to work in both the fixed and mobile worlds and, more importantly, bridge the gap between them. Whether the subscriber is using a mobile phone or a PC client to communicate, the same presence and group list functions in IMS will be used. Additions or changes to a buddy list will directly be reflected on any terminal that the user logs on to.

2.4.2.8 Access-Aware Networks

Different services have different requirements. Some services demand high bandwidths, some demand low latency and others demand high processing power in the device. This means that in order for different services to be executed properly, the network has to be aware of the different characteristics of the access methods.

Multi-access functionality is inherent in the IMS architecture. If this is extended with access-aware control and service logic for multimedia services, IMS offers a way for fixed and mobile operators finally to deliver true fixed–mobile convergence. This will enable the delivered service to be adapted to the characteristics and capabilities of the currently selected device and its network access method. We will expand later in this book on access awareness.

2.4.2.9 Device Types for Converged Networks

One traditional distinction between fixed and mobile calls is that with a mobile call one calls a person, whereas one calls a location using the wireline network. With personal SIP addresses, fixed calls can become personal as well – according to user needs.

In this converged world, the device type will become more important than the underlying network architecture.

2.4.2.10 Safe Communication

Reliable, secure communication is a top priority for both users and operators. With IMS, operators can implement end-to-end communications services built around a number of IMS security and network architecture cornerstones.

These include the fundamental IMS attribute that *operator-controlled services* are provided to *authenticated users*. The originating operator has end-to-end responsibility in the operator community: no services are delivered to anonymous or untrustworthy end users, and no service requests are relayed from anonymous and untrustworthy operators and enterprises. The chain of responsibility is based around the following: IMS authentication, controlled IMS services that provide service to authorized users, interoperator agreements mandating responsibility, etc. and secure network interconnect.

In addition, payload (primarily non-voice and video) can be checked for viruses.

Access domain security is provided through user authentication and single sign-on. Network domain security is provided through site security for hosted solutions, node hardening, virus protection and audit logging.

2.4.2.11 Scalability

IMS services are primarily intended to address a mass market, with telecom grade quality of service. Using the IMS network architecture to provide converged services is advantageous as it is designed to scale independently of the traffic. This means that CSCF capacity can grow in proportion to the number of subscribers and that the number of application servers can grow in proportion to utilization of the different services.

2.4.3 Conclusion on IMS

Evolving the networks which currently deliver traditional voice and data services into adaptable, cost-efficient all-IP networks that secure future revenue growth presents a number of technical challenges. The main goal is to design a cost-efficient IP-based network which meets customers' expectations for high quality of services, availability and reliability, while still offering network operators the flexibility they need to adapt quickly to new revenue-generating market opportunities and unexpected new demands.

Achieving the above-mentioned main goal requires the realization of convergence in all its forms. Device convergence allows multiple applications to be run on a single device, reusing the same functions for identification and authorization. Furthermore, mobile devices upport more and more functions in addition to telephony using several common access types (CDMA 2000, WCDMA, GSM, fixed broadband and WLAN). A multitude of services (person-to-person, person-to-content and content-to-person) can be provided to the same user over different access networks and to different devices.

IMS is a key component of a multi-service layered network architecture for delivering attractive, easy-to-use, reliable and profitable multimedia services. Thus, it is the cornerstone of a converged network, providing different user services, with telecom-grade QoS, to several access types with an emphasis on operator cost efficiency.

2.5 Towards Ambient Networks

Having seen the philosophy and requirements behind IMS, which will become a market reality soon due to its offered benefits, we should now take a look at the network itself from an IMS perspective. What kind of control interface would the IMS ideally find and what abstract view of the connectivity is necessary? Furthermore, what will other non-IMS-based service platforms expect like the DLNA (Digital Living Network Alliance [16])? The new service platforms on top of a connectivity layer based on Internet technology highlight the importance of a clearly defined control interface to this connectivity layer in order to efficiently realize the service features envisaged for these platforms. Ambient Networks is an approach to respond to these and other new requirements as outlined below.

2.5.1 New Requirements

The mentioned developments towards all-IP networks will enable the design and deployment of converged networks in a multitude of flavours. However, is it enough to put mobile networks and Internet technologies together to enable widespread converged networks? We think that in addition new capabilities are needed to reflect the *scale* of the networks from very small personal area up to large operator networks. Also, the *user side* of networks has changed dramatically since the turn of the millennium. Users increasingly own networking equipment and set up small edge networks in their homes, in small enterprises and around their persons. They need a better, more responsible and equal representation in the network model. And finally, converged networks require a *common control framework* to avoid a multitude of feature interaction problems in a dynamically changing end-to-end delivery chain. These requirements lead to a new approach which we call *Ambient Networks.*

2.5.2 Why Discussing the Internet Architecture?

We will first have a brief look at two related research and development efforts.

Research in the Internet community on future network architectures is mainly influenced by the discovered deficiencies of the current Internet, where mobile networks and mobility aspects are treated with comparatively low priority. Furthermore, there are few aspects of end-to-end system management and security handled in a consistent and satisfactory manner. The motivation for reconsidering the design of the Internet is often focused only on its current prevalence and its impact on social, economic and political aspects [13]. Efforts are made to define a common set of architectural principles and tenets, which will guide the development of a new Internet architecture. In parallel to these efforts, which adopt a top-down analysis, several technology-focused activities exist working on solutions for new naming and addressing schemes, QoS and mobility support, security, manageability and routing scalability (e.g. analysis in accompanying AN deliverables and [14]).

Interestingly, one of the findings made within these research activities is the concept of separating the Internet into domains or realms, which are interconnected on a higher logical level to build an Internet. The separate domains are defined to encompass networks deploying different technologies from the ones used in the existing Internet. This approach is supposed to ease technological advances, as a localized introduction would be made possible. However, the definition of the required border functions, translating between different domains, has not been provided yet.

A second approach of the Internet research community is the work on overlay networks, for example, introduced to improve network reliability or to derive a network structure enabling new routing paradigms (see e.g. [26] for the discussion of an overlay architecture specifically suitable for content delivery).

Although all of these research efforts are essential to build upon, we believe that the emerging needs of future wireless and mobile networks are not fully satisfied. Issues concerning the (potentially automated) control of larger networks are not conclusively addressed, e.g. how to combine smaller networks into larger ones and how to support mobility or heterogeneity. The following sections will analyse these needs and the solutions proposed by Ambient Networking.

2.5.3 All-IP Evolution

Ambient Networks are based on the vision known as 'all-IP networks' and can be regarded as the outcome of a continued adoption of Internet design principles. 'All-IP' based mobile networks can be characterized by – among other aspects – a clear separation between transport and control related tasks. The functions concerned with either of these tasks are grouped in two distinct layers to ease the independent development in both areas.

The Ambient Networks concept adapts these tenets and assumes the presence of a layer to ensure basic connectivity between different networks. The control functions already identified and isolated by 'all-IP' networks are extended to the 'Ambient Control Space', which embraces an extensible set of control functions required to guarantee the cooperation between networks. The control space is thus no longer concerned only with the management of the transport functions within its own network domain, but is additionally responsible for the establishment of agreements with neighbouring networks to provide end-to-end services through the set-up of a service chain involving several networks.

2.6 Motivation for a New Approach

In today's wired and wireless networks, the trends in networking technology very much point to a dominance of Internet technology in all its flavours. IP is the common 'lingua franca' to enable the exchange of data across various networks. There is, however, an increasing divergence in the network control layer: Different control environments are established to facilitate services like VPNs, security, mobility, QoS, NAT, multicast, etc. For a multitude of services, uniform Internet networking might still handle data, but the control of such services is becoming increasingly fragmented. Increasingly, the network as a whole therefore diverts from the pure end-to-end view of the Internet. Furthermore, applications would like to rely on enhanced and consistent support from the network for the complete delivery chain.

This lack of a common control layer for joining the services (in a wide sense of service) of multiple networks represents a crucial challenge both technically and from a user perspective. Usage scenarios that should be realizable in the mid-term future include utilization of multiple devices, multiple networks and multiple access technologies in an integrated fashion.

This is not easily controllable or manageable with today's technologies. These usage scenarios partly evolve from the vision of 'Ambient Intelligence', developed and popularized by various research efforts. The core notion of this vision is to make technology invisible and useable, whenever necessary. This notion holds in particular for networking technologies – networking should be available whenever required in the most appropriate, accessible and affordable form.

Our approach, Ambient Networking, aims at providing a domain-structured, edge-to-edge view of the network control. In this way, an Ambient Network is expected to embrace the heterogeneity arising from the different network control technologies so that it appears homogeneous to the potential users of the network services. The vision is to allow the agreement for cooperation between networks on demand, transparently and ultimately without the need of pre-configuration or offline negotiation between network operators. End users are increasingly not just owners of a terminal or a PC, they own and effectively operate a network of devices in their homes, offices and around the body. Consequently, they are included in this network of cooperation and are treated as operators of special, low-complexity networks. This approach generalizes to different kinds of networks that are currently appearing, such as in-vehicle networks, body area networks or sensor networks: By modelling devices as networks, the network is the primitive building block of our architecture, allowing all types of networks to be integrated into a larger system.

This approach is confined to the network layer as such. Session layer signalling is for instance not included, but it is open to various schemes such as SIP and the IMS mentioned above. Similarly and in good Internet tradition, no specific assumptions of the infrastructure network technologies are made. The internetworking architecture is based on a control plane that coordinates the network state for features such as mobility, QoS and security. The internetworking control plane is designed to control not only IPv4 and IPv6 infrastructure networks, but also networks based on other types of emerging and legacy technologies. A key feature of the architecture is an API that allows for backwards compatibility with IP-based applications. Also, efficient migration from legacy IP networks is a design goal for the architecture. See Chapter 7 for more details.

To motivate the architectural requirements outlined below, consider a brief scenario which both matches the strategic objectives and illustrates some of the corresponding technical challenges. It is derived from one of the reference scenarios developed in the project, which tells the story of a rock band on tour in 2015.

We follow a rock band, while they tour Europe using a special "rock train" by which they travel between gigs, use it as stage when having concerts and also to provide some exclusive interviews and materials to special guests and fans paying for travelling with the band between stations.

There will be multiple ANs set up between different actors on-board the trains as well as between actors on and off the train. There will also be temporary ANs set up at the concerts when the rock group plays, to facilitate information sharing and content distribution between the band and the audience, among the audience, as well as between the audience and some of their friends not being at the concert. Potentially, all these ANs will be using different access technologies, and end users will only minimally notice when their connection is transferred from one access technology to another or when new traffic is added to already restricted resources. At the same time, ANs will allow for capacity being dynamically built up and configured with minimum human interaction following a network plug-and-play paradigm.

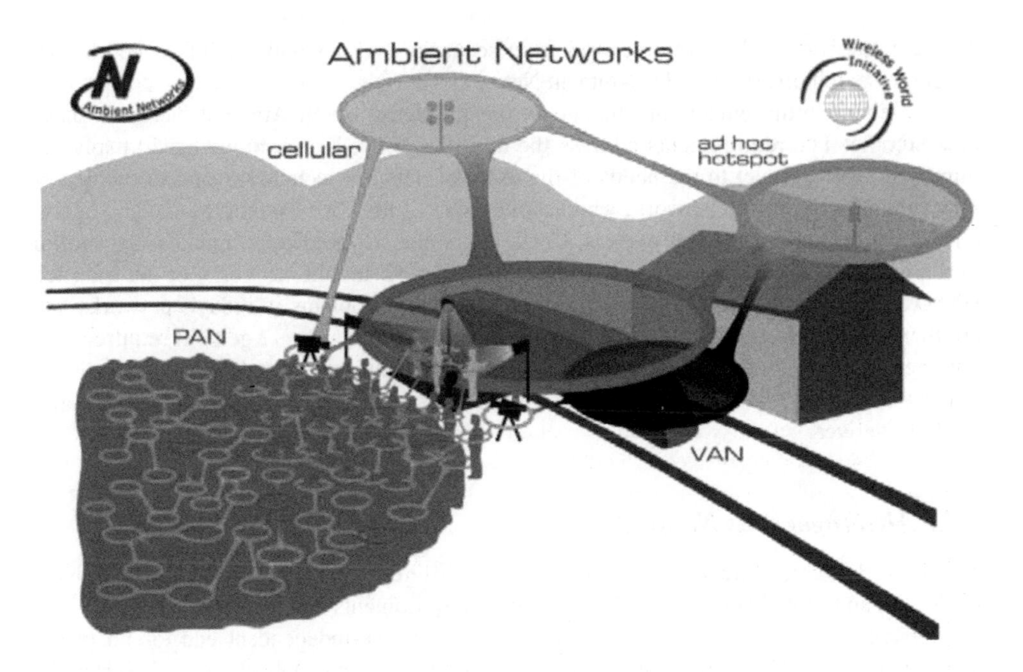

Figure 2.6 The Ambient Rock Express – a scenario

The scenario shows many different networks and actors, ranging from personal area networks (PANs) of the band members and fans to a wide area cellular network; there are also short-range ('hotspot') access networks providing high-speed coverage at particular locations (the concert venues and on the train providing transport between them), and finally networks hosting mobile services, which are not directly associated with any specific wireless access at all. Each of these networks has been provided by a scenario participant for their own different applications and needs, which has led to a huge diversity of scale and technology (type of radio access and the level of network functionality beyond pure data transport), see Figure 2.6.

Some of these networks may even be useful in isolation. However, the true potential of the scenario is obtained only when they become interconnected in a way which allows their resources to be shared and new communication patterns to be established between their users and services. All of this must be done while respecting the different business motivations behind the individual components and at the same time achieved automatically because the dynamic nature of the set of network–network relationships rules out time-consuming or complex manual configuration.

2.7 Architectural Requirements for Ambient Networks

The architecture of an Ambient Network is a result of a structured design process starting with the identification of the main requirements posed on an Ambient Network and its intrinsic architecture. This section gives an overview of the 12 main general requirements on

Ambient Networks. They were derived from scenario analyses and are influenced by the general research direction set for Ambient Networks.

The general requirements are related to the problems which Ambient Networks have to overcome. The requirements address the question 'what issues do we have to solve or improve?' with respect to the needs of the users of network technology and to the necessary functionality of the network, whereas the users of network technology include operators, end users and service providers. Certainly, some of these requirements are similar to requirements already found in networks like GSM/3G mobile networks or the Internet. However, the requirements address or outline unsolved problems in today's networks and are therefore necessary to be fulfilled for Ambient Networks. The 12 general requirements together as a set – and not necessarily per individual requirement – describe what distinguishes Ambient Networks from earlier technologies and what Ambient Networks really intent to deliver.

2.7.1 Heterogeneous Networks

The network architecture shall efficiently support different kinds of technologies to provide users with an optimal connectivity based on their requirements and preferences.

Ambient Networks must enable application- and service-independent end-to-end reachability in the global network environment. Network services and application services shall be provisioned independently of each other.

Although not new, the explicit objective to provide a network architecture embracing technology heterogeneity is one of the main properties of Ambient Networks. Furthermore, Ambient Networks are expected to hide this heterogeneity, which becomes especially visible in the different current (radio) access technologies, from the application process. Besides addressing heterogeneity of access and network technologies, the concept of Ambient Networks also addresses the business heterogeneity of the service delivery chain by not posing any restrictions to the business models that can be implemented in Ambient Networks.

2.7.2 Mobility

Ambient Networks must support mobility management schemes for user, service, session, terminal and network mobility.

The mobility management mechanisms should be able to locate and update the current location of the user and to support seamless mobility. Ambient Networks should be capable to support both existing and new mobility mechanisms that enable terminals and networks to move around without being closely tied to so-called 'home' networks. Real-time transfer of information flows must be supported.

The need for mobility across different heterogeneous networks imposes some constraints on the design of the network architecture. Specifically, networks must be able to move to different physical or logical locations at any time. This requirement impacts the naming and addressing convention for Ambient Networks. A specific control function in Ambient Networks is needed to deal with allocation, registration and if necessary translation of names as well as resolving them to physical addresses (locations and end-point IDs).

2.7.3 *Composition*

The network architecture shall support mechanisms that achieve on-the-fly negotiations and agreements across different administration domains.

Composition is a new feature introduced for Ambient Networks. The control space of a network shall be responsible for providing and conveying the information required by the various networks to compose with other networks. With this requirement the concept of real-time automated negotiation and agreement of services between networks is introduced. A network can be composed out of many smaller networks, which look as 'one' to the outside world. Still, the individual components of the composed network will not loose their identity or capability to connect to other networks with their original identity.

2.7.4 *Security and Privacy*

Ambient Networks must provide a seamless, comprehensive and flexible security scheme that operates consistently across a dynamically changing environment of constituent heterogeneous networks, component entities and services. This embedded network security shall cover a multiple network operator/service provider environment and be characterized by

- user friendliness and helpfulness, while remaining as far as possible invisible to the user;
- smooth transition between different accesses and services;
- trustworthy operation;
- robustness and resilience under attack and mishap;
- ease of security management;
- protection and privacy of user and network information and assets;
- protection and privacy of identity and location;
- accountability.

Security must take regulatory and law-enforcement requirements into account. It also must contribute to the overall availability and dependability of networks and services.

Feasible and suitable security procedures should be inherent to each Ambient Network and built according to one security architecture such that all the communications and negotiations take place in a secure manner. It is also very important to establish a trust relationship between the various networks that interact. Three different types of trust relations have been identified and should be supported. These are direct trust, brokered trust and no trust at all.

Security together with management and testing is most often overlooked during the system architecture design phase. This leads to an 'add-on' treatment and often results in poor, compromised or very rigid security solutions. The design of the Ambient Networks architecture therefore has to consider security from the very beginning. The ultimate goal is to make Ambient Networks immune against security threads.

2.7.5 *Backward Compatibility and Migration*

Ambient Networks must support migration paths from existing networks and mobile terminals, e.g. allow reuse of infrastructure of current 2G/3G systems including their evolution as well as the Internet and non-cellular access systems, such as xDSL and WLAN. Legacy applications (e.g. designed for not only UMTS, GSM but also IPv4) must be able to run within the Ambient Network environment.

Ambient Networks have to interact with legacy networks. As these do not provide the services expected from an Ambient Network, advanced features like network composition are not possible. However, Ambient Networks may interact with them through an appropriate adaptation function providing the missing functionality. The provisioning of services from legacy networks to Ambient Network aware applications can be accomplished similarly.

2.7.6 *Network Robustness and Fault Tolerance*

The network architecture and network management must allow for building Ambient Networks that are scalable, cost effective, robust, reliable, with high availability and survivability across heterogeneous networks in dynamically changing environments. It must also allow building small and affordable Ambient Networks that do not possess some or any of these capabilities.

The Ambient Network architecture should avoid single point of failures. The design should allow partial Ambient Networks to function even when isolated from the rest of the network.

2.7.7 *Quality of Service*

Ambient Networks must offer multiple QoS classes for end-to-end services when needed.

The QoS mechanisms employed must be independent from link-specific technologies, but provide a consistent QoS coordination across multiple access technologies. It must also provide means to renegotiate a QoS class either from the user or the network side.

The QoS agreed with an application or another network shall be guaranteed by the Ambient Network even when composing with other Ambient Networks or interfacing to legacy networks. The Ambient Network Interfaces shall be able to communicate the parameters needed for the establishment and maintenance of QoS-aware data services across network boundaries.

2.7.8 *Multi Domain Support*

The network architecture must transparently support network functionality spanning multiple administrative domains, i.e. areas operated and managed by different authorities.

Ambient Networks must support various existing and future network business models, and the security and trust framework within the Ambient Network architecture shall support this requirement on demand and in real time. It should be possible to negotiate with multiple Ambient Networks simultaneously and form agreements for the need of one or multiple applications as well.

2.7.9 *Accountability*

The network architecture must support mechanisms that enable auditability of single entities and subsequent enforcement when appropriate.

In particular, Ambient Networks must provide an efficient, reliable and secure way to collect and manage accounting independent of the business models employed. Ambient

Networks must also be flexible to interact with legacy accounting systems and new compensation schemes.

When Ambient Networks compose, a trust relationship is created. This includes especially authentication, authorization and accounting. Accountability denotes the need for the various interfaces that the Ambient Network supports, to be capable of conveying or collecting accounting information so that users can be made accountable for the communication they participate in.

2.7.10 Context Awareness

Ambient Networks must support a common framework for context awareness across all functions in the Ambient Control Space in order to automatically adapt service availability and delivery to heterogeneous networks and dynamically changing environments. The framework shall include support for collection, processing, management and dissemination of context information enhanced with context-level agreement negotiations and support for conflict resolution.

User-specific information such as personal preferences shall be easily communicated across networks in order to always achieve the required action with a minimum user involvement. This will result in a very user-friendly and adaptable communication system built in a user-centric manner. In this respect, the controlled distribution of user context information is important, so that user privacy concerns are considered. These concerns are mainly focusing on specifying limitations about the spread and use of context information and about ensuring anonymity of users if they demand for it. This holds especially for information gathered in one domain and forwarded into another domain in which the originating user has no control over the information anymore.

2.7.11 Plug-and-Play Extensibility of the Network Services Provided

Ambient Networks must support extensibility of the control layer in order to be able to adapt to different network scenarios demanding a more or less feature-rich control layer. Extensibility of the control layer shall be organized in a plug-and-play manner requiring little to no configuration effort.

In this respect, Ambient Networks provide an evolution of the current IMS framework, which deploys application servers to implement services, which extend the basic service set of IMS, and already now provides certain flexibility in the network services portfolio.

2.7.12 Application Innovation and Usability

Ambient Networks shall support short innovation and deployment cycles for applications. The architecture must be able to attract application developers by offering a stable, rich and migration-friendly API. Usability of this API is of paramount importance for the success of Ambient Networks.

It is important to offer application developers the support and performance of network services they demand, ranging from simple connectivity up to media delivery overlays.

2.8 Summary

The trend for converging services delivered via a multitude of channels based on IP transport has posed new requirements and challenges on the network layer. The all-IP and IMS approaches are first answers to these challenges. Furthermore, the issue of the network control layer requires a more fundamental approach, also due to the multitude of protocols and technologies which have come to the fairly simple IP platform since the commercialization of the Internet.

Ambient Networks are set out to define a universal control plane for current and future IP-based networks, especially those featuring a radio access link. New network layer capabilities are composition, the greatly enhanced treatment of the user domain, embedded security, increased availability of network accesses, context information and media delivery, and the increased affordability for the user as well as the business value of the network for the operator. These business perspectives are being elaborated in the following chapter.

3

The Business Environment for Ambient Networks

Acknowledgements

This chapter is based on the joint experiences and efforts of the researchers in the first phase of the AN project and particularly the following people listed as contributors and authors (i.e. in alphabetical order): Bengt Ahlgren (Swedish Institute of Computer Science), Antonio Alves (Critical Software SA), Ulrich Barth (Alcatel), Hendrik Berndt (DoCoMo), Marcus Brunner (NEC), Bryan Busropan (TNO Telecom), Lars Eggert (NEC), Svante Ekelin (Ericsson EAB), Anders Eriksson (Ericsson EAB), Hannu Flinck (Nokia), Robert Hancock (Siemens (RMR)), Frank Hartung (Ericsson EED), Eiko Heuer (Ericsson EED), Geert Kleinhuis (TNO Telecom), Takashi Koshimizu (DoCoMo), Lars Lundgren (Ericsson EAB), David Moro (Telefonica Investigación y Desarrollo SA Unipersonal), Luis Munoz (University of Cantabria), Norbert Niebert (Ericsson EED), Gunnar Nilsson (Ericsson EAB), Toon Norp (TNO Telecom), Borje Ohlman (Ericsson EAB), Manuel Quadros (Critical Software SA), Juergen Quittek (NEC), Jarno Rajahalme (Nokia), Simone Ruffino (Telecom Italia Lab), Andreas Schieder (Ericsson EED), Mikhail Smirnov (Fraunhofer FOKUS), Michael Soellner (Lucent), Heiner Stuettgen (NEC) and Olle Viktorsson (Ericsson EAB). This chapter has been edited by Irena Grgic Gjerde and Bryan Busropan.

3.1 Business Drivers and Benefits

A business motivation for the introduction of the features which the Ambient Networks concept offers to different actors is required to justify the costs for adopting a new technology. In order to get a clearer picture of the AN business environment, it is essential to investigate how the AN concept would affect today's telecom business environment – actors, roles they are playing and their relationships. Also, its effects on the new roles introduced and the rearrangement of actors to adapt to the situation are of interest. Figure 3.1 depicts the AN business environment. Both traditional roles like end users, service providers and network providers and new roles like aggregators, brokers and ID providers are included. Different actors may

Ambient Networks: Co-operative Mobile Networking for the Wireless World Norbert Niebert (Ericsson GmbH), Andreas Schieder (Ericsson GmbH), Jens Zander and Robert Hancock

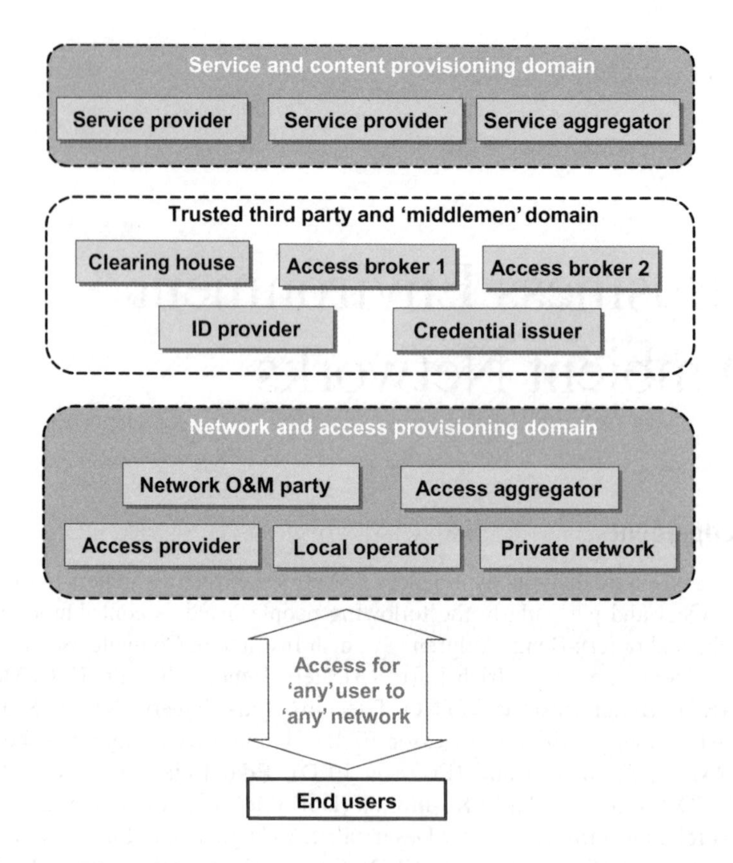

Figure 3.1 Overview of business roles in Ambient Networks

play different roles, and often a single business entity plays several roles to deliver (a set of) products to its customers. For example, a large operator may combine the roles of a service and content provider and a fixed and mobile network operator.

Business drivers to adopt the AN concept are naturally different for different actors, depending on the roles they play, their current situation, current market, etc. For example, end users would benefit from an increased range of potentially exploitable networks and a general ease of operation where less technical skills are required to get everyday tasks done. Also, Ambient Networks concepts may lower entry thresholds to implement the role of a network operator, opening the possibility for the former end users to become 'operators' of their own private networks. This may extend the range of services offered by operators to ease the end user's 'operator' role as well as provide cost benefits by automating configuration of edge networks which often cause problems in end-to-end service delivery. As end users, in general, have no interests in the technical details of operating these private networks, the simplicity of operation of an Ambient Network is a requirement for mass-market adoption of the proposed concept. A simplification of the operation of telecommunication technology for end users also leads to benefits for other parties. Removing complexity reduces obstacles for the adoption of telecommunication services and technology such that operators, service providers and vendors can benefit from an increase of the overall telecommunications market.

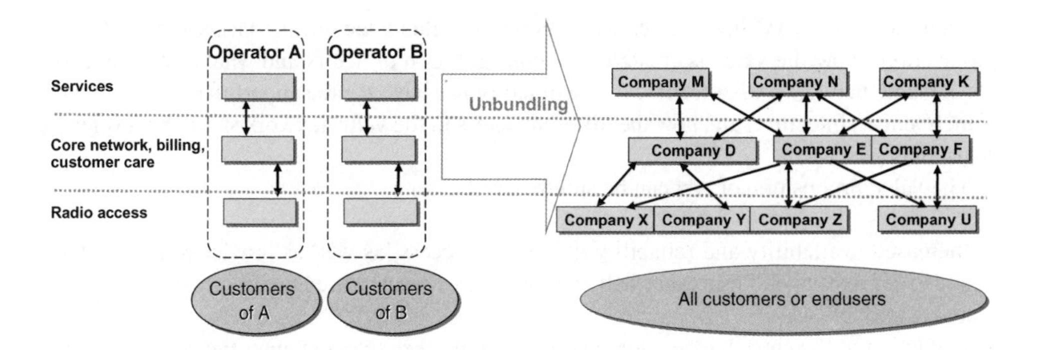

Figure 3.2 Schematic representation of the disintegration of the current market structure into a fragmented market structure

As a specific example, mobile network operators (MNOs) provide a majority of mobile services and access to customers based on vertically integrated value chains. Customers are usually attached to a particular operator's network and given the access to certain services via a postpaid or prepaid subscription. Currently, the communication and information industry is facing large structural changes driven by new regulatory frameworks pushing disintegration or unbundling as well as driven by internal restructuring like data–voice and fixed–mobile convergence. Moreover, in the future new networking technologies will be introduced. Hence, one may expect a multitude of new ways to deploy and operate networks, as well as new business roles and relations, and new forms of cooperation.

New types of actors can emerge as a result of disintegration of the vertically integrated value chains typical for today's MNOs, as illustrated in Figure 3.2. Cooperation between operators and local networks deployed by users in homes or offices will also result in the need for new types of roles, either as a part of the traditional MNO or as an independent actor. MNOs are believed to be well suited to fulfil many of the new roles in the disintegrated market, as well as being able to find new ways for interoperator cooperation.

In this chapter, we will describe the business roles of some of the more prominent market actors and show how the Ambient Networks technology and concepts are capable of enabling new forms of cooperation and competition. With access to virtually any network, the users will experience a higher service level with increased availability and reliability as well as increased quality. The AN functionalities for multi-radio and multi-operator access ensure that services can be offered at a lower price through more efficient deployment and use of resources. For cooperating access providers, the operational cost and the investment risk can be lowered further through a shared use of network infrastructure and reuse of user-deployed networks resulting in a reduced need for over-provisioning. Furthermore, cooperating providers will also get access to a larger potential customer base [17].

An evaluation of the AN user and operator benefits has been performed using a Business Blueprint Method (BBM), which offers a structured approach for analysing technical and economical interrelations. The BBM approach is mainly qualitative, but a part of the financial component has been analysed through a detailed quantitative cost comparison for a selection of scenarios. The BBM investigates what economic value a technical concept creates (value proposition), how the value is created for a configuration of actors and what role each actor plays (value network), what technical functionality is necessary to support the value network (functional architecture) and how the different actors in the value network share the revenues, costs and risks.

The value proposition of AN can be summarized as

- Increased availability and reliability of network access, as well as provisioning of value-added services to end users that will experience this as increased quality in an easy-to-use and affordable way.
- Reduction of the entry barriers for new potential market actors to enter the wireless arena.

In order to discuss the new roles in more detail, the following sections provide an overview of all roles and the value network. After that we look at two elements of the AN concept from a business point of view in more detail – a composition function and a compensation function.

The discussion of new business roles will focus on the final migration stage called 'native Ambient Networks'. However, it is assumed that the migration to such a native AN takes place in several logical steps from current networks and market structures to this native stage. Looking at the final migration stage means that the market structure will be adjusted to reap the benefits of the AN technology. So probably, it is a flexible and unbundled market in which small local access providers will for instance cooperate with big international service providers.

3.2 Business Actors

Through the new capabilities that AN offers not only end-user services will change but also by what kind of actors and what means these services are delivered. As services and access will be independent and every network will be accessible, other roles become more important than at present. So we will analyse such roles and how AN influences the way they conduct business with the end user as well as amongst themselves. After describing these roles in this section, we will in the next section place them in an integrated and a disintegrated value network and discuss aspects of relations like cooperation and competition, transaction costs and some other financial aspects.

3.2.1 Local Access Provider (LAP)

A local access provider is a business role offering local access/local network resources for local communication. Combining local resources with nationwide networks will possibly reduce production and operational cost of network access achieved through more efficient resource management and through 'reuse' of local infrastructure and user-deployed equipment using AN technology. This, for example, means that a chain of shops with a wireless access point in each shop can act as a local access provider for its clientele. Local access providers will most likely always be combined with the role of local service provider as they offer access services and probably some services that use the access to target the local customers. In general, local service providers offer content and services for a local market and/or local context. Entry barriers will be lowered as AN supports a multitude of network providers, i.e.

many channels to the customers. Local service providers that will offer content and services for a local market and/or local context can emerge supported by the advertising and service discovery functionality of AN.

3.2.2 Access Aggregator

Although there will exist LAPs with national presences (e.g. food store chains), we believe that most LAPs will be truly 'local'. This means that there will be hundreds or thousands of LAPs in a large town or city. To handle this multitude of local providers, the type of actor exemplified by the access aggregator will likely emerge. This is a business role that aggregates and bundles the different radio access services bought (or supported) from (local) access providers. This access will thereafter be sold either to service aggregators or to access brokers offering an intermediate step. The role of access aggregator can exist within a MNO and also as an independent actor.

3.2.3 Access Broker

An access broker represents a particular type of capacity broker, dealing with 'brokerage' of access capacity. It offers and manages end-to-end connections between the service aggregators, service providers, access aggregators and access providers. An access broker could also have billing relationships with customers (making him an actor playing already two roles); hence, we can distinguish between two different types of brokers: (i) the retail access broker, offering capacity directly to the end users, and (ii) the wholesale access broker, offering the capacity to other players like MNOs, MVNOs or service aggregators.

3.2.4 Service Aggregator

This role provides an aggregated service to customers by bundling services. This is mainly a marketing and customer relations role providing a one-stop shop to customers for a variety of services. Today there exist a few MNOs in every country, each with many millions of 'well-known' users. Most users have a subscription with one MNO only. In the envisioned scenario, a user may use access and end-user services from 10–100 different local network providers or LAPs during one day. In total, a user may have used thousands of LAPs during one day. To make the daily life easier, service aggregators and retail access brokers can act on behalf of the users. From the LAP point of view, hundreds or thousands of different users may access the network every day; as mentioned before, the agreements will in most cases not be prenegotiated.

Higher availability of services allows for better economy of scale as it is exposed to a wider user audience in more locations. Furthermore, the seamless combination of networks combined with the possibility of aggregating services has the potential to offer new kinds of service bundles.

3.2.5 Trusted Third Party (TTP)

This entity facilitates interactions between other parties who all trust this entity. By using this trust, TTPs offer services like authorization, ID management and payments. For example, a

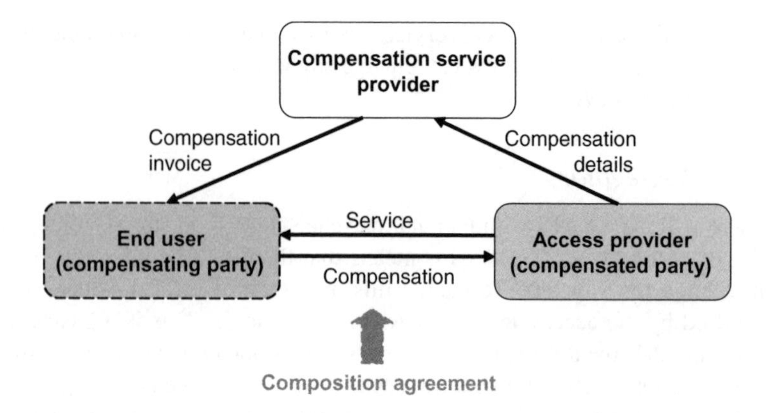

Figure 3.3 Representation of the business entity: compensation service provider

TTP may fulfil a notary function by authenticating entities or signatures on agreements. Note that traditional operators looking for new business opportunities and a crucial position in new value webs could play this role. Also, strong security authorities like billing and payment service providers and financial institutes like banks or credit card companies could naturally take this role.

3.2.6 Compensation Service Provider (CSP)

This is a business role offering compensation services to parties who have to compensate each other for delivering services. In this respect, the compensation service provider is an entity that helps two parties getting the compensation done if they themselves do not have the adequate compensation functionalities. One may expect that big traditional parties like MNO play this role themselves. We note that here we describe compensation service providers, depicted in Figure 3.3, offering retail compensation services. In this view, the clearing house entity described below could be seen as a wholesale compensation service provider. In fact, a dedicated party in the market could play both roles together moreover as wholesale compensations are aggregations of retail compensation details.

3.2.7 Clearing House

This entity provides a clearing solution to parties who have to pay each other due money as their own customers have used services of the other parties. Hence, a clearing house, entity depicted in Figure 3.4, settles wholesale demands including agreed revenue sharing. This role could be played by a specialist party in the business-to-business market. Another possibility is that, for example, an aggregator who bundles access or services for customers also fulfils this role.

The roles discussed above are interrelated in many different ways and all contribute to the value creation, distribution and consumption, and thus build a value network, which is explained next.

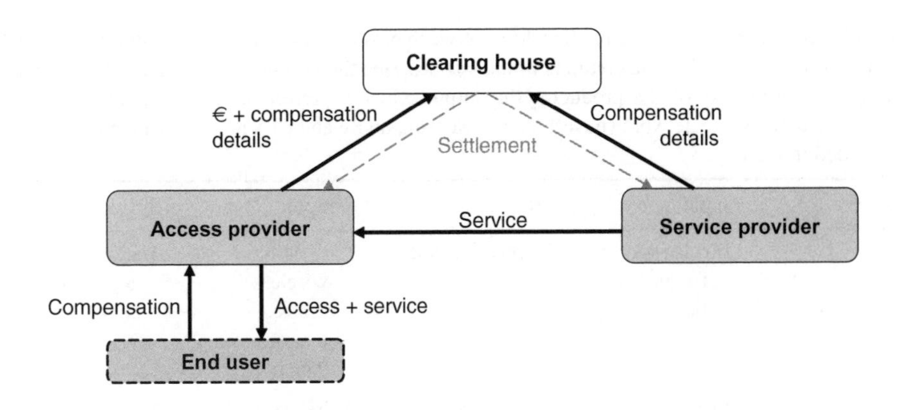

Figure 3.4 Business entity clearing house

3.3 The AN Business Proposition: The Value Network

An example of the envisioned roles and their relations is presented in Figure 3.5. Even though the figure includes only the most important roles of the value network, it becomes visible that the market structure is highly challenging and numerous different value configurations summarized in Table 3.1 are plausible.

One of the challenging aspects within fragmented market structures is, as can be seen from Figure 3.5, that there will be numerous relationships between existing roles. The actual level of fragmentation of the market will to a large extent be determined by transaction costs related to the establishment and maintenance of relations to other actors. If AN fails to reduce

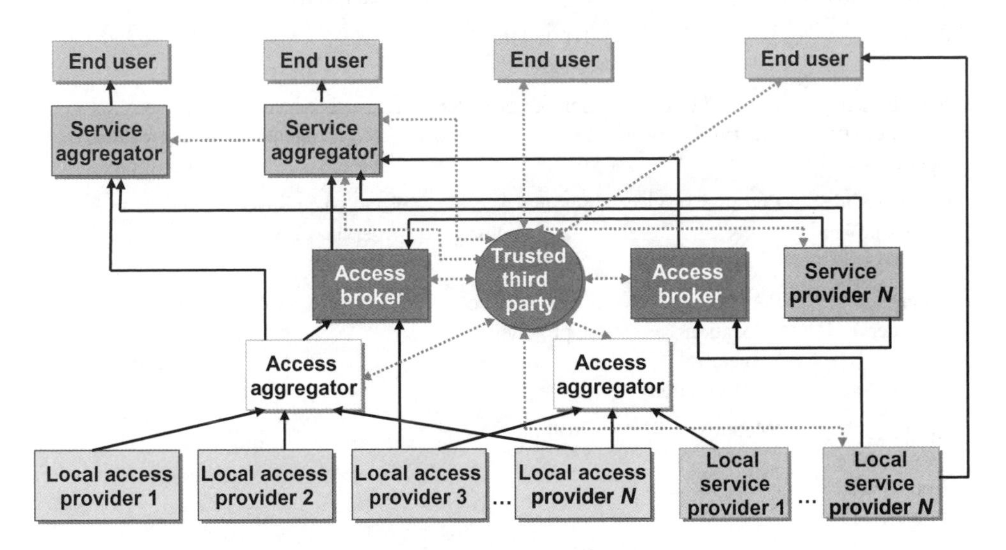

Figure 3.5 Value network for a fragmented market structure. The solid lines indicate services flowing in the direction of the arrow, implying that a monetary compensation is required in the opposite direction. The dotted lines indicate trust relationships. Clearing house and CSP are not presented in this picture to reduce the complexity

Table 3.1 Summary of possible relationships between business roles (middle column) present in the value network (Figure 3.5). The columns to the left describe the producers and the production factors used by the investigated role for producing the output, which is presented to the right and provided to the next role in the value network (buyer). Again clearing house and CPS are not included as they were not included in Figure 3.5

Producer	Input	Roles	Output	Buyer
	Infrastructure, frequency license	Access provider	QoS assured wireless access, measures on data usage	Access aggregator, access broker
	Infrastructure	Local access provider	Wireless access	Access aggregator, access broker, LSP
Access provider, LAP	Different types of wireless access (licensed, licensed exempt)	Access aggregator	Bundled RA services with E2E connectivity	Access broker
Access aggregator, access provider, LAP	Bundled (or unbundled) RA services	Access broker	Bundled RA service including E2E connectivity	Service provider or aggregator, consumer
All		Trusted third party		All
Access broker and aggregator, access provider	Bundled RA services	Service provider	End-user services (e.g. content based)	Service aggregator, consumer
Service provider, access broker	End-user service	Service aggregator	Bundled end-user services(service portals)	Consumer
(Content providers)	RA services, location-specific information	Local service provider	End-user services	Service aggregators, consumers
Service aggregator, (retail) access broker, TTP	'Monthly' bill	Customer		
Service aggregator, retail access broker	Bundled services	End user		

the transaction costs considerably – compared with the ones in current market structure – a consolidated version of value networks where all actors encompass several different roles will develop. Naturally, this will reduce the fragmentation (and consequently also the competition) of the market structure, in the extreme rendering a similar structure as today. Thus, for this scenario composition constitutes a crucial functionality and apart from simply providing the technical solutions the functionality also has to be designed such that the amount of business-sensitive information is kept at a minimum level. We believe that in many cases, and especially within the fragmented market structure, one has to outrank the technical merits (e.g. the capacity) of the solution. The fact that consumers would like to have only one billing (or compensation) relation will most likely consolidate the number of actors and we envision that the role of the access broker and trusted third party in many cases can be combined.

Access and service providers can choose to cooperate horizontally or vertically. It is obvious that AN increases the opportunities and incentives for horizontal cooperation. Cooperation is needed to hand over end users seamlessly from one network to another, but this will be done only if both networks benefit from that. From the end-user perspective, 'continuous connectivity' is a kind of service that represents an added value and hence could be paid for. From the service provider perspective, it would also be more beneficial to cooperate with access providers that are willing to contribute to 'continuous connectivity'.

For a disintegrated market scenario, we envision a market structure characterized by 'fierce' competition between different access and service providers. At the same time, some actors will cooperate on long-term basis, e.g. form alliances depending on their corporate strategy and their position in the value network. It can be argued that the large number of both access and service providers will result in a need for aggregators that act as agents for other actors by managing the complexity of the business environment. Examples are the service aggregator acting on behalf of end users and the access aggregator acting on behalf of LAPs.

Assuming a migration path with increasing disintegration of vertically integrated MNOs, a certain development of roles can be seen. Starting at the vertical integrated market with little cooperation, it is plausible that the role of access aggregator within an operator can be identified. In this role, the multiple access networks of the operator are composed to work as one seamless network. Additionally, the MNO will incorporate several local networks if it is efficient at that point in time to use the available infrastructure instead of building it itself (Figure 3.6). In the same way, the role of service aggregator is played to aggregate and integrate the own access and services into one seamless bundle and handle all customer relations. The service bundle will be complemented with services from service providers like banks or music distribution companies. The actors and their mutual business relations in a market with vertically integrated value chains with a high degree of cooperation with independent access and service providers are illustrated in Figure 3.6.

The next step in cooperation is reached when access aggregators start sharing resources with each other. This might be buying extra capacity when needed and selling spare resources. It would be the task of an access broker to organize this market. AN technology will help to make this a seamless experience for the customer and automate large parts of the trading. In addition, the access broker can provide bargaining power as well as a one-stop shop, as it can act on behalf of LAPs. Furthermore, the access broker can use aggregated access of a vertically integrated operator as a wholesale offer. A LAP can generate extra revenue when it joins an access broker by opening up their networks for others. LAPs can also sell valuable context information – like location – that they have of the local users.

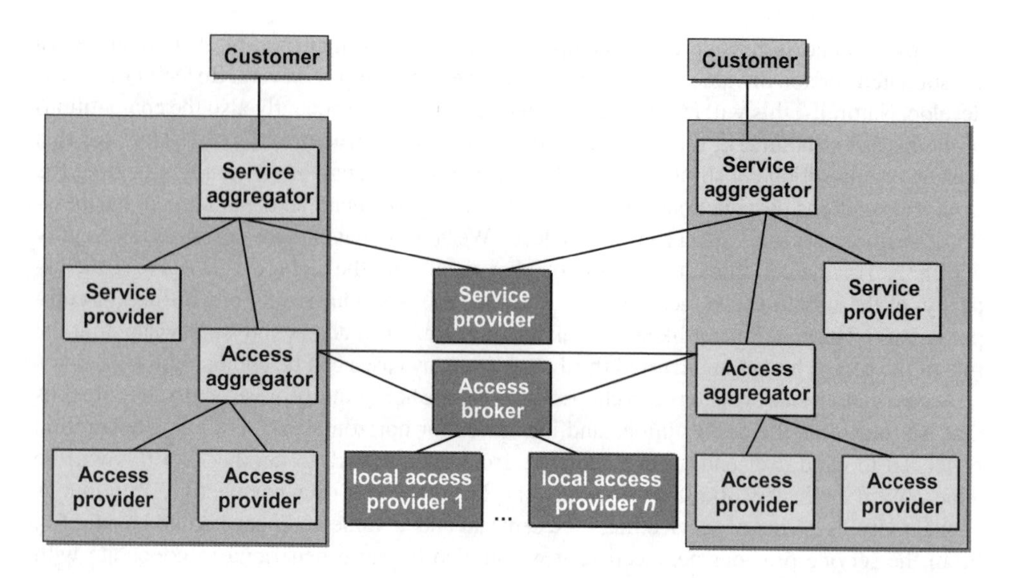

Figure 3.6 Actors and roles in a market with vertically integrated value chains with a lot of cooperation and their mutual business relations

At this point, services are available from independent service providers and access is available from access brokers. This means that it is possible to take up a role as independent service aggregator (not being a part of a MNO) bundling services and access and target this bundle at a certain customer group.

The multitude of independent roles and the multitude of actors that can play a role will make it necessary to have one more intermediate role. Instead of passing identity, security, authorization, profiles, preferences and payment information from one provider to another, it will be better for all to have a central party that manages this, the TTP. These roles and relations are visualized on the right-hand side of Figure 3.7.

For the customers, it will be possible to buy access independent of services and services independent of access. In principle, the service aggregator is not really necessary because the customer can take up this role for himself, but most customers will appreciate a party taking all the burden of organizing, negotiating and buying all separate items of a complete service bundle. In the unbundled market scenario, the differences over time can shift dramatically in terms of quality, costs, access and services available as ad hoc negotiations and network composition cannot be predicted. The financial risk involved can be taken by an intermediate role as a service aggregator because it spreads the risk over a large number of customers. An individual end user cannot spread the risk in this way.

It has to be realized that in the real world all possibilities will exist next to each other. There will always be vertically integrated operators offering a nice and integrated bundle of services and next to that will be independent providers of services and access. Some will cooperate and some will not. This will make a much more complex market as in one of the separate figures. In order to make the AN world work, relations and negotiations, resulting in sharing of resources, are a focus point. If there is an intermediate role in the middle or not, it is important that all information is available to check if for instance composing networks is possible and

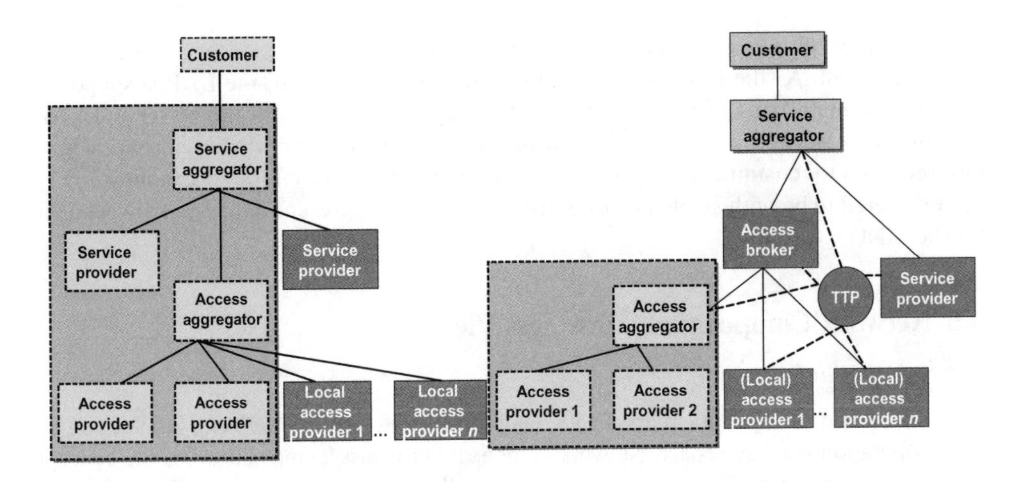

Figure 3.7 Cooperation between access aggregators sharing common resources. On the left-hand side, the operator represented with dashed rectangles acts as an access as well as a service aggregator. On the right-hand side, the independent service aggregator, illustrated with solid rectangles and black text (acting on behalf of the user), cooperates with an access broker and a TTP; here the operator, represented with dashed rectangles, is only an access aggregator

if services can use the available access and at what costs while maintaining a certain amount of security. All this information should be available, compared and agreed upon without the customer noticing or influencing the seamless service experience.

3.4 Financial Aspects

An analysis of cost per user for amount of data transferred during one month is presented in [189] or the case where a MNO cooperates with LAPs and uses a part of their capacity. One conclusion that can be drawn from this analysis is that there is potential for significantly lowering the investment size, risk level and overhead costs if specialized local network operators exist and are willing to share their assets. This is enforced if these assets were purchased initially for other purposes. Moreover, a partnership with strong brand companies will reduce the marketing costs and open up more financial resources for network development.

One of the main points of AN implementation is to provide lower transaction costs among the multitude of players willing to cooperate in the market. As was suggested above, cooperation among horizontally specialized players is a must if they want to be independent of MNOs.

From a traditional MNO perspective, the investment risk in venturing into a new wireless service with unknown demand is considerable. This is a good enough reason to look for opportunities of minimizing the initial investment by sharing and reusing existing assets belonging to LAPs as well as marketing competence and brand of large corporations. By deploying their networks in selected areas with high population density, LAPs can develop a small-scale but healthy business with low investments. One possible outcome may be that these LAPs can drive the deployment of broadband transmission lines for Internet connection to the areas where they deploy the infrastructure.

From the MNO point of view, the cooperation with LAPs can present a more or less safe investment. As the infrastructure is used by the LAP himself, the LAP is supposed to make the major part of the investment. If the agreement between the MNO and LAP also includes 'access to all traffic', both internal and public, the MNO will experience lowered costs for customer acquisition. For companies, the internal traffic can most likely be considered to be both stable and established; thus, this represents a low-risk investment for the MNO.

3.5 Network Composition – Business View

3.5.1 Introduction

In an unbundled and dynamic market with many access and service providers, flexible and ad hoc cooperation is a key aspect. Network composition (or just 'composition') is the concept for how Ambient Networks interconnect dynamically in order to make use of services and resources available in or via interconnecting Ambient Networks [78]. Composition applies on a global scale and is possible and supported for any level or size of networking. Through the inherent support of dynamic networking, it also means that business agreements between interconnecting networks are supported as run-time operations. This together with that composition is defined generically for any type of network (and of any size) and provides for a platform of very efficient and scalable deployment of new technologies, applications and services.

Network composition can support many different use cases, e.g. a personal area network interconnecting with a residential network, a train network with a cellular network or carrier networks connecting to offer dynamic service levels. Already mentioned are the dynamic roaming agreements, where neither the user nor the visited network needs to have any prior arrangements with a home network. The necessary agreements can be established at the time the user roams into the visited network. Moreover, they could make it possible that the user does not need any prior arrangements with any network/service provider, i.e. the user can be seen as 'homeless'. Another example is a community network, even a residential network, composing with a MNO to provide extended coverage and capacity.

3.5.2 The Composition Agreement (CA)

Ambient Networking is about cooperation among networks, and this implies that there are a lot of different issues that need to be arranged between the networks involved. When establishing an internetwork agreement today (e.g. for interconnect or roaming), typically the non-technical aspects take most time to agree. As the goal of composition is to enable networks to cooperate easily, it implies that the establishment of the composition should be done 'on the fly', therefore speeding up the process of negotiating and execution – and eventual release – of cooperation. Therefore, the aim of composition agreements is to agree on technical and business aspects to some degree automatically. It is also necessary that these agreements are established in a secure, trustworthy way.

The wide scope of a composition agreement implies that a huge number of parameters, values and options need to be agreed. Negotiating all these parameters and values may

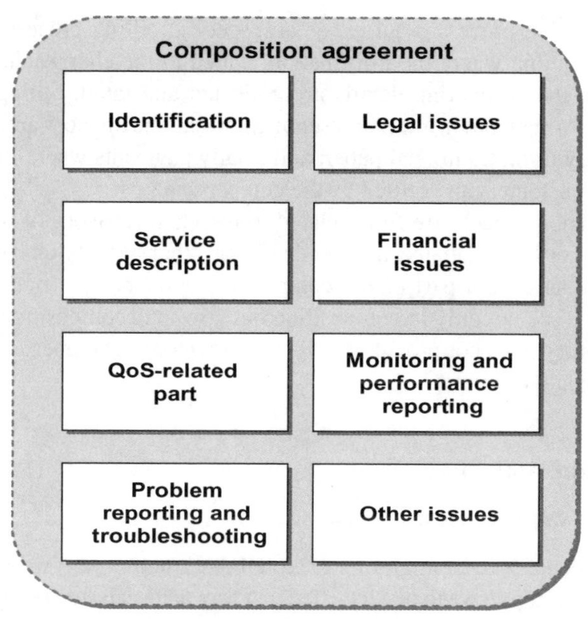

Figure 3.8 Aspects of composition agreement (based on the SLA template developed in [22])

require a lengthy process which conflicts with the idea to negotiate CAs on the fly. Fortunately, some more generic elements of these negotiations can be done in advance using composition agreement templates. This template predefines parameters, values and options for a CA. For specific use cases, the parameters that should be agreed are generally the same. By using an applicable composition agreement template in a composition process, one may reduce the amount of negotiation and time needed. Standardization bodies could define composition agreement templates. An example is the GSM Association that in a way defines a template for roaming agreements. A CA template could include the following items: ID, service description, QoS requirements/guarantees, legal issues and financial issues, monitoring and performance reporting, and problem and failure reporting as depicted in Figure 3.8.

3.5.3 Compensation

To ensure commercial viability for all the different business roles, the composition process needs to deal adequately with charging and billing. We term this compensation, i.e. basically, the charging party provides some service for which it gets compensated. Figure 3.9 describes this basic compensation case.

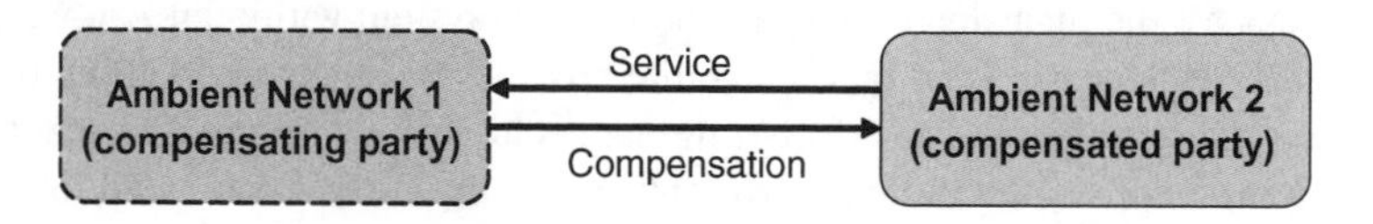

Figure 3.9 Basic compensation case

The compensation process requires, as any specific billing process, the usual functionalities like accounting where the information related to the chargeable events is generated, charging where the accounting details are collected and rated, billing where the charged events are transformed into invoices, presentment where the invoices are presented to customers electronically or on traditional paper, and finally payments where the presented invoices are paid by the customers and settled by the compensated party.

As compensation aspects are fully related to the characteristics of the services being offered, they cannot be negotiated independently of other aspects of the service. Therefore, compensation issues are a part of the composition agreement. It must be noted that compensation is an issue not only in composition but also in decomposition. For example, if the compensated party does not provide the service anymore, recompensation – as long as it is manageable – could be attempted.

3.6 Migration Aspects

3.6.1 Towards Native Ambient Networks

If the business benefits from migrating to an 'all-AN' (native AN) world are proven, still the various migration steps have to be identified in a way that each step brings business advantages by itself and requires only moderate investment. Here we consider the migration roadmap process of moving from today's legacy systems to Ambient Networks with full functionality. We identify two main phases in the roadmap towards AN: (1) overlay and (2) interworking. In the overlay phase, an 'AN island' connects to and composes with a distant AN over the legacy environment, which is used as a simple transport mechanism. In the interworking phase, an interworking function (IWF) is introduced, which translates the signalling between the ACS and the control plane functions of a legacy network.

We have identified four steps for a smooth migration path towards a native AN environment in the future:

- Phase 1a: Immediate use of existing layer 2 or layer 3 transport bearer of the legacy network, which is more or less the starting point.
- Phase 1b: Implementation of an abstraction layer, which adopts the ARI to the control functionality of the legacy network.
- Phase 2a: Installation of IWFs to 'emulate' an AN behaviour in the legacy network in order to appear as an AN from the perspective of other ANs and as a legacy network from the existing environment.
- Phase 2b: Add AN functionality in a stepwise fashion to the legacy network to become a real AN with full ACS capabilities and all AN interfaces.

The different steps described together with a possible time frame are depicted in Figure 3.10. These four steps for the inclusion of a legacy network or system within an Ambient Network are not mutually exclusive, and in fact, it is quite possible that they could all be deployed as intermediate phases on the road to a fully migrated solution. An important aspect of the incremental migration presented in this deliverable is that the steps do not need to be followed in the exact sequence presented in Figure 3.10. They may take place in parallel, depending on the legacy network from which the process starts or from the strategy pursued by the

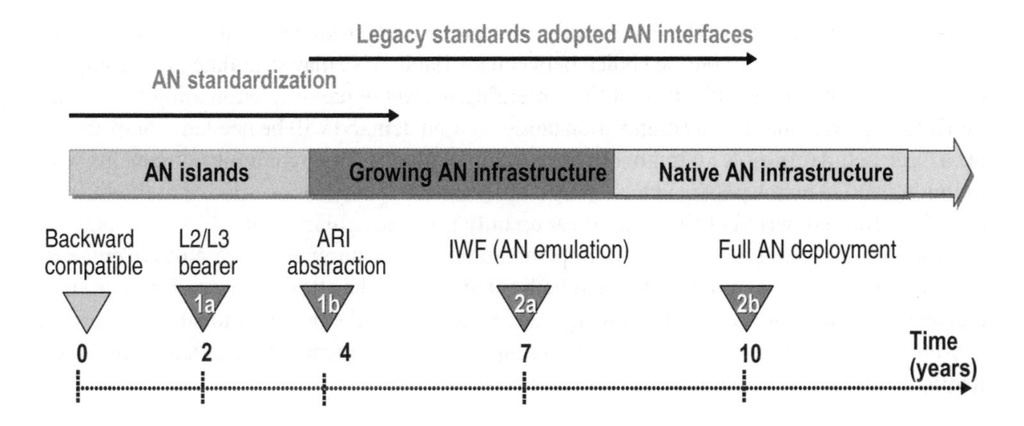

Figure 3.10 Roadmap plan for migrating to a native AN

individual operator. The aim is to be flexible enough to accommodate as many different business cases as possible. Consequently, based on this assumption a stakeholder can, for example, go straight to the deployment of a full native AN environment.

In addition to the four migration steps on the operator network, an important step is the emergence of AN-capable end systems – either existing user equipment upgraded to AN or new AN native devices. This step will often be a part of the migration phases. Indeed, it is most likely that migration towards AN will start with end-user equipment, as it demonstrates the benefits to the end users who can easily communicate among each other and use new as well as legacy networks over any type of network.

3.6.2 Business Deployment Conditions

Ambient Networks is all about cooperation among networks, and this implies that there is some agreement in place.

In the future, both the number and diversity ranging from access network operators (visited networks) as well as operators who mainly act as identity and trust providers (AAA operators or home operators) are expected to increase. Examples of current trends or developments include that WLAN hotspots become more widespread due to the ease of use and relative low cost of deployment. In such a situation of many independent small and large networks, the AN resources need to be advertised (both the existence and the offers) from different ANs (AN advertisement), to look for and to evaluate network offers (AN discovery) and to select the 'best' network. The 'best' network can be selected with respect to performance, user preferences, cost (AN selection) and service requirements (e.g. media processing requirements). Furthermore, several networks may be selected at the same time which would allow choosing between different routes in order to perform load balancing, improve the reliability of an ongoing connection or react to topology changes. Benefits of Ambient Networks become most apparent when it comes to rapid network deployment, cost efficiency, increasing network dynamics, creation of new business models and foreseen technological improvements, e.g. multi-hop communications.

Relying on pre-established and manually configured composition agreements is not practical if we envision that the online choice between available ANs may stimulate price competition. If we want to take advantage of this emerging market dynamics, automating the formation and implementation of a composition agreement on demand will be needed. For example, for a large cellular operator it might not be worth to manually set up roaming agreements with individual WLAN hotspot providers. This is relevant especially for hotspots that might exist only for a limited period of time, e.g. if set up only for a particular event.

However, if the agreement and its implementation were automated, the attractiveness of forming a composition agreement would be limited only by the attractiveness of the commercial terms and technical capabilities being offered and not by the organizational and technical overhead of creating and implementing a composition agreement. The challenge for establishing a CA between operators has two aspects:

- Commercial – What are the terms and conditions of the CA?
- Technical – How to achieve network interworking on the fly?

The basic concept behind AN is easy composition among networks that will benefit the user as he gains a wider choice of available networks. Networks that compose need to establish a CA in which the terms and conditions of compositions are described.

3.7 Summary

Ambient Networks is a new technology which offers technical benefits and creates new business opportunities. Today's telecom business environment may be impacted by the AN concept of network composition, as it increases the flexibility for cooperation between the different actors in the telecommunications value chain. The potential changes in the market structure as well as the opportunities for the creation of new business roles were discussed in this chapter. It was highlighted that the increased flexibility in setting up business relationships leads to an increased demand for aggregators bundling access options and services such that useful and stable configurations emanate. The AN concept will be implemented only if clear benefits for the different actors in the value chain exist. This chapter has elaborated the business motivations for the different actors and it was deduced that end users, operators and vendors are main beneficiaries from the Ambient Networks concept. The migration process from legacy systems to Ambient Networks with full Ambient Control Space functionality is a project priority from the beginning and was outlined in this chapter. We described two phases in the roadmap towards AN, using overlay and interworking functionality, respectively.

4

Architecture and Components

Acknowledgements

This chapter is based on the joint experiences and efforts of the researchers in the first phase of the AN project and particularly the following people listed as contributors and authors (i.e. in alphabetical order): Bengt Ahlgren (Swedish Institute of Computer Science), Antonio Alves (Critical Software SA), Ulrich Barth (Alcatel), Hendrik Berndt (DoCoMo), Marcus Brunner (NEC), Bryan Busropan (TNO Telecom), Lars Eggert (NEC), Svante Ekelin (Ericsson EAB), Anders Eriksson (Ericsson EAB), Hannu Flinck (Nokia), Robert Hancock (Siemens (RMR)), Frank Hartung (Ericsson EED), Eiko Heuer (Ericsson EED), Geert Kleinhuis (TNO Telecom), Takashi Koshimizu (DoCoMo), Lars Lundgren (Ericsson EAB), David Moro (Telefonica Investigación y Desarrollo SA Unipersonal), Luis Munoz (University of Cantabria), Norbert Niebert (Ericsson EED), Gunnar Nilsson (Ericsson EAB), Toon Norp (TNO Telecom), Börje Ohlman (Ericsson EAB), Manuel Quadros (Critical Software SA), Jürgen Quittek (NEC), Jarno Rajahalme (Nokia), Simone Ruffino (Telecom Italia Lab), Andreas Schieder (Ericsson EED), Mikhail Smirnov (Fraunhofer FOKUS), Michael Söllner (Lucent), Heiner Stüttgen (NEC) and Olle Viktorsson (Ericsson EAB).

4.1 Introduction

Analysing recent work on future networking scenarios reveals that two key issues are emphasized by the different approaches and initiatives such as [19–21]. First, networks are expected to become more technologically heterogeneous, accommodating old and new access systems as well as various kinds of applications and services – indeed, migration and feature rollout will not be a 'one-off' but a constant activity. Second, networks will become organizationally heterogeneous, as it is expected that today's cellular systems will be complemented by a diverse mixture of other network types operated in different environments such as personal, vehicular, sensor and hotspot environments. Ambient Networks is addressing the need to dynamically form and re-form these diverse networks in response to changing conditions. Besides Ambient Networks, an increasing number of research initiatives come to similar conclusions and compatible assessments of the future. The result

Ambient Networks: Co-operative Mobile Networking for the Wireless World Norbert Niebert (Ericsson GmbH), Andreas Schieder (Ericsson GmbH), Jens Zander and Robert Hancock

is a rising number of research initiatives focusing on the design of new network architectures such as the recently created NSF NeTS research area Future Internet Network Design (FIND) [22]. This makes the Ambient Networks approach increasingly creditable and acknowledges the approach taken.

The requirement to address the identified needs for future networking defines an extremely broad problem space, too large for any single activity to consider. Therefore, the Ambient Networks project concentrates on networking aspects only, leaving the development of a new air interface and a service platform to sister projects embedded in the same overall framework of the Wireless World Initiative (WWI [10]) [23–25]. This chapter presents the new network architecture that addresses the problem space discussed above and shows how it provides a basis for the realization of the Ambient Networks vision.

The key constituents of the Ambient Networks architecture are the Ambient Layer Model, the Ambient Control Space (ACS), a set of security principles and the network composition procedure.

The Ambient Layer Model defines the upper and lower boundaries of the networking functionality embraced in Ambient Networks and delimited by a pair of interlayer interfaces. At the upper boundary, the Ambient Service Interface (ASI) provides uniform access to the Ambient Networks functionality from higher layers, and a bearer-level abstraction has been developed to represent end-to-end communications services. Bearer capabilities range from simple best-effort data transfer to special-purpose media handling services autonomously providing a means to publish multimedia content (described in Chapter 10). The ASI and the bearer abstractions are key enablers for a universal deployment of new services and applications and the transparent and incremental deployment of value-added networking functionality to support them.

At the lower boundary, the Ambient Resource Interface (ARI) encapsulates the capabilities of the underlying infrastructure, and a flow-level abstraction provides the basic building block of data transfer between addresses. Because these abstractions are not tied to any specific network technology, the AN control functions are portable between different network types. This decoupling enables a simple migration path from current systems to the Ambient Networks approach.

The Ambient Control Space is the environment within which a set of modular control functions can coexist and cooperate. The environment includes plug-and-play concepts that allow the ACS to bootstrap and discover the set of present functions dynamically. Further, a naming structure and registration mechanisms are defined to ensure that new functionality can be developed and integrated without impacting the overall system design and implementation.

4.2 The Ambient Network Approach

In the process of turning the general requirements discussed in Chapter 2 into offered functionality of Ambient Networks, a set of common design principles is derived from the requirements. These principles commonly define the overall approach taken for Ambient Networks and clearly highlight the key aspects of the Ambient Networks approach. This intermediate step towards the Ambient Networks architecture also ensures that the general requirements are considered appropriately in the design of the Ambient Networks architecture.

Principle #1: Ambient Networks Build Upon Open Networking Functions

One basically new way of defining networking in Ambient Networks is to remove architectural restrictions on 'who or what can connect to what'. Compared to existing internetworking, the goal is to enable all networking services, e.g. quality of service or media delivery support, for connected networks instead of connected nodes. This is motivated by the observation that current, node-centric designs fail for many scenarios including PANs, moving networks or sensor networks, when connecting such networks to other networks. In general, we will always assume a network at the end of the communication flow. Hence, we talk about 'end environments' rather than end nodes or terminals.

The Ambient Networks concept defines a set of support functions required to satisfy the business needs of the operators of such end environments, where operators can be commercial entities as well as end users. For such an end environment, capability offerings are broken down to their nucleus, which is the end-to-end relationship between control functions. The challenge is to provide suitable mechanisms to enable any such relationship regardless of whether the partners reside in an operator's network or an end user's terminal.

Principle #2: Ambient Networks are Based on Composition and Self-Management

Although composing networks is an easy task if only packet forwarding is concerned, advanced end-to-end control functions like QoS and security as well as mobility are currently very difficult to establish across network boundaries. Ambient Networks treat network composition and reconfiguration in a self-managed way as the guiding design principles for the research work. The goal is to use network composition and self-management as basic, locally founded building blocks of a networking architecture. This includes composition across different business domains and a wide-ranging support for policy-based networking. These features will broaden the business case for the operator and enable a fast introduction of new services in all connected networks.

Principle #3: Ambient Networks Functions can be Added to Existing Networks

In today's networks, there is a large degree of homogeneity in the very basic connectivity/packet forwarding functions, but the network control is distributed over multiple layers and is specific to network technology, operator and even implementation. As a starting point for Ambient Networks, the connectivity and control level are logically separated. The control level, referred to as the Ambient Control Space (see Section 4.3), can enhance existing technologies with distinct control functions, which are compatible across all parts of an Ambient Network.

Principle #4: Ambient Networks Secures and Simplifies Network Control

Network control was very simple in the early days of the Internet, when the network had only simple tasks to fulfil, i.e. delivery of packets between fixed network endpoints at best-effort condition, often within a single administrative domain. Similarly, the first digital mobile network according to the GSM standards initially had a relatively simple single service centred architecture. Adding constant innovation to both networking approaches and finally converging their architectures has now led to a quite complex networking architecture, with a very large number of interfaces.

Ambient Networks takes a radical approach using overlay techniques to simplify the overall networking control architecture to enable future growth of functionality in a cost-efficient manner. The three basic reference points of Ambient Networks described later in Section 4.3.1.2 are the cornerstones for the inherent simplicity of the approach.

Similarly, security was not a major threat in the early days of the Internet. Security add-ons on all layers have been added since the commercialization of the Internet. Still, security fixes barely cope with the threats. Ambient Networks build security into all network control functionality as a generic feature. All functions are being analysed for their security implications and the external reference points are being protected by conceptual measures.

Principle #5: ANs Scale Between PANs and WANs Using the Same Protocols

Telecommunication networks traditionally differentiate between network internal and user interfaces, limiting the recursivity and scalability of the architecture. Ambient Networks do not make this difference and treat the user domain like a small-scale operator. In conjunction with the very small minimal functionality and the wide-ranging self-configuration, this approach will cater for the required scalability. A part of this principle is to allow for disconnected operation as not all parts of a network will be interconnected all the time. Especially, smaller networks like PANs will often be disconnected from their composition partners as well as from their central components and still have to work consistent to their policies.

This approach yields the following advantages for the users of Ambient Networks technology:

- *Network operators* can decide what level of support they want to give to users and business partners, based on flexible sets of network control and composition functions. Affordable and simple as well as service-rich networks can be built without restricting the integration into the global connectivity. The approach can be extended to incorporate the infrastructure of existing networks, providing protection of investment and a path for operators to unify their management, security and service offerings, and furthermore new technologies can be introduced in a coherent way.
- *End user's* ability to select between different control functions on their devices or personal networks provides the feature-rich service environment they like to use. Innovation is nurtured by the opportunity to develop any service on top of the basic connectivity with as few barriers as possible because of the simple but extendable connectivity interfaces.
- *Service providers* can easily address large numbers of users and deliver their specific services to them without having to worry about their users' access or network capabilities.

4.3 The Ambient Control Space Concept

Following the design principles and requirements discussed above, the notion of the *Ambient Control Space* is introduced to encompass all control functions in a certain network domain. The Ambient Control Space together with a (possibly legacy) connectivity network is called an Ambient Network. The main characteristics of such an Ambient Network are

- It provides well-defined reference points to other ANs and to service platforms or applications.

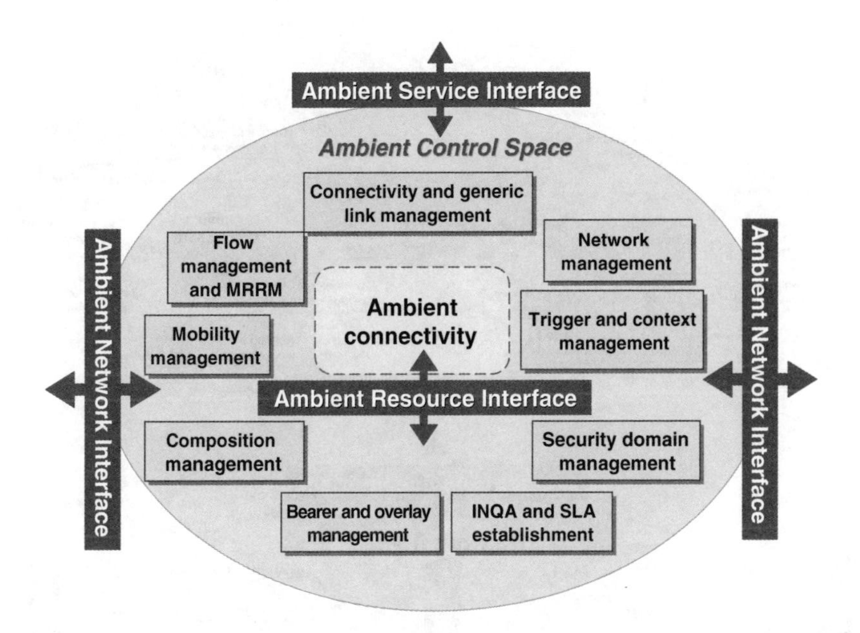

Figure 4.1 Schematic representation of ACS, its components and reference points to external entities

- It provides (at least an essential subset of) the Ambient Control Space functions.
- It can be dynamically composed with several other ANs to form a new AN.

Examples of functions hosted by the Ambient Control Space (see Figure 4.1) include the support for mobility, network management, network composition, QoS control or multi-access and also more abstract functions like the provisioning of context information or support for media delivery overlays. All functions are working according to common security principles explained in more detail in Section 5.3.

Ambient Networks could potentially belong to separate administrative or economic entities. Hence, ANs can collaborate in both a cooperative and a competitive way. When ANs and their control functions are composed, care must be taken that the responsibilities of each individual function are clearly defined such that concurrency is avoided.

Ambient Network reference points

Control spaces spanning several domains are required, as users will roam across several network domains and want to exploit the capabilities of as many domains as possible. Still, they will assume their personal services to be at their disposal regardless of their location and physical connections utilized. The intrinsic complexity of domains might vary and economic considerations will decide about the feasibility of different approaches. To allow cooperation across different Ambient Networks, the *Ambient Network Interface (ANI)* is provided. It is a reference point offering standardized means to connect the functions of an Ambient Control Space with functions of another domain (see Figure 4.2). It also advertises the presence of control space functions in adjacent domains to allow for a dynamic set of control functions in each.

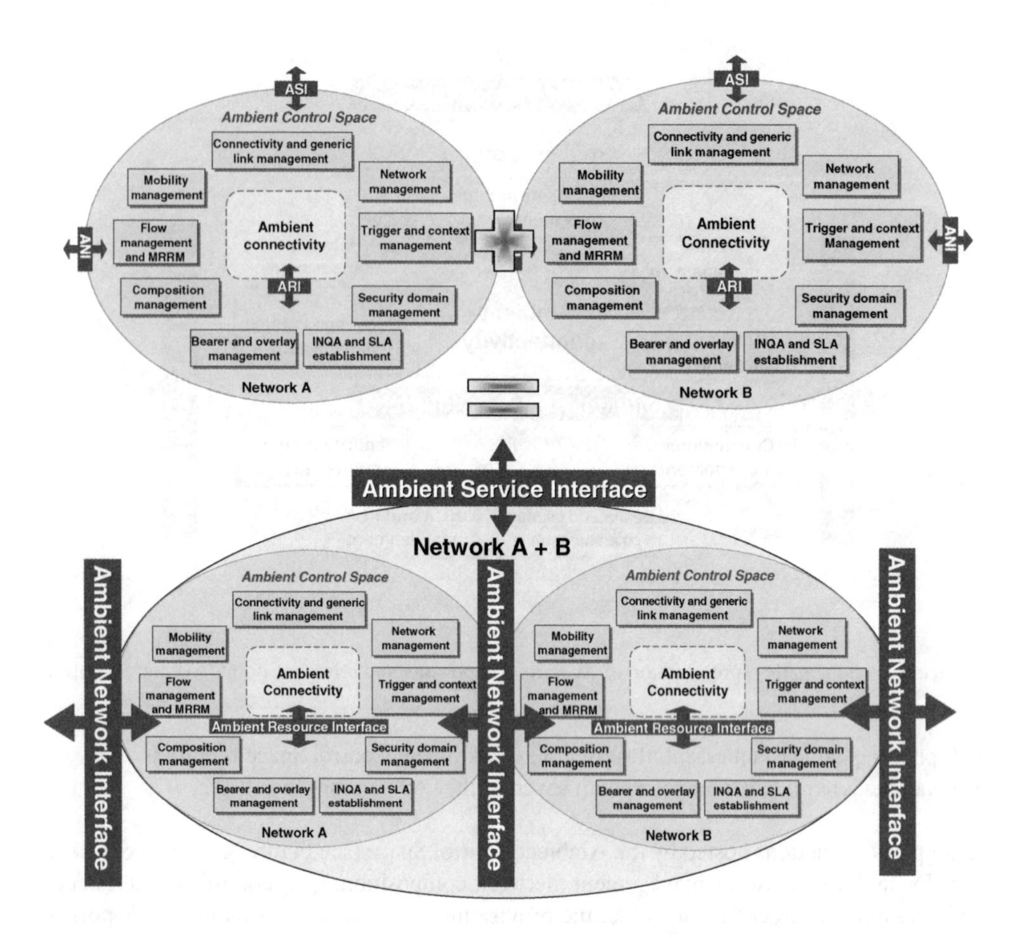

Figure 4.2 Cooperating ACSs appear as a single domain to external service users

Accessing the services of an Ambient Network happens via the *Ambient Service Interface*. Even in a composed Ambient Network this reference point ensures that only a single homogeneous control space becomes visible to external entities. An application or service, making use of the functions in the control space, should find the same environment regardless of which Ambient Service Interface it is currently connected to. The mechanisms implemented in the Ambient Control Space and the Ambient Network Interface will support this.

Finally, the control space operates on an abstraction of the physical infrastructure in a way that only those resources are visible that are actually controlled by the ACS (see Section 4.4 for further details). The resources provide access to the ACS via a common reference point. This reference point is called *Ambient Resource Interface*.

A particular issue is to ensure scalability of these reference points to be applicable for small personal networks up to large-scale networks. A related issue is the universality of the approach covering all types of networks.

To summarize, the three Ambient Network reference points jointly

- offer a standard functionality independent of the nature of the underlying AN;
- support a simple plug-and-play connection between ANs;
- enable network reconfigurability; and
- provide a single reference point to the outside.

4.3.1 Architecture and Interfaces of the Ambient Control Space

The Ambient Control Space introduced above constitutes the logical control plane of an Ambient Network. Figure 4.1 shows a simplified overview of the logical structure of the Ambient Control Space. It illustrates that an AN consists of two major, distinct components. The first component is the Ambient Connectivity, an abstraction of the existing network infrastructure that is controlled by the second component, the Ambient Control Space. The ACS again is comprised of two subcomponents: the control functions (indicated by rectangles in Figure 4.1) and the control space framework functions, which hosts all functions necessary to allow new control functions to plug into and interact with an already established control space, execute their control tasks and coordinate with other control functions.

4.3.1.1 Control Space Structure

The ACS consists of a number of control functions that cooperate to implement the overall control functionality. The key features of the ACS architecture are its distributed nature and modularity. Although a small number of control functions must be present in any ACS to make a network 'ambient' (i.e. establish a 'minimum ACS'), additional control functions can be added to or removed from the control space during regular operation of the network.

The control space can be thought of as having two halves representing the two main areas of functionality that the ACS provides. The first half sets up, maintains and moves user plane connectivity and the second half deals with network and security domain management, policies and composition (see Figure 4.3).

The part of the ACS that supports and manages user plane connectivity embraces the following main areas of control functionality:

- The *bearer and overlay management functions* offer end-to-end bearer services to applications through the ASI. Overlay management provides both basic end-to-end bearers and advanced bearers that are aware of the semantics of the media streams they convey, which allows manipulation of the media streams along the delivery path. These advanced bearers, called service-specific overlay networks, are constructed through interconnected media processing nodes called media ports [26], which implement transcoding, caching and synchronization functions. For a more detailed description of the Ambient Network bearer, refer to Section 4.4.
- The *flow (and multi-radio resource) management functions* configure and establish flows both in the wired and in the wireless part of the network. MRRM is the subclass of flow management that is responsible for handover decisions for the radio access parts. The access selection procedure decides (based on cost, QoS, policies, etc.) which of the available access flows should be used to provide a requested bearer.

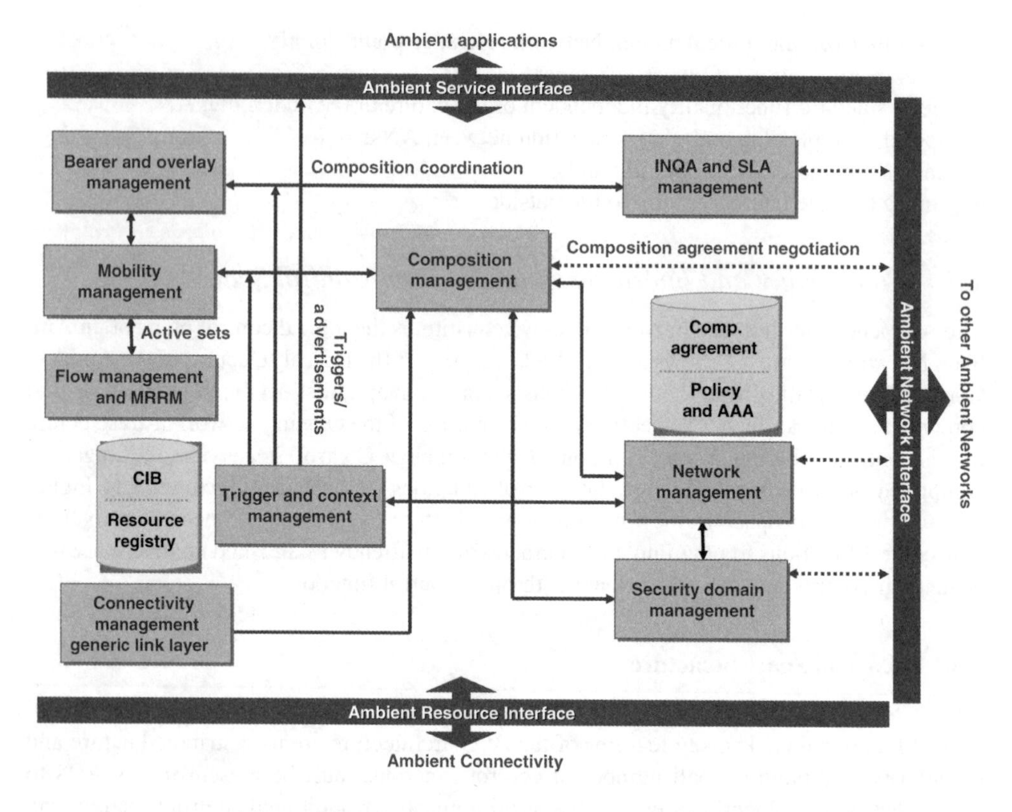

Figure 4.3 Illustration of the high-level ACS structure

- The *connectivity management (and generic link layer) functions* detect new links and initiate the establishment of connectivity between ANs utilizing them. Once new connectivity is detected, the secure attachment procedure is initiated. The generic link layer is a subclass of connectivity management, which is responsible for managing access links. Topology information is provided to the overlay and bearer management FEs to support them in their bearer set-up decisions. Flow configuration and set-up requests are received from flow and multi-radio resource management and the links are then configured accordingly via the ARI. Links are observed by receiving measurements from the connectivity layer.
- The *mobility management functions* ensure that bearers remain independent and unaware of connectivity changes or movement events that occur in the underlying connectivity which consists of flows that are bound to locators (the current point of attachment to a network). The bearers are bound to higher-level objects in the naming framework which have nonchanging identifiers, such as application points of attachment or cryptographic node IDs. The key role of mobility management is to maintain the mapping between the current locator and the identifier, thus enabling communication to take place regardless of the current network point of attachment. It also supports mobility for groups of nodes. The mobility management FE bases its handover decisions on triggers, context information and policies.

- The *triggering and context management functions* coordinate the storage, access control, dissemination and aggregation of context information. Context information, such as triggers, user preferences or network status, is available in the distributed context information base. Each object is identified by a unique identifier and is registered with the context coordinator. Context clients either retrieve context information directly from the triggering and context management function or are directed to the context source, which is able to satisfy the request. It is also possible to subscribe to continuous updates of context information. Context information can only be accessed according to explicit policies, to ensure privacy and protect business information.

The second part of the ACS manages administrative domains including security domains. It ensures that only actions that are compatible with current policies are performed. It also provides the means for ANs to cooperate and share resources in an automated way through automatic negotiation of service and composition agreements. The main areas of this part of the ACS are

- The *security domain management functions* manage membership and policies for security domains, which are groupings of resources with a common management object according to certain security policies. They allow secure interactions within the group or with other objects in a common way. Security domain management provides information to composition management about what resources are available and about associated policies and receives information about agreements made.
- The *composition management functions* govern the composition process through coordination with other control functions, negotiate with composing ANs and realize the established composition agreements. One specific type of composition agreement is a dynamic roaming agreement which can be used to establish trust between two operators with no previous business relationship and the automation of both the technical and commercial realization of the roaming agreement. This aims to remove the barriers often associated with establishing bilateral roaming agreements between network operators today, e.g. cost, time and technical knowledge.
- The *network management* is realized through two entities. The plug-and-play management deals with the processes related to autoconfiguration, like attaching a new element, component or network. The domain management accesses the policy DB in order to support the composition process in the management plane. It reads the policy DB before a composition process and runs a negotiation process to decide on management overlay types. Finally, it updates the networks policies based on the outcome of the composition.
- The *INQA*[1] *and SLA management functions* establish and maintain service level agreements on connectivity resources.

4.3.1.2 ACS Reference Points

The services and functions of the ACS are accessible through three reference points, which are also employed to implement the ACS's control tasks. This section provides a summary of the design and structure of these three reference points. It also presents the application of the

[1]Dynamic control of internetwork QoS agreements.

three reference points for the implementation of specific AN functions in order to visualize their roles in the overall service provision. The three reference points discussed are the Ambient Network Interface, the Ambient Service Interface and the Ambient Resource Interface.

The Ambient Network Interface connects the ACSs of different ANs (inter-ACS communication) as well as it might be used to facilitate communication between functions residing in the same ACS (intra-ACS communication). When interworking between ANs that belong to different authorities is to be ensured, conformance testing on well-defined interfaces is eventually required. Conformance testing is prepared during the design of systems by specifying potential conformance testing points, named reference points. The ANI is a reference point located either between different ANs or between functions belonging to the same ACS. In the former case, this reference point is made out of a set of interworking interfaces, which will be selectively used based on the purpose of the interworking between the ANs. When used between domains, the reference point is prescriptive, i.e. it mandates the way in which the objects providing the interfaces are built. When used intradomain, then the ANI can be declared to be descriptive not constraining any implementation.

The Ambient Service Interface encapsulates the connectivity and control functions for use by upper layer applications and services within a node operating in an Ambient Network. It allows applications and services to issue requests to the ACS concerning the establishment, maintenance and termination of end-to-end connectivity between functional instances connecting to the ASI. The ASI also includes management capabilities and means to make network context information available to the applications.

The Ambient Resource Interface is located inside a node between the ACS and the connectivity layer. It offers control mechanisms that the ACS can use to manage the resources residing in the connectivity plane. These resources can be routers, switches and radio elements (terminals, relays, access points), and also media transcoders, filters and proxies. The resources are accessible though an abstraction layer, which allows the ACS functions operating on them to stay technology independent.

Design and structure of the ACS interfaces

The ACS interfaces play a major role in the implementation of the Ambient Network services, which often are provided in cooperation between functions residing in different ACSs and networks. Figure 4.4 provides a typical example of application/service layer objects requesting a service from the ACS through the ASI. Depending on which functionality is requested, different functions of the ACS are invoked to implement it. The services requesting the functionality are though assumed to be unaware of the ACS structure. In the example depicted in Figure 4.4, the functionality requested requires the local ACS to contact functions residing in a second ACS. This network-to-network communication is facilitated by the ANI. Finally, the functions in the ACS use the ARI to control resources in their network to implement the requested functionality, e.g. the set-up of an end-to-end user plane connection.

The scenario in Figure 4.4 should be regarded as an example clarifying the role of the three different interfaces. The sequence of actions presented is though not binding and many other conceivable options exist. Functions in the ACS need not, for instance, be triggered by service requests received through the ASI. One can construct equally valid scenarios in which functions are invoked due to a message received through the ARI, e.g. indicating changes in the connectivity resources, such as the loss of a radio link.

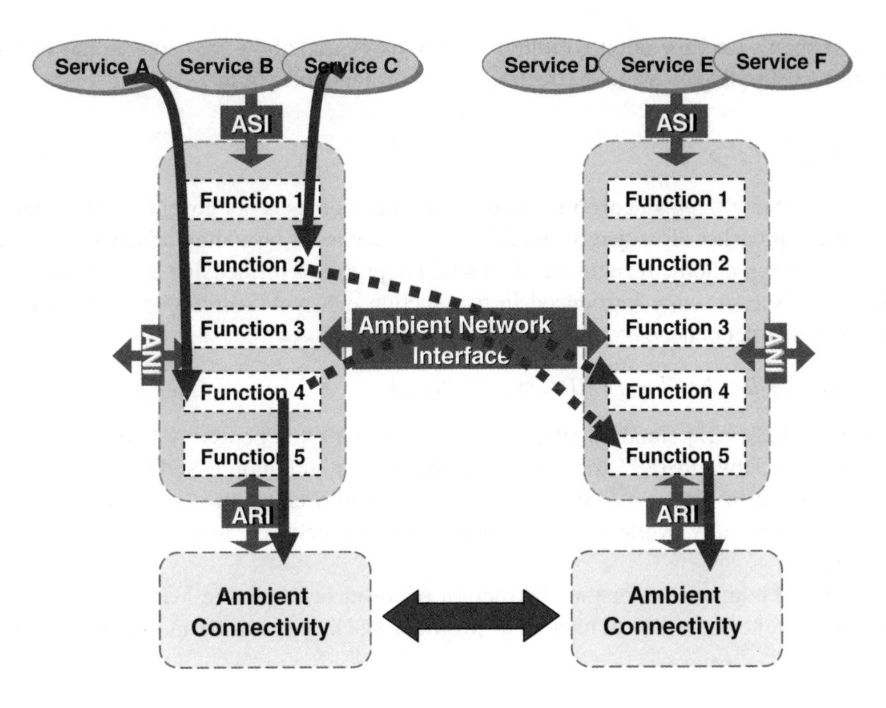

Figure 4.4 Role of the three AN interfaces in the provisioning of AN services

Apart from the basic roles these interface fulfil, a modular structure is intrinsic to all of these interfaces. As the ACS is supposed to be constructed out of cooperating control modules serving different control purposes, the AN interfaces are structured according to the control task division. The AN interfaces are thus to be regarded as reference points, each implementing a set of interfaces serving different control purposes.

It is envisaged that each particular implementation of an AN reference point may include a different set of interfaces depending on the functionality present in the ACS to which the reference point belongs. One can distinguish between mandatory interfaces that need to be present and optional ones that are implemented when required. The set of mandatory interfaces has not yet been exhaustively studied, but a few elements in this set have been identified. An interface offering basic security mechanisms to allow the secure attachment of network elements, including the support for a basic identity management, is mandatory. Also, an interface offering plug-and-play mechanisms to determine the set of optional interfaces implemented in a particular ANI is mandatory.

4.4 The Ambient Layer Model

The Ambient Control Space needs to interact with user plane resources, which might deploy various technologies and are accessible through diverse protocols. It is though one key requirement of Ambient Networks that the control functions gathered in the ACS remain as independent of specific network technologies as possible. To fulfil this requirement, Ambient Network has developed the Ambient Layer Model, which is discussed in this section. The goal is to enable interoperation with legacy technology and allow easy integration of future networking technologies without affecting the control modules in the ACS.

4.4.1 State of the Art

The idea of generalizing and abstracting user plane resources is not novel as such. A number of existing approaches already rely on this principle, but focus on comparatively narrow fields of application only. Some recognized abstraction approaches influencing the solution chosen by Ambient Networks are introduced in this section, whereas Section 4.4.2 introduces the Ambient Network connectivity abstractions.

Media-Independent handover (MIH) (IEEE 802.21)

The IEEE 802.21 [27] standard supports handover and interoperability between heterogeneous wireless networks (802 and non-802 networks) for both mobile and stationary users.

As illustrated in Figure 4.5, the standard defines an architecture that enables transparent service continuity while a mobile node switches between heterogeneous link layer technologies.

The central component for the abstraction of lower layers is the MIH function. It is a shim layer in the mobility management protocol stack of both the mobile node and the

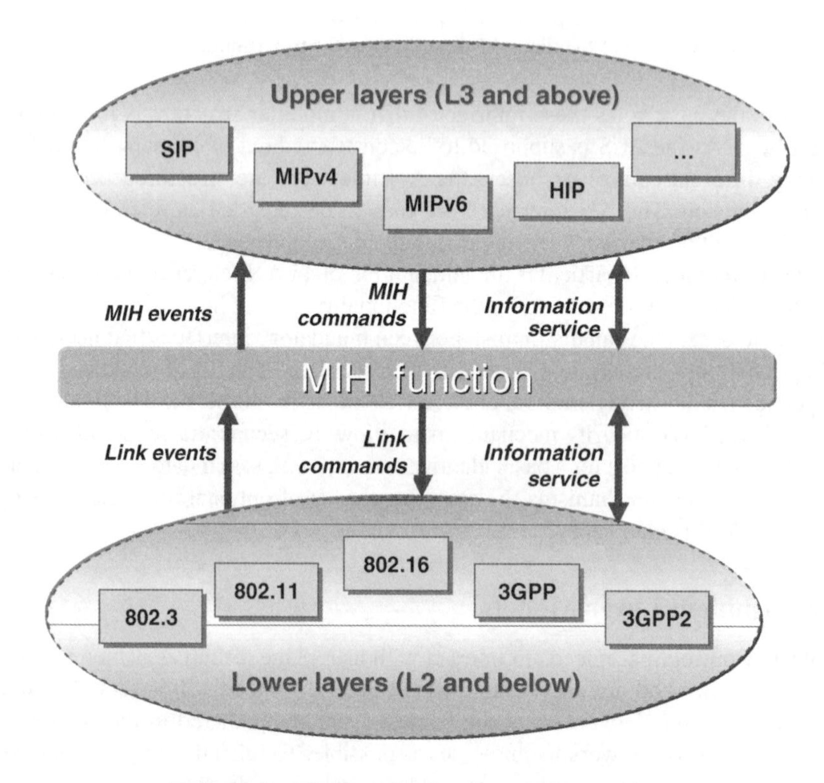

Figure 4.5 MIH function location and key services

networks. Upper layers select links and make handover decisions based on inputs and context received from the MIH function. It offers a unified interface to the upper layers and provides them with abstracted services. The MIH Event Service translates link events that originate from event sources below the MIH function to MIH events and propagates them to upper layers. The MIH Command Service translates MIH commands to link commands that can be understood by the lower layers (MAC and PHY). Examples for local commands are MIH Poll, MIH Switch, MIH Configure and MIH Scan; remote MIH commands could, for example, be MIH Handover Initiate, MIH Handover Prepare and MIH Handover Commit. In addition, the bidirectional MIH Information Service is used to propagate information needed to facilitate handovers across heterogeneous networks (for more information see [27]). The MIH function communicates with the lower layers of the mobility management protocol stack through technology-specific interfaces, which are already defined as service access points (SAPs) within the standards for the respective technologies.

H.248: Gateway Control Protocol

The ITU-T H.248 [28] (Gateway Control Protocol) and its IETF counterpart Megaco [29] are signalling protocols that enable switching of voice, fax and multimedia calls in the PSTN and in IP networks. The protocol defines a connection model, which describes the logical entities within a media gateway (device that interconnects two possibly disparate networks) that can be controlled by a media gateway controller. The main abstractions used are terminations and contexts.

A termination is a logical entity that sources and/or sinks media and/or control streams. It is described by a number of characterizing properties, which are grouped in a set of descriptors that are included in commands. The link between two terminations is not explicitly modelled; it is implicitly described by the properties of the terminations.

A context is an association between a certain a number of terminations. If more than two terminations are involved in the association, it also describes the topology and the media mixing and/or switching parameters. In other words, contexts are used to connect terminations within a media gateway. Terminations can be added to contexts using the Add command, they can be removed from a context using the Subtract command or they may be moved from one context to another with the Move command.

Generic functional architecture of transport networks (ITU-T recommendation G.805)

The generic functional architecture of transport networks [30] applies a technology-independent method of describing a transport network from the viewpoint of the information transfer capability. The abstract network description is provided using a small number of architectural components and applying layering and partitioning.

The architectural components are defined by the function they perform in information processing terms or by the relationships they describe between other architectural components. They are classified as topological components (layer network, subnetwork, link and access group), transport entities (connections and trails), transport processing functions (adaptation and trail termination functions) and reference points. Some diagrammatic conventions have been developed to support the descriptions.

With link partitioning, any subnetwork may be partitioned into a number of smaller (contained) subnetworks interconnected by links. A link is constructed by bundling a set of link connections which are the smallest unit of manageable capacity. Links may be partitioned into a set of parallel links (or link connections) or into a serial arrangement of link connection–subnetwork–link connection. The partitioned links may themselves be further partitioned recursively.

In addition, G.805 defines a layer network, which describes the generation, transport and termination of particular characteristic information. The transport network can be decomposed into a number of independent layer networks with a client–server relationship between adjacent layer networks. The layer networks identified in the transport network functional model offer the same service using a specific protocol (the characteristic information). Thus, they differ from the layers of the OSI model, where an OSI layer offers a specific service using one protocol among different protocols.

Functional architecture of connectionless layer networks (ITU-T recommendation G.809)

The ITU-T recommendation G.809 [31] describes the functional and structural architecture of connectionless layer networks independently from network technology. It is an extension of the methodology defined in G.805 providing a common framework for a connection-oriented layer network and a connectionless layer network.

The methodology described by G.809 uses a small number of architectural components and applies layering and partitioning (as classified in G.805). One of the main differences between the connection-oriented layer networks in G.805 and the connectionless layer networks in G.809 is that the former assumes a bidirectional transmission, whereas the latter assumes a unidirectional transfer of information only. ITU-T recommendation G.809 uses topological components (layer network, flow domain, flow point pool link and access group), transport entities (flows and connectionless trails), transport processing functions (adaptation and flow termination functions) and reference points. The associations of trail terminations (connection-oriented network) are defined by processes and depend on their changing connectivity. The associations of the flow terminations (connectionless networks) are defined on datagram basis.

Conclusions

The presented abstraction mechanisms are solutions for different application areas; they are tailored to the specific purpose they were developed for. The abstraction layer introduced for IEEE 802.21 is very focused on mobility issues. H.248 focuses on terminations in switching equipment rather than on the links connecting them. The G.805 and G.809 standards devised by the ITU focus on a functional description of transport networks, but they do not model the dynamic interaction and configuration of resources. For the application in Ambient Networks, all these existing approaches are not comprehensive enough.

This motivated the development of a new, more generic model for the use within Ambient Networks. As described later in Section 4.4.5, we do explicitly foresee that existing mechanisms are at least partly reused, in line with the Ambient Network vision of bridging and harmonizing access networks. How exactly this happens however depends on the particular case and is within the freedom of implementation for the abstraction layer.

4.4.2 Ambient Networks Abstraction Approach

Ambient Networks fulfil the requirement of remaining independent of specific network technologies by a generic abstraction of the user plane. Applications shall not be aware of all details of the underlying connectivity. Through the use of an abstraction mechanism, it can be avoided that they need to implement different mechanisms to set up and maintain end-to-end connections in different connectivity technologies. Applications simply create an abstract connection entity that fulfils their given communication requirements. ACS functions are then responsible for setting up and maintaining this end-to-end connection across any given connectivity technology. The abstraction model also ensures that the implementations of control space functions themselves can remain as generic as possible.

The Ambient Networks connectivity abstractions define different views on the underlying physical connectivity that provide different levels of detail. They present a generic, technology-independent view of the underlying network connectivity to ACS control functions. The control space functions interact with network resources through the technology-independent ARI without knowing the implementation details when they, for example, set up or configure network resources.

The actual transfer of data over these resources is referred to as flow. In addition to that, the abstraction model in Ambient Networks provides a higher-level view of the connectivity to the application, service or user interacting with the ACS via the ASI: The bearer is the communication primitive provided on this level. It hides the implementation of the connectivity and provides only an end-to-end view.

The following sections describe the different abstractions and the relation to the Ambient Networks naming framework. In addition, implementation issues are explained and examples of applying the abstraction model are provided.

4.4.2.1 Flow Abstraction

Following a very generic approach, a network can be modelled as a set of nodes that is interconnected by links. Almost all types of networks, regardless if they use circuits, packets or other data transport mechanisms, can be described in this general way. The sequence of nodes and links that data is traversing when it is transmitted between two nodes is denoted as a path. This generic view of a network forms the basis for the abstraction model used in Ambient Networks.

The control functions of the ACS, however, do not operate on the detailed level of links, nodes and paths. The flow abstraction exposes only a subset of these entities. As illustrated in Figure 4.6, only some of the entities from the connectivity level are shown on the flow level. The control space functions operate on a common representation of the underlying network that is visible through the ARI and can also be influenced through the ARI. The connectivity graph resulting from that (illustrated by the cylinders and dark lines in the bottom level of Figure 4.6) forms the basis for all connectivity-related operations. Obviously, not all of the connectivity resources need to be used at a given time, which is illustrated by the chequered cylinders not currently carrying a flow.

A flow is the transfer of data between two flow endpoints. It is constrained to a single network technology and defined as being unidirectional. Flows transfer data transparently with certain performance characteristics, which may include the required level of integrity

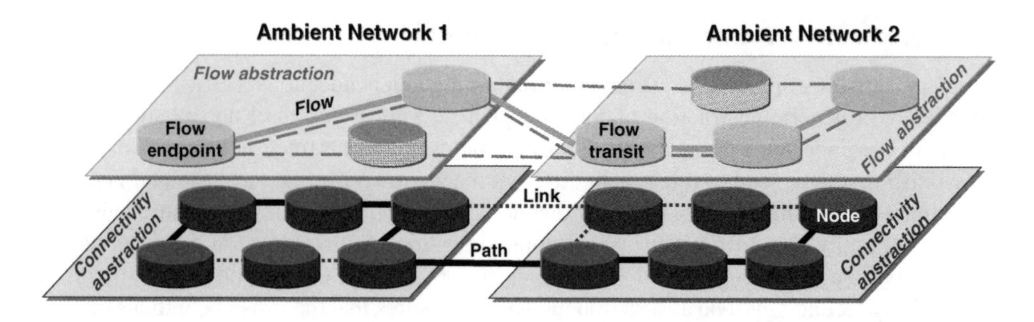

Figure 4.6 Illustration of the flow abstraction and connectivity plane

of the data. A flow is at least associated with a specific source locator and destination locator that are tied to the flow endpoints. In certain cases, an additional description that picks out a subset of the data with a given source/destination locator pair may be needed. This could, for example, be a QoS classification or a UMTS bearer ID.

This unambiguous identification is required in order to ensure that the ACS functions have a common means of referring to flows. For communication between different functions referring to this common flow naming is mandatory, whereas the internal mechanisms of the functions may also deviate from this naming where needed. Whether an additional, entirely technology-independent flow ID is required is left for further study.

A flow may pass through intermediate resources ('flow transit' in Figure 4.6), which are not explicitly tied to the flow, but which can be controlled through the ARI. The set of intermediates may change over the lifetime of the flow without changing the flow itself. The flow may also pass other nodes not visible, and thus not controllable, through the ARI.

For some types of network technologies (such as connection-oriented technologies), a flow may require a connection set-up, but for other types (such as packet switched technologies) that may not be necessary. The control space may use the control and configuration capabilities of the ARI to request certain treatment of the flow by the connectivity resources. Flows transfer data transparently with certain performance characteristics, which may include the level of integrity of the data. Mobility of a data transfer requires a flow to be modified (as locators are modified) or a new flow to be created and the old one deleted.

4.4.2.2 Bearer Abstraction

The applications or services operating on top of the ACS are however not aware of details like the flows that are used to transport the data. They operate on an even higher-level view of the connectivity referred to as bearer. The bearer is an end-to-end communication mechanism that is offered at the ASI. It hides the implementation of the connectivity and provides the end-to-end transport service applications require.

The bearer endpoints, unlike flows, are not bound to locators, but to higher-level objects in the naming framework. The control space functions manage the mapping of bearers to flows and potential updates of this mapping, e.g. in the case of a mobility event. They are capable of constructing a bearer from a set of concatenated flows using the name resolution capability of the ACS naming framework. Multiple bearers may be mapped to the same flow, i.e. flows

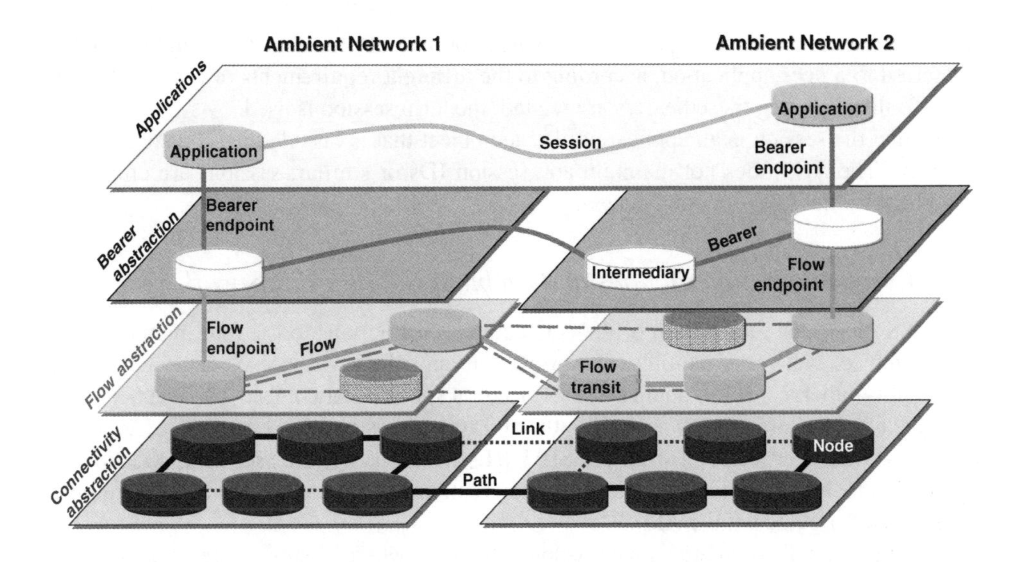

Figure 4.7 Complete illustration of the connectivity abstractions

can multiplex bearers. In turn, a bearer can also be distributed among multiple parallel flows, e.g. in a multi-access scenario.

In a simple case, a bearer provides little more than what a flow provides, which is often sufficient for relatively basic services like a best-effort file transfer. For other applications, like the service specific overlay networks (see Chapter 10) developed in Ambient Network, a bearer can also be more sophisticated. In such a case, additional functionality provided by the ACS functions can be dynamically included to enhance the bearer, e.g. by providing QoS reservation, mobility or customized media manipulation and routing capabilities. Nodes implementing media manipulation functions as well as other entities (e.g. a function providing VPN encryption/decryption or spam filtering) the bearer passes through are then modelled as bearer intermediaries (see Figure 4.7).

When special treatment is needed at a bearer intermediary, the bearer is mapped to at least two flows – one from the source to the intermediary and the other from the intermediary to the destination. The name resolution support of the naming framework controls this mapping of a bearer to a set of concatenated flows between the intermediaries.

The service specific overlay network function is currently the only type of bearer defined in Ambient Networks. Work on an additional, simple best-effort bearer is planned for phase 2 of the project. Please also note that the bearer is currently defined only for unicast; bearer types supporting multicast or broadcast are for further study.

4.4.2.3 Application Sessions

On top of the bearer abstraction, applications may define sessions that combine multiple bearers into application-defined transport entities. For example, an application could decide to request different bearers for data (e.g. file sharing or an FTP transfer), voice and video data transferred to a peer application, according to the different requirements of each of the data types. To illustrate that these bearers are related, the term session is used.

Note that the session is an application-defined object that is outside the scope of Ambient Networks. The ACS does not maintain any session IDs or similar; sessions are only mentioned here for the sake of completeness.

4.4.3 Connectivity Abstractions and Ambient Naming Framework

Figure 4.8 illustrates the relation of the connectivity abstractions to the four main elements of the Ambient Networks naming framework. An application service or data object is an entity that is either a specific application service or a specific data object. The identity of the object persists over time and is not tied to the end system hosting the service/data. Examples of names in this layer include HTTP and SIP URLs. These names are used by applications as references which are resolved and result in the creation of application sessions utilizing the bearer service provided at the ASI.

The point of application attachment denotes the point where clients can reach an application program that implements parts of an application service. This point is located at the ASI and can be compared to a standard TCP/IP socket API. A bearer endpoint is instantiated at a point of application attachment when a bearer set-up is completed.

The network topology independent identification of a node in an Ambient Network is referred to as host end system, which does not necessarily need to be a physical entity – it may also be a logical entity that can move between different physical entities. The host end system is the entity 'hosting' the ASI. Cryptographically generated identifiers present an example of secure, self-assigned host end-system identities. The host end-system abstraction makes address/protocol translations on the locator level invisible to the point of application attachment.

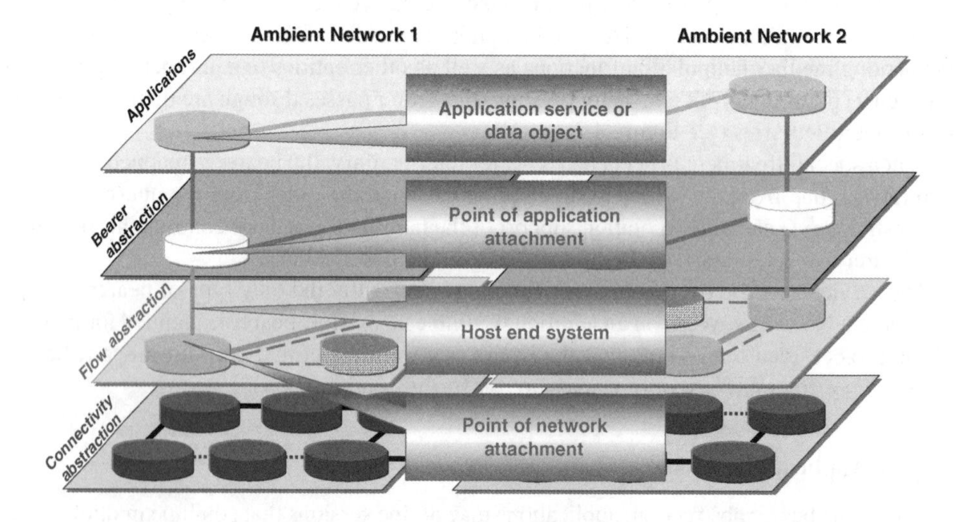

Figure 4.8 ANs connectivity abstractions and their relations to the naming objects

Finally, the point of network attachment defines a location in the network. The location is identified by a locator, which is a routable address in the underlying network technology, such as an IP address. This point is exposed to the ACS via the ARI. A flow runs between two network points of attachment. As already pointed out, each flow is restricted to a single locator domain. Thus, bridging or translation may be needed at the locator domain boundary in order to support communication between the nodes in the Ambient Network. This translation is best managed by mapping of host end-system identities to locators in each locator domain passed.

The requirement to support bridging raises the issue of how to handle the routing between networks. This routing function requires the identification of Ambient Networks as first class objects, similar to the autonomous system numbers used in the Border Gateway Protocol (BGP).

The Ambient Networks naming framework defines dynamic bindings between the identifiers used at the different levels of the connectivity abstractions. With such dynamic bindings at multiple levels, names of objects become location independent and support for mobility can be provided at all levels.

It is the task of the Ambient Control Space to manage these bindings and to provide resolution mechanisms that map a name of an object in an upper layer into a lower layer identifier of the corresponding object. For example, the procedures for updating the bindings for ongoing communications are defined by the handover and locator management functional entities in the ACS.

The naming framework supports the notion of indirection or delegation. It is an extension to the dynamic bindings that enables more advanced mobility schemes and the explicit control of so-called middleboxes, such as network address translators (NATs), firewalls and transcoders.

The concept of indirection also includes the possibility to let the location of an object be an object of the same kind. That is, it does not restrict the binding to an object one level below, and also allows bindings horizontally within one level. The upper dashed arrow in Figure 4.9 illustrates the use of indirection. One important application is to enable efficient mobility mechanisms for moving networks. A node in the moving network binds its location to a designated gateway node. Only the latter node needs to update its binding to a new network location as the moving network moves.

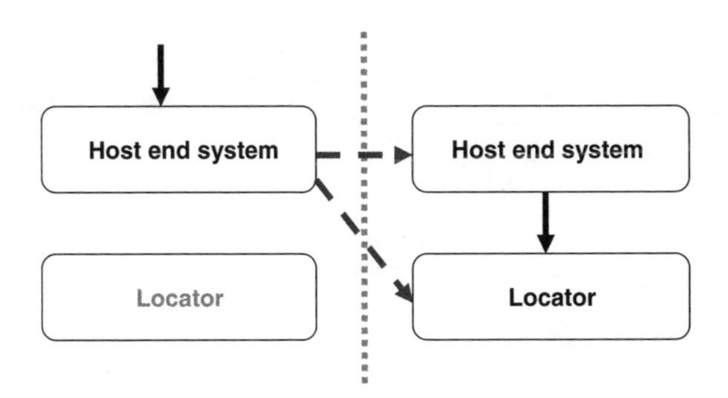

Figure 4.9 Indirection and delegation

4.4.4 Interaction Between Control and User Planes

To further illustrate the relation of control and user plane, this section introduces the outline of a typical bearer set-up process. The bearer set-up is initiated by the application through a bearer set-up request made at the ASI. In a simple case, this request could contain only a destination address. In more advanced scenarios, additional details like bandwidth requirements or the request for a special media processing can be included in the request. The control space functions use the abstract topology and capability information maintained in the ACS to determine which flows could be reused or would need to be set up in order to construct the bearer, also taking into account certain intermediaries the bearer needs to traverse, e.g. for media processing. As illustrated in Figure 4.10, commands issued through the ARI are used to configure the connectivity resources when needed. In addition, the overlay and bearer management may issue commands to bearer intermediaries over the ARI, e.g. to configure a transcoder. If a requested functionality is not available in the own Ambient Networks, the ACS functions can also use the ANI to negotiate with other networks to make additional resources like transmission bandwidth or more advanced processing facilities available.

Once all these steps are completed, a bearer endpoint is instantiated at the point of application attachment (e.g. an extended 'socket'), which can now be used by the application or service to transmit data to its peer entity over the user plane. It is important to note that application-level signalling traffic, such as RTSP, is also transmitted through the bearer.

4.4.5 Implementation Considerations

The way the abstraction from the actual network topology is implemented plays a crucial role for the proper operation of the ACS. Although the implications of different abstraction

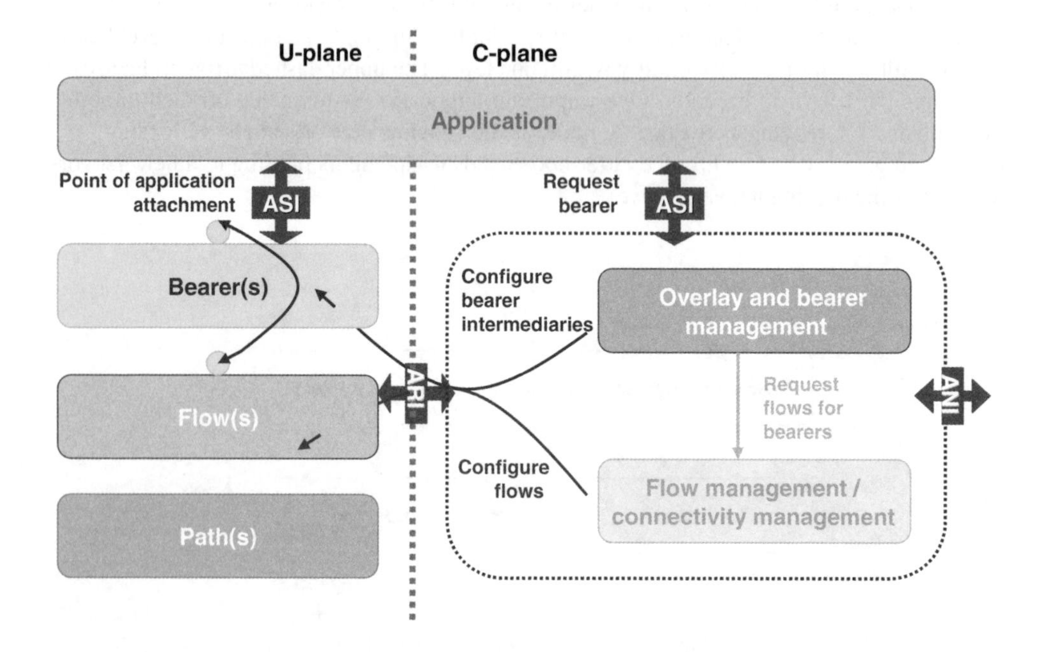

Figure 4.10 Illustration of the interactions between control (right) and user (left) planes

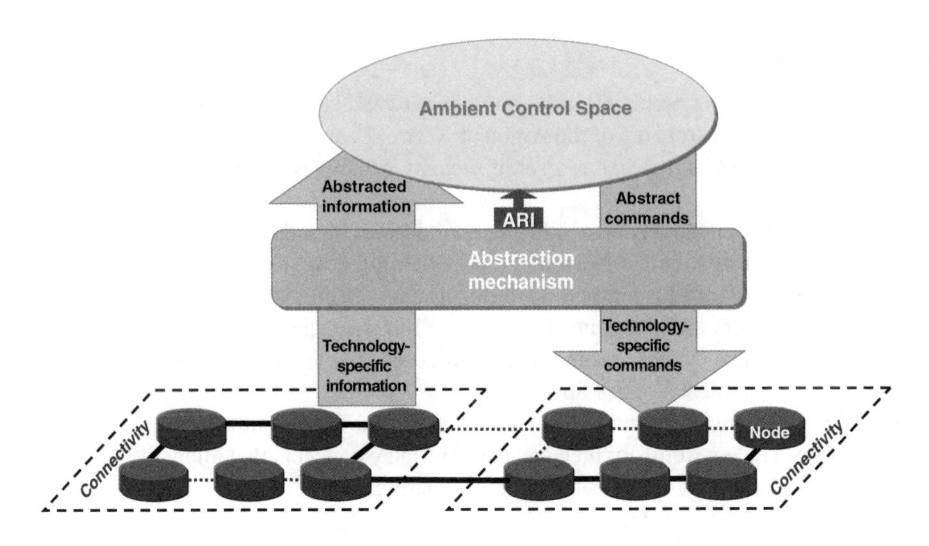

Figure 4.11 Illustration of abstraction principles

mechanisms have not been studied in detail during phase 1 of the project, a number of general characteristics and requirements can be stated.

As illustrated in Figure 4.11, the abstraction must work in two directions. It needs to be capable of providing data about the network topology and capabilities of the network elements, e.g. using a description based on graph theory. In the other direction, the abstract commands issued by the control space functions have to be translated into concrete, technology-specific commands and procedures. This may also include the interaction with legacy control mechanisms in the user plane.

Thus, the abstraction mechanism is very specific to the underlying network technology and therefore needs to be developed or at least adapted for each type of network. As pointed out in Section 4.4.2, we foresee that these specific abstraction layers build on and extend existing mechanisms wherever possible. This approach guarantees that Ambient Networks are able to support each existing future network technology as long as it fits the very general assumptions on the properties of nodes and links.

Different control space functions may require different levels of detail on the flow level. Therefore, the current assumption is that a recursive modelling of flows may need to be considered when implementing the abstraction mechanism. In such a model, a flow can also more easily be defined as a combination or concatenation of other flows. A situation where the Ambient Network multi-radio access is used could be an example where this modelling can prove useful.

One example that would allow for different modelling approaches could be the case of two applications using, e.g. the RTP protocol (Real-Time Transport Protocol) for the exchange of data over an IP network. Although one application is working on an IPv4 network, the peer application is working on an IPv6 network. Obviously, a protocol translation between the networks is required. In some cases that may be entirely transparent to the ACS (e.g. encapsulation of the IPv4 address in the IPv6 address), in other cases this would need to be explicitly controlled and configured over the ARI. In the former case, the point at which the protocol translation takes place would likely be modelled as a flow transit or not shown at all in the abstraction, whereas in the latter case this would likely be regarded as a flow endpoint.

4.5 Summary

The Ambient Networks architecture targets the network layer of future telecommunication networks. The key constituents of the Ambient Networks architecture are the Ambient Layer Model, the Ambient Control Space, a set of security principles and the network composition procedure.

The Ambient Layer Model defines the upper and lower boundaries of the networking functionality embraced in Ambient Networks and is delimited by a pair of interlayer interfaces, the Ambient Service Interface providing uniform access to the Ambient Networks functionality from higher layers and the Ambient Resource Interface encapsulating the capabilities of the underlying infrastructure and making them accessible to the network control functions embraced in the Ambient Control Space.

The ACS is the environment within which a set of modular control functions can coexist and cooperate. The environment includes plug-and-play concepts that allow the ACS to bootstrap and discover the set of present functions dynamically.

5

Security in Ambient Networks

Acknowledgements

This chapter is based on the joint experiences and efforts of the researchers in the first phase of the AN project and particularly the following people listed as contributors and authors (i.e. in alphabetical order): Jari Arkko (Oy LM Ericsson AB), Pasi Eronen (Nokia Corporation), Rainer Falk (Siemens AG), Michael Georgiades (University of Surrey), Seppo Heikkinen (Elisa Corp.), Ian Herwono (British Telecom Plc.), Günther Horn (Siemens AG), Keith Howker (Vodafone Group Services Limited), Mattias Johansson (Ericsson AB), Rieks Joosten (KPN/TNO), Olavi Karasti (Elisa Corp.), Geert Kleinhuis (KPN/TNO), Florian Kohlmayer (Siemens AG), Mika Kousa (Nokia Corporation), Julien Laganier (DoCoMo Euro-Labs), Tim Leinmüller (DaimlerChrysler), András Méhes (Ericsson AB), Daniel Migault (France Telecom SA), Anand Prasad (DoCoMo Euro-Labs), Mark Priestley (Vodafone), Peter Schoo (DoCoMo Euro-Labs), Göran Selander (Ericsson AB), Kristian Slavov (Oy LM Ericsson AB), Hannes Tschofenig (Siemens AG), Tseno Tsenov (Siemens AG) and Alf Zugenmaier (DoCoMo Euro-Labs). This chapter has been edited by Alf Zugenmaier, Michael Georgiades and Peter Schoo.

5.1 Introduction

Developing a future mobile/wireless communications system that will be heterogeneous in nature and work in a seamless manner induces to consider several security problems. The vision of the Ambient Networks project is a world where many different types of networks can work together in a seamless manner to provide networking services to users. In order to realize this vision, it is important and essential to consider different types of business and interaction models, including traditional operator models as well as cooperating individuals and ad hoc networks. This perspective reveals some of those assets that are to be protected with appropriate security solutions. It is also necessary to support communication with individual hosts as well as entire networks. A heterogeneous link layer technology shall be used, which enables communications to continue across changes in the link layer or location at the IP network. Solutions should provide dynamic capability negotiation and resource allocation and allow applications to be aware of their context. Finally, the solutions should be efficient in terms of their bandwidth usage and latency.

Ambient Networks: Co-operative Mobile Networking for the Wireless World Norbert Niebert (Ericsson GmbH), Andreas Schieder (Ericsson GmbH), Jens Zander and Robert Hancock

These scenario characteristics throw light on the challenges for the security work within the project. Besides the fact that some of the technologies used are not yet available, traditional measures are applicable only to a limited extent: First, technology-specific threat analysis methods cannot be applied, as technical details are yet missing during the research phases, which is to be balanced with technical know-how, expertise and experiences. Second, security measures are often set in relation to the assets that are to be protected, aiming to ensure that the trade-off between the efforts spent and the remaining risks is well balanced. In research work, such effort estimations can often not be applied, as application fields are innovative or values that help selecting suitable solutions do not yet exist. Again, it is only expertise and experience that can provide some guidance.

Two main principles were followed throughout the development of the results presented in this chapter. Based on past experience, it has been shown – all too often – that retrofitting security leads to suboptimal results. Accordingly to this, the Ambient Networks project agreed to seek for adequate security solutions in all major elements that define the Ambient Networks architecture. The second main principle followed is that existing technologies were preferred over those solutions that will require new development and employment. Only if a clear need is identified, which cannot be satisfied with existing technology, then the design of new security mechanisms was to commence.

A subset of the problems considered during the first phase of the project work, together with the related solutions that were elaborated, is presented in this chapter. Comprehensive and more detailed descriptions for all can be found in the publicly available project documentation [33].

There are many challenging security problems, related to managing the risks associated with the Ambient Networking technology, some of which are addressed in this chapter. Hence, the following contains solutions and results from key problem areas of security, privacy and security (trust) relation management using two complementary approaches: top-down and bottom-up. In addition to securing the new and attractive features provided by Ambient Networks, a major objective is also to avoid the known security vulnerabilities of present architectures, thus making Ambient Networking a competitive communications technology from a security perspective.

Therefore, the top-down approach produced a set of guidelines and common building blocks. These were derived from a comprehensive requirements analysis and the guidelines should be followed by the specific technical solutions. The four identified major concepts are identifiers, security (trust) relations, authorization and security by default. The latter includes a security toolbox which includes commonly used and existing mechanisms. Ambient Networks should have cryptographic identifiers. These cryptographic identifiers could be ephemeral, to enable privacy, or long-lived, where privacy is not a concern. These identifiers should integrate with legacy technology where necessary. The need for cryptographic identifiers comes from the requirement to be able to bind specific authorization rights to an identifier. Authorization rights should be granted explicitly and at a fine granularity [32]. The process for granting authorization rights fundamentally depends on the security (trust) relations between the involved entities, i.e. direct, brokered and no security relation. To address the requirement of zero configuration, the concept security by default is introduced. It means that security needs to be incorporated right from beginning in design and deployment, as well as being always present, automatic and activated during operations. This is a fundamental part of the security architecture because previous approaches have shown that attacks on a system that exploit basic principles of the system architecture cannot in general be mitigated by applying security *a posteriori* to an already completed system design, less than by redesigning the whole system. To ensure the appropriate architectural principles for Ambient Networks, the AN security architecture is an integrated part of and developed in parallel with the AN architecture. The presented principles and guidelines were applied to and validated for particular key problems.

The bottom-up approach consisted of the identification of key technical security problems, which needed to be solved for an Ambient Network. The key security problems discernible at this point in time were tackled and extensively investigated. The developed solutions for these key problems ranged from the analysis of state of the art, identification of future work, investigation of the new dynamic Ambient Network scenarios, highlighting of suitable existing protocols (and combinations thereof) and the development of new protocols that advances the state of the art when needed.

The structure of this chapter is as follows: Section 5.2 gives an insight to the problem scope addressed. Section 5.3 explains the architectural aspects of security in Ambient Networks. Finally, Section 5.4 presents the key problems for future mobile and wireless communication systems that were identified, i.e. a new and improving network attachment procedure, multi-hop access solution, specific cases that can support composition but have not been discussed elsewhere in this book, secure session handover and mobility procedures, GANS security solutions and finally generic link layer security. This chapter ends with the conclusions and some outlook.

5.2 Security Problem Space in Ambient Networks

Security problems that have been considered while developing the main concepts for a future mobile/wireless communications system are discussed in this section, along with some basic assumptions that have guided the work. The security architecture presented here attempts to ensure that these requirements can be met. In particular, the results [33] concern the security of the following functions:

- Link layer communication, including signalling related to control of radios, etc.
- Network attachment procedures.
- IP attachment procedures.
- Procedures related to multi-hop ad hoc networks.
- Composition and roaming procedures.
- Internetworking across multiple administrative and technology domains.
- Quality of service negotiation.
- Session handover and mobility mechanisms.
- Secure and authorized access to context information.
- Secure group communication.
- Management of authorized use of resources in Ambient Network.
- Resistance against attacks and coordination of response.
- Backward compatibility with existing networks.

This covers essentially all functionality of the link and network layers, as well as some support for context awareness of higher layers. This is consistent with the approach to provide a complete system-level view to the functionality, security and efficiency of Ambient Networks. Transport and application layer issues are considered out of scope for this project, however.

By link layer communication, all the procedures associated with the establishment and control of radio communications between our devices are understood, as well as the actual payload packet protection over the wireless media. This function covers all scanning, beacon, link layer connection establishment, transmission power control and handover procedures. Note that unlike in traditional wireless communication networks, Ambient Network assumes the ability to use heterogeneous technology as well as new types of business models and participating entities. This implies, among other things, the ability to perform operations such as power control or handovers across technology and administrative borders [34].

By network attachment procedures the procedures required for implementing network access control are meant, such as those currently present in UMTS [35] or 802.11/1X/11i standards [36]. In contrast to traditional technology, however, Ambient Networks attempt to provide a more flexible set of business and cooperation models. This implies, for instance, the ability to provide more information about the available networks for network selection decisions or the use of nonsubscription-based control models, such as credit cards [37].

In order to be able to communicate, the parties involved in network attachment need to establish IP communications. In current networks, this occurs typically after network attachment procedures, but it is desirable to perform this during the network attachment procedure, as this makes it possible to perform all signalling over IP. The establishment of IP communications involves finding suitable routers, network prefixes, etc. These are existing procedures that currently have (some) security issues that may deserve to be addressed. More importantly, the existing procedures have a significant negative impact on handover latency, which implies new procedures would be desirable.

Traditional network attachment procedures have also concentrated on providing access to a single host. In Ambient Networks, the term composition procedure has been used to generalize this into complete networks attaching to each other for some purpose. For instance, a personal area network (PAN) can attach to a vehicular area network (VAN) or a vehicular network can attach to a public access point. This is similar to traditional network attachment, but needs to consider the different roles of participants, such as attaching to a network in order to use its services and attaching to it in order to allow others to use those services through you [38,39]. One part of the composition is roaming procedures and especially the automated procedures for setting up roaming have been worked on.

Another difference to traditional attachment procedures involves multi-hop ad hoc networks as extensions to public access networks. Here, a set of willing relay nodes needs to be found in order to gain access to the public network. Depending on the scenario, this may involve finding a compensation for the services of the relay nodes, even when the client and the relay nodes have no prior relationship.

Session handover and mobility mechanisms ensure that communications can be continued even across change of topological location at the IP layer [40,45]. This also includes the ability to use several attachment points simultaneously. The primary security problems with these mechanisms are related to the ability of attackers to make illegitimate requests to move someone else's communications to them or the use of these mechanisms to bomb other parties with amplified traffic streams [46]. Existing mobility mechanisms can prevent these attacks, but often suffer from performance problems related to the necessary security procedures [47].

The Internet consists of independent networks that belong to different administrative domains. Within the last ten years, fundamental discrepancies have developed between different domains, including the use of address translation and other mechanisms that impede end-to-end communications. Many of the existing attempts at improving the architecture have been either small, incremental improvements or patches that do not fit very well in the overall architecture or larger attempts, as in the case of IPv6, that have not been successful or that imply new barriers in end-to-end communication, such as address translation. In the Ambient internetworking architecture, multiple technologies, address domains and various middleboxes are first-class components of the architecture and not just nuisances that need to be worked around.

Networks can perform quality-of-service signalling in order to establish priorities and reservations suitable for the particular user and application. This signalling takes place either on the physical wireless link or through a network path. Securing the latter type of signalling is hard, as it involves communication with a number of previously unseen entities such as routers in the path.

Context awareness procedures help to ensure that participating Ambient Networks, nodes and applications, are aware of the type of communication services available. For instance, a transcoding device may assist a movement to a lower-bandwidth wireless interface. This device shrinks a given stream to a more suitable size [48]. In addition, the broadcast and multicast functions offered by the AN architecture need to be secured. The work on group security has been concentrating on broadcast encryption, and applications of this in the Smart Multimedia Routing and Transport (SMART) work.

Ambient Networks need to be resistant to various forms of harmful traffic and attack in the Internet. Most interesting from an internetworking perspective are denial-of-service attacks on individual hosts, networks or distributed infrastructure. Making authorization a key concept, Ambient Networks also need to address the management of authorization of the access and usage of the devices included within them.

The ability to work with existing networks in a backward-compatible manner is also necessary. This implies an ability to support legacy techniques for communicating between Ambient and other networks. As an example, the work done regarding mobility and moving networks in the Ambient Networks project expects that several different mobility mechanisms need to be supported. Another implication is that architecture and security models need to be able to have the flexibility to support such mechanisms. For instance, a flexible authentication framework is able to support both legacy credentials, such as SIM cards, along with newer types of credentials.

This leads to the summary of the underlying assumptions that were mainly based on items like resources of AN nodes, network layer protocols, network structures, etc. [49]. They can be categorized into two subsets, namely security-related assumptions and general assumptions. The latter cover basic topics such as the availability of resources or the used network layer. The general assumptions are

- Regarding network organization, two different types of networks are assumed: managed networks and ad hoc networks. Both types are coexisting and are able to cooperate, i.e. communicate with the other network type.
- Given today's view on networks, it is assumed that the network layer is IP based (IPv4 or IPv6).
- Furthermore, in order to deploy security mechanisms, a certain amount of bandwidth overhead is acceptable, but on the contrary, round trip times are expensive, especially to the home network, and should therefore be kept as small as possible.

Further, security-related assumptions are

- Cryptography is inexpensive with respect to computation, storage and power consumption. However
  - Asymmetric cryptography is a venue for denial-of-service (DoS) attacks.
  - Cryptography might restrict the range of devices, i.e. sensor networks or legacy networks are represented by their gateways; routers and middleboxes must be able to handle high throughput with low latency.
- A basic set of security functionality should be enabled by default and it should be working automatically, e.g. without user interaction.
- Nesting of cryptography is acceptable, i.e. transferred information and data might be encrypted several times by different entities.
- In case there is no better option, opportunistic security is better than no security at all (leap of faith).
- It is always possible to add authentication to anonymous entities, the reverse is hard (monotony of identification); hence, design for privacy is to add authentication.
- There is no single global authentication authority, i.e. PKI.

The next section presents the security aspects in Ambient Networks from an architectural point of view.

5.3 Security Architecture

Ambient Networks take a systems view on networks. As such, the security of Ambient Networks also has to consider the whole system. This is different from the approach taken by a lot of security work dealing with security of one particular aspect only. The approach taken by this work is two pronged: a top-down approach defines basic elements of the security architecture – the rules and principles that should be adhered to in designing Ambient Networks. In addition, a bottom-up approach shows how individual problems are solved securely applying these principles.

Architecture is defined as a set of rules, or principles, that characterize the structure of a system and the interrelationships of its parts. In this section, we study AN security architecture principles, i.e. the subset of rules that directly relate to security. This section is an abridged version of the security architecture considerations that can be found in [33].

Ambient Networks takes the approach that the security architecture needs to be an integral part of the overall AN architecture. This way it should be possible to avoid mistakes and clumsy workarounds that may be the result of adding security as an afterthought. However, defining security without knowing the full spectrum of functionality and limitations provided by Ambient Networks turns to be very difficult and can lead to shopping cart mentality as described in [40]. Therefore, instead of defining a functional architecture, we propose general principles. These make it easier to secure functionality if the design of this functionality follows these principles. Just as AN functionality is evolving, requirements for the security architecture may also evolve. We assume that the principles we propose here will remain valid.

However, further principles may be required. The principles proposed here are an outcome of our work in the first phase of the AN project summarizing our experience from the top-down approach.

The main principles that build up the security architecture relate to secure identification, security relationships, authorization and security by default. They are discussed in the following subsections.

5.3.1 Secure Identification

This section provides a background on identifiers, in particular cryptographic identifiers, and analyse where they should be used.

5.3.1.1 Identifiers

Entities and identities are both real-world concepts, which exist independent of AN. Identifiers, on the contrary, can and will be defined in AN to provide a name for a particular entity in a particular context. An entity can have multiple identifiers, not necessarily of different types, and some of these identifiers can be ephemeral (i.e. used/valid only for a short time) to allow privacy.

Within the Ambient Networks concept, an identifier is simply a name used to identify an object; thus, the words identifier and name are used as synonyms. The syntax and format rules of (a type of) names are called a namespace. A name system extends the namespace by usage rules, allocation rules and a resolution mechanism/infrastructure, the latter for mapping names in the name system to names in other name systems.

Namespace

Although the security consequences of the syntax and format of names may not be immediately apparent, their privacy aspects, cryptographic properties and uniqueness deserve attention.

As for cryptographic properties, some identifiers, such as public keys, have a well-defined intrinsic mathematical structure useful for cryptography, whereas most others do not and need to be augmented by additional, usually secret, data to enable cryptographic operations.

Allocation rules

Security-wise two main types of name systems must be distinguished, namely those with and without a designated naming authority. In name systems with a designated naming authority, the authority can be considered to own all names in the system, and an authorization scheme can be used to transfer ownership, delegate usage rights or otherwise grant permission to an entity (or group of entities) to use a particular name. Essentially, the naming authority acts as a trusted third party (TTP).

In a name system where entities are allowed (or required) to allocate their own names, only statistical uniqueness can be guaranteed and the proof of ownership must rely on the names themselves. In other words, the names should have intrinsic cryptographic properties, and a collision detection/recovery procedure also needs to be defined.

Usage rules

It is recommended to apply fixed rules for Ambient Networks that generally apply to all names and a naming authority whose jurisdiction is restricted to authorizing the use of names pursuant to these predefined rules. This has the benefit of a unified approach to the different name systems, as in systems, where the identifiers are independently self-allocated, the general usage rules can only be predefined. In either case, only enforceable usage rules should be included.

Resolution mechanism

Resolution mechanisms concern the interactions between different name systems, providing a 'translation' between the respective names. Thus, they are not strictly a component of any single name system. As different name systems generally name different types of entities, most resolution mechanisms map from one entity to another entity, rather than between different identifiers of the same entity. For example, the Domain Name Service (DNS) resolves a fully qualified domain name (FQDN) of a host to an IP address, the name of its network location, and the Address Resolution Protocol (ARP) in turn resolves the resulting IP address to an L2/MAC address, the name of the host's physical interface.

A secure resolution mechanism requires proofs of ownership for all the names involved and/or a statement authorizing the binding of these names. How this authorization is produced will depend on the name systems involved and the exact nature of their interactions.

Connection to authorization

Authorizations by a naming authority beyond providing a means to prove ownership of a name are possible, but undesirable. Proof of ownership simply establishes the fact that the entity presenting the name does actually 'own' its identifier. Further details on authorization are available in Section 5.3.2.

5.3.1.2 Cryptographic Identifiers

The majority of typical identifiers in widespread use today do not come with any guarantees, e.g. IP/MAC addresses may be spoofed, usernames replaced, etc. To get some assurance from an entity presenting a particular identifier, one could require the claiming entity to also present some sort of proof of ownership, typically involving cryptography.

5.3.1.3 Identifiers in Ambient Networks

Name system syntax and semantics may dictate the form and usage of certain identifiers, e.g. an IPv4 address is 32 bits long and implies routability, whereas a HIT is 128 bits long and comes with built-in cryptographic properties, but lacks any structure that would facilitate routing. Human-readable identifiers are also important and necessary for user-centric communications. Apart from the above constraints, cryptographic identifiers are not only suitable, but also necessary to provide adequate levels of security. A number of secure identification schemes where a TTP guarantees the binding between an identifier and its associated secret are generally applicable, as they do not restrict the form and interpretation of the identifiers used. Cryptographically generated identifiers (CGI) can be applied directly in machine-to-machine or even application-to-application contexts, but they may need to be adapted to meet the constraints specified above. On the contrary, the need for such an adaptation is arguably minimal, as human-readable (or otherwise restricted) names can be securely bound to cryptographically generated identifiers. Starting from the lower layers then, we apply the above analysis in the following subsections. As a conclusion of our treatment of identifiers, we note that secure dynamic bindings between the different identifiers are a prerequisite for many of the promises of Ambient Networks. For example, automated dynamic agreements, various forms of mobility and also privacy all benefit from cryptographic identifiers.

Interfaces

Interface identifiers (i.e. L2/MAC addresses) have only local significance and no user-friendliness requirement, and hence could easily be replaced by hashes of public keys if desired (cf. CGI). Doing so allows nodes both proof of ownership and a degree of privacy by periodically replacing the identifier by a newly generated one.

Network points of attachment

Under the assumption that the identifiers for network points of attachment will continue to be IP addresses,

- these identifiers are administratively assigned, and thus, could be certified by the assigning party (binding to an identifier logically above, below or both);
- these identifiers must preserve their routing semantics; consequently, the fact that traffic is actually routed to/from a attachment point provides a sense of 'authorization', although this can be verified only if an assurance of the sameness is provided (e.g. via a cryptographic binding to said attachment point's identifier).

Host end systems

As described in [51], public keys have a number of useful features as identifiers for host end systems: they can be self-generated; they come with intrinsic cryptographic properties; they are statistically unique; they can be certified directly by TTPs or other hosts (i.e. public keys); and they can themselves act both as issuers and subjects of authorization statements. With the advent of trusted computing [42], the secure storage of private keys in trusted host end systems may finally make this type of cryptographic identifiers meaningful in practice.

Ambient Networks

The lifecycle of Ambient Networks includes a variety of common operations from creation through composition (merging, splitting and different kinds of interworking) to termination. ANs need to be able to communicate with each other, possibly through other networks, and to negotiate and execute agreements that define their interactions. In most of these operations, we need to be able to identify an AN, or a suitable subset of thereof, and indicate which nodes belong to it. The purpose of an Ambient Network identifier would be to facilitate this, in other words, to identify a group of nodes as opposed to just individual nodes (even if a network may contain only a single node at times).

In Annex 3, we present two natural grouping concepts: the administrative domain and the security domain, roughly characterized by ownership and interaction, respectively. Similar to host end systems, cryptographically generated identifiers are used to identify both types of domains. This has the advantage that it removes the need for an external party to authorize particular nodes to act on behalf of the network. A domain public key (often self-generated) is also used as the identifier for the controlling object of the domain, and the corresponding private key is used to exercise authority over the domain. In the administrative domain, the keys are used to manage ownership; in the security domain, they manage interactions within or between domains. It is for further study to see if the administrative domain, the security domain, a combination of both or some other grouping concept will serve as the best model for AN identification. For the time being, when AN identifiers are mentioned, we think of some sort of domain identifiers as modelled in Annex 3 of [33].

To summarize then, cryptographically generated identifiers are also the top candidates for AN or domain identifiers. Some domain identifiers will obviously have bindings to host identifiers (e.g. for hosts proving membership in a domain, and potentially in several domains at the same time), and they may (and sometimes should) also be bound to the unique identifiers of users/organizations that 'own'/operate them.

Application points of attachment

Traditionally, identifiers for application points of attachment (such as port numbers) have been severely size constrained; thus, they are less suitable to carry much cryptographic significance. Depending on the final namespace semantics, some of these identifiers may even be predefined (cf. 'well-known ports'), otherwise it may be possible to compute them as a function of the host/domain identifiers 'below' and the application identifiers 'above'. In either case, cryptographic bindings both above and below can be created with relative ease (assuming, for the 'above' part, cryptographic identifiers on the application layer as well).

Application services and data objects

On the one hand, identifiers for application services and data objects must be human readable; on the other hand, the need for – or, at least, the benefit from – cryptographic identifiers is undisputable. The current practice used for web servers achieves this by binding a human-readable name and a public key by means of a certificate issued by a TTP. As the example shows, this approach works well in specific application domains, but the viability of independent global certification authorities for the rather wide scope of Ambient Networks is questionable. Nevertheless, the basic model of binding human-readable names and public keys combined with delegation and federation concepts may provide a workable solution.

Users and other legal entities

The Ambient Networks architecture is expected to enable agreement/contract negotiation in an automatic and business model independent manner. Although automation implies that these agreements will be negotiated and executed by network elements, there is always a person or organization behind each contract (whether by direct involvement or configuration). On the one hand, accountability and (when compensation is desired) chargeability require unique identifiers for all legal entities involved; on the other hand, privacy calls for owner-controlled exposure of all unique identifiers. The proposed AN security architecture achieves a balance between accountability/chargeability and privacy as it requires explicit authorization for every action, but enables authorization without revealing the authorized parties' unique identifiers by allowing ephemeral identifiers in most protocols and even in contracts, as long as a binding is guaranteed between the ephemeral identifier and the unique identifier. Naturally, the nature of this guarantee must also be recorded (e.g. if the guarantor is a trusted party, their identifier can be included in the agreement). Protection for unique identifiers can be provided by a TTP or, in a more limited form, by opportunistic security based on the ephemeral identifiers.

Note that uniqueness of an identifier can be guaranteed only for a given naming authority, usually called an identity provider (IdP) in a user identification setting. Traditionally, at least in the cellular context, the IdP has been the PLMN (public land mobile network) operator, but in many of the AN scenarios other players such as banks, credit card companies and corporations may take this role (cf. alternative payment solutions). Different IdP-unique identifiers can be 'federated' as specified by the Liberty Alliance [41] (again, using ephemeral identifiers).

Also note that the additional level of indirection introduced by the binding between ephemeral and unique identifiers allows the integration of legacy identification/authentication mechanisms to the AN framework in a relatively painless fashion. Each TTP/IdP is free to choose their own (cryptographic) namespace of unique identifiers and bind them to ephemeral identifiers that conform to the AN-defined requirements. Thus, identifiers for users and other legal entities in AN may follow arbitrary legacy solutions (specific to each IdP) as long as bindings to, say, (possibly ephemeral) host and/or AN identifiers can be guaranteed.

5.3.2 Authorization

The goals of information security, such as confidentiality, integrity and availability, are usually defined in terms of authorized and unauthorized entities: for example, something should not be '... made available or disclosed to unauthorized individuals, entities or processes' [48]. Making and enforcing authorization decisions, either permitting or denying some actions, is at the core of many (or even most) security mechanisms – much more than, for example, authentication or encryption.

Often the authorized entities are defined by referring to their identifiers. Thus, secure identification as discussed in Section 5.3.1 is an important prerequisite. This section contains recommendations for approaching authorization in Ambient Networks.

5.3.2.1 Consider Authorization in Architecture Documents

When designing the architecture of a system, it is quite common to treat authorization as something that 'just happens' at some step and assume that the authorization decision is based only on the authenticated identity. In the worst case, the document can simply say something like 'AN1 and AN2 authenticate and authorize each other', and that is assumed to take care of the authorization issues. However, this is highly inaccurate and misleading and can easily lead to systems that do not properly take security issues into account.

Whenever an authorization decision needs to be made, architecture documentation should, in the very minimum,

- Identify the logical entities involved in making the authorization decision and their roles in it. Eventually, these entities will need some kind of protocols, so they can be involved in the decision.
- Identify an example or some examples of what kind of policies each of the entities are likely to use. In most cases, different entities will have totally different policies: For instance, in roaming network access the home network's policy is based on the subscription's details and possibly account balance, whereas the visited network's decision is based on receiving a 'promise to pay' message (usually implied by a simple 'success' message) from a trusted roaming partner.

Some examples of different ways of doing authorization:

- Authorization based on cryptographically generated identifiers. For instance, when updating HIP bindings during mobility, knowing the corresponding private key gives an implicit authorization to update the bindings for the corresponding HIT.
- Authorization based on tokens created by an offline third party, such as certificates or SAML assertions. To take another mobility example, an authorization certificate could assert that the owner of a particular public key is allowed to update mobility bindings for a certain home address. (However, Mobile IPv6 decided not to use this approach for route optimization.)
- Authorization based on a protocol exchange with an online third party. This is the approach used in, for example, IETF RADIUS/Diameter work. The protocol design may have to consider the need for several round trips to the third party (for instance, EAP methods).
- No authorization is needed: For instance, a WLAN could provide access for free or at least allow access to local portal pages without any authentication or authorization.
- Authorization certificates are one example of delegation: The issuer of the certificate gives the subject a permission to do something and possibly the right to further delegate the permission to some other entity. The concept of delegation is especially powerful [49] when delegations can be chained: For instance, if a mobile node is authorized to update its mobility bindings, it could delegate this permission to a local access router to improve performance.
- Identify whether the authorization decision is 'one time only' or may be changed or revoked during the session. In particular, if the decision is based on information that changes as a result of the session (e.g. prepaid account balance), this usually has to be taken into account in the protocols. Revoking an authorization also requires a way to contact those nodes that are currently using the authorization decision.
- When it is likely that policies will be based on some kind of payment, it is usually more accurate to describe the situations in terms of payment, rather than just abstract 'authorization'. This approach usually takes better into account the dynamic nature of the situation (e.g. need to update prepaid account balance) and the different roles of the home and visited networks.

The examples in this subsection have dealt primarily with network access, but authorization decisions occur in all contexts. For instance, is this request to hand over the video stream authorized? Is this access point allowed to advertise this service? Or is this composition agreement in line with my policies?

5.3.2.2 Decouple Network Access Authentication/Authorization from Services

Authentication, authorization and payment for higher-layer services should not be tied to network access authentication unless it can be safely assumed that the higher-layer service and network access will always be provided by the same entity (or closely cooperating entities). Even if the services are provided by the same entity, the user might prefer to use separate accounts or payment methods for them (e.g. even if the user's employer pays for the network access, the user might prefer to pay for some other services from his or her own account).

5.3.3 Types of Security Relations

Security relations are important to describe the relationship entities have with each other in respect of security relevant functions. The term security relation in this document refers to technical issues such as knowledge about encryption keys or authentication information. We believe that security relations between any two AN entities fall into one of the following three categories:

- *Direct security relation*: We generally refer to direct security relation between two entities if they already have a preconfigured security association (e.g. encryption key) before proceeding with the first phase of communication.
- *Brokered security relation*: Two entities without direct security relation between themselves (e.g. entity A and entity B) may build a brokered security relation from segments of direct security relations. Network roaming is a good example wherein the home network operator acts as a security relation broker between the user and the visited network operator. It is important to keep in mind the role of the broker and how this maps onto potential business models.
- *No security relation*: We assume that no prior security relationship exists between two entities if neither direct nor brokered security relations (as defined above) can be established between them. However, it is possible and desirable to set up an opportunistic security relation, which can ensure that communication is secured after an initial set-up phase.

5.3.4 Security by Default

The concept of security by default is important to achieve an adequate security level in practice. AN security should be built in from the start and applied by default as early as possible. In many cases, there should not even be an option to turn it off, as 'optional security' often means 'no security'. Having options increases the complexity, and too many options make a system difficult to configure and use. Both these issues are in themselves security risks.

In this section, we present by means of additional examples what security by default is all about.

5.3.4.1 Security Being Easy to Configure and Use

Ambient Networks are designed with the end user in mind. Considering Ambient Networks set up by end users themselves, the user-friendly, i.e. easy and understandable, configuration of security measures is an important issue. The user should have to be exposed to these administrative things as little as possible, and the needed configuration as seen by the end user should be simple and self-explaining. As many end users may have only a rudimentary understanding of security, security should be set up automatically hiding the complicated technical issues like creating and configuring cryptographic keys and defining policies behind the scenes. It should not be assumed that ordinary users take correct security decisions as shows experience with web browsers where some users tend to override security-related warning messages without really reading them and understanding the consequences.

Easy set-up can be achieved in particular in the following ways:

- Providing secure default configurations.
- Users should not be tempted to just 'Click OK' overriding security: Security should be maintained without just 'switching it off' or 'accept everything' in case problems occur.
- Complexity of security configuration is hidden from end users by providing predefined configuration templates with meaningful names (as security level 'low/medium/high') or for typical usage environments (home network, public hotspot) that the user has only to select from.

Hiding the set-up of security parameters (e.g. cryptographic keys, access control policies) behind high-level configuration decisions has an intuitive meaning for end users (I want to communicate with that network X that I just discovered in the mode 'unknown PAN').

5.3.4.2 Default Security Mechanisms – Security Toolbox

The building blocks described in this paragraph apply in particular to signalling between Ambient Network Control Space functions within and between administrative domains. The HIP base exchange is proposed here modified to allow explicit authorisation (see Figure 5.1). The implications of these toolbox mechanisms and whether they are the most appropriate ones for Ambient Networks need further investigation.

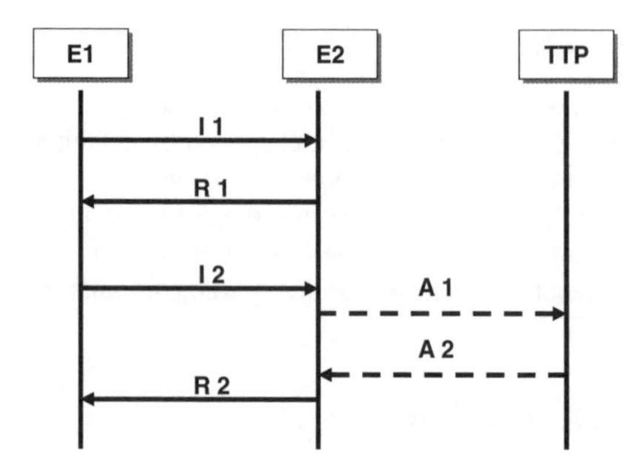

Figure 5.1 Association protocol based on HIP base exchange

5.4 Key Problems and Solutions

This section presents a number of security solutions proposed in AN which have resulted from a 'bottom-up' approach as well as continuous feedback from the related or affected working areas in the AN framework. The solutions also aimed at the same time to meet certain common requirements and complement the security architecture principles proposed in Section 5.3. Each of the proposed solutions contains a description of the state of the art in that area, a problem statement, proposed solution(s) and how the general security requirements and concepts are met. A more detailed discussion about each solution, their relationship with the AN security architecture as well as their impact on the AN functional entities and AN interfaces is provided in [33].

5.4.1 Network Attachment Procedures

Network attachment is the initial procedure performed by a device or network (as in AN) connecting to a network on wake-up or otherwise. Current attachment procedures involve several steps such as layer 2 parameter agreement, access control, authentication and network layer configuration. These steps result in large latencies as an attachment typically requires over 20 link and IP layer messages (see Figure 5.2). In addition, there are other limitations in current protocols like denial-of-service vulnerabilities, difficulties in trusting a set of access nodes distributed to physically insecure locations and so on.

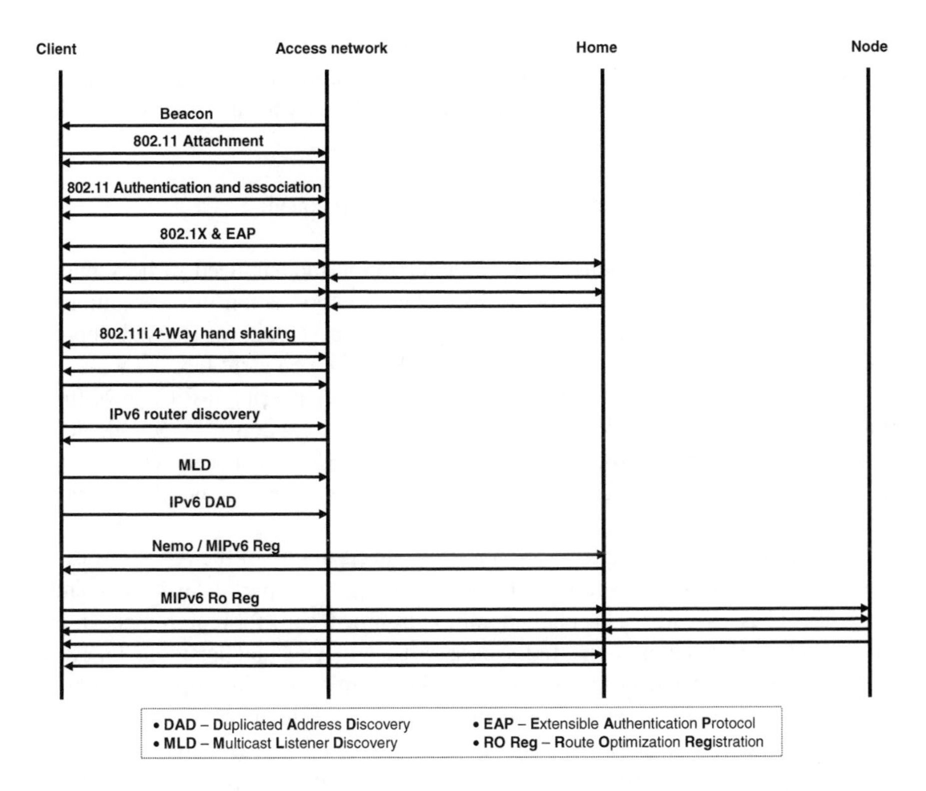

Figure 5.2 Example of today's network attachment

The heterogeneity of radio access technologies and network structure that the AN project encompasses makes network attachment ever challenging. Besides the technology involved there are also different business models to be considered in AN. In addition, an efficiency improvement is necessary in order to achieve seamless operation in heterogeneous environments. To tackle this problem, we take a system-level view that comprises the following principles:

- *Naming*: Entities are explicitly named with cryptographically generated identifiers, if possible.
- *Backward compatibility*: Legacy authentication mechanisms need to be supported in order to ease the deployment of the system.
- *Authorization granularity*: Increased hotspot and other such deployments lead to complex roaming infrastructure. As a consequence, users might not trust the network in the same way as today to ensure proper accounting and charging of consumed resources. High granularity of authorization is employed in the proposed solution together with mechanisms to authenticate not just home networks but also other network nodes. Together these make it possible to run ANs without assuming a uniformly trusted network infrastructure. This applies to all types of network nodes. For instance, the AAA infrastructure needs to be built in a manner that minimizes the need to place trust into its individual parts. This is achieved through the use of high granularity in authorization and the identification of all involved parties.
- *Deployment*: Security mechanisms for network attachment must allow incremental deployment providing minimal security also without configuration using opportunistic security. Enhancing authorization and security properties must be easy to enrol to offer solutions for home networks and ad hoc type networks.

In addition, the following specific design principles are employed:

- *Authorization models*: Different types of authorization models need to be supported to provide a higher degree of flexibility. This includes authorization using traditional two/three-party models, opportunistic models and token or credit card based approaches. Authorization can be based on roles and traits and not only on identities.
- *Delegation mechanisms*: These should be used whenever possible as it allows functions being executed by the network on behalf of the end host (or a network). Mobile networks, as an example, can benefit from this approach to achieve higher security combined with a performance benefit. For instance, hosts can delegate tasks such as mobility signalling to access networks.
- *Robustness*: The strong dependency on the real-time capability of the AAA infrastructure impacts the robustness of the entire network architecture and particularly when end hosts are considered as networks. This challenging environment calls for requirements regarding security mechanisms being robust in the case of disconnected and partitioned networks.
- *Network capabilities*: The properties of networks (roaming relationships, cost, QoS capabilities, privacy statements, available security protocol mechanisms, etc.) need to be disclosed to the end host (or network). This allows to make more reasonable selections by an end host when multiple networks are available or to instruct other protocols in the protocol stack in order to adapt its behaviour.

- *Business relationship*: A user and an end host need to be aware of the characteristics and the business relationships of networks when selecting an identity for authentication and authorization. A user might additionally influence AAA routing because of cost, policy reasons and personal preferences. To enhance network selection functionality available today, the following primitives need to be offered by a network attachment protocol that aims to interact with the AAA infrastructure:
  - source route;
  - name resolution request/response;
  - roaming request/advertisement;
  - service query/response.

The information offered by these functions can be more powerful than today in order to assert properties of networks.

The proposed solution, as a side effect, creates a secure communication channel that can be used to convey various kinds of control information like connectivity configuration and AN specific negotiations. Figure 5.3 depicts the possible message exchange occurring between different parties using the proposed solution. Compared to Figure 5.2, the amount of messages is considerably decreased especially on the first leg that most likely is the scarcest with respect to the available resources.

5.4.1.1 Multi-Hop Access

Multi-hop access to an infrastructure network is an important technology for future mobile communication systems. It can be employed in different variants to increase coverage and throughput, and it can enable new business opportunities for intermediates setting up multi-hop extensions offering access to an infrastructure network. This technology is also investigated in [49] and named nonconventional and low-cost access. The access security for multi-hop access to a public network should be based on the same preconditions (trust relations, existing security material) as the case where the public network is accessed directly without using intermediate nodes. However, additional security requirements have to be met to ensure that the multi-hop extension operates correctly and that it cannot be misused for unintended purposes. This multi-hop access can appear in different variants which have significantly different security properties. These justify handling them to some degree separately as they lead to different security solutions. Therefore, we define the following three variants:

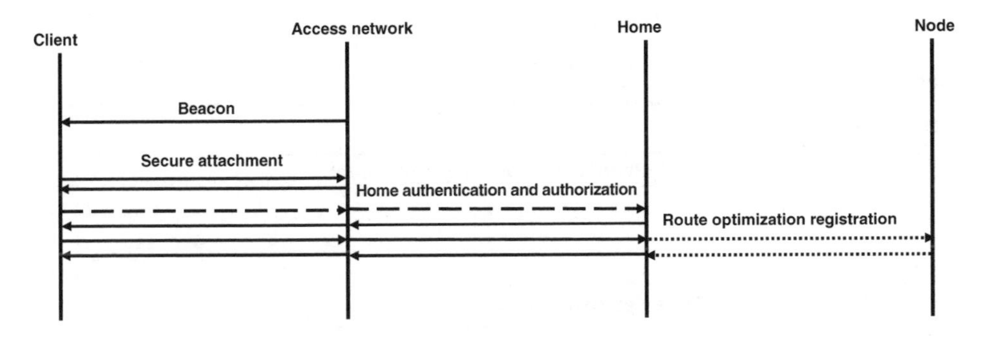

Figure 5.3 Example of message flow in the proposed architecture

- infrastructure mode;
- cooperative mode;
- user compensation for multi-hop network access.

5.4.1.2 Infrastructure Mode

Relays are set up by the operator of the infrastructure network. They are typically fixed or stationary. The relays are a part of the infrastructure network domain. In Figure 5.4, an example of infrastructure access with relays owned by an operator is shown. In the first step, the relays are authenticated against the network and a group key is established between the relays to protect the relay traffic. In the second step, the user authentication takes place and the tunnel between the user and the gateway is established.

Highly mobile devices, for instance vehicles, provide multi-hop communication. Only a very short term relation exists between such mobile relays. In Figure 5.5, an example of the cooperative scenario is shown. As already mentioned before, relay nodes *R* are at the same time user terminals UT in this scenario. In the ad hoc network part, beacon messages are exchanged on a regular basis, in order to communicate terminal identifiers, routing information and gateway announcements. It is important to note that there is no exchange of security credentials between relays and the gateway GW in the set-up phase. The operational phase then is similar to the infrastructure access, as shown in Figure 5.4.

The third multi-hop case with focus on compensation can be found in Section 5.4.1. Next steps are securing the routing protocol used in the cooperative mode, investigation and

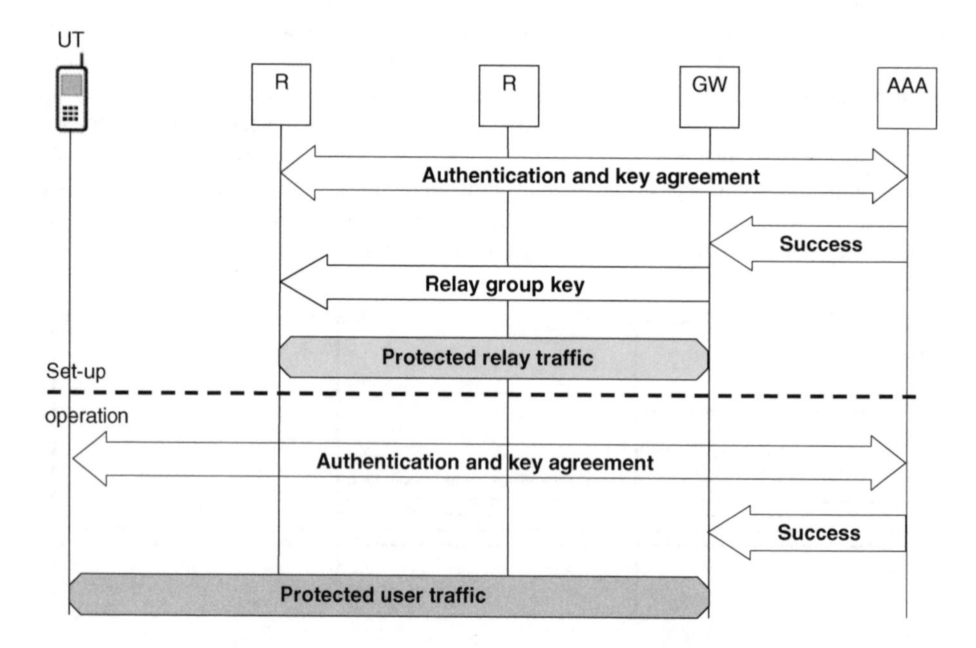

Figure 5.4 Example solution for the infrastructure case

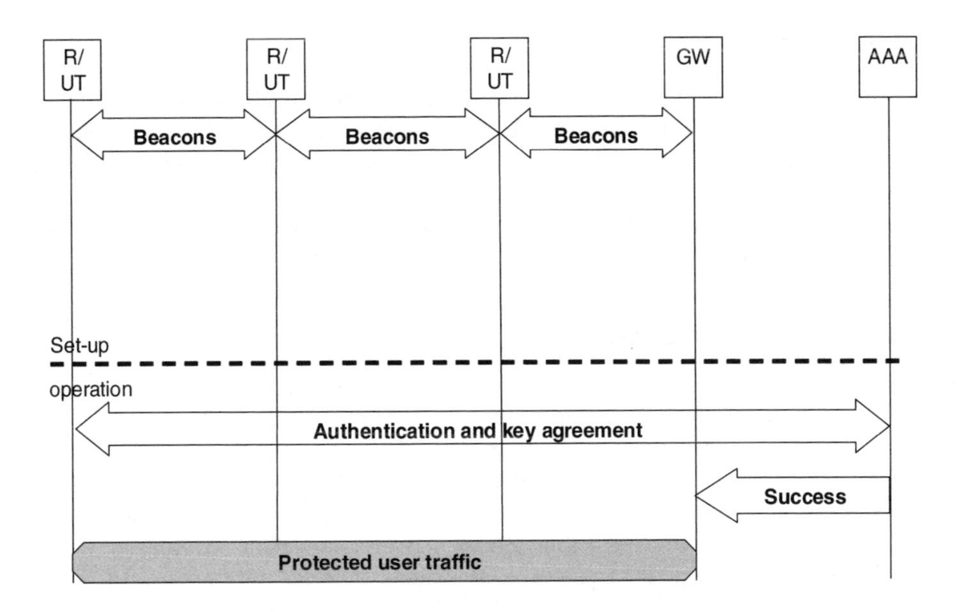

Figure 5.5 Example solution for the cooperative case

development of mechanisms for encouraging node cooperation. In the infrastructure case, future security work depends on the design choices about how the multi-hop extension shall be realized, i.e. which protocols on which layers are used, to map the described security concepts to concrete security technology that can be applied to the chosen technology.

5.4.1.3 User Compensation for Multi-Hop Network Access

Public access to wireless network is today offered primarily by network operators, their partners and large institutions such as universities. Such public networks also typically offer global roaming with the help of network infrastructure, such as AAA [65]. Currently, public access to wireless networks happens through wide area cellular networks as well as a smaller number of high-speed 'hotspots', typically employing wireless LAN technology. However, the number of actually installed high-speed local networks is much larger as enterprises, private persons and various communities have already deployed networks for their own use. Such private networks may not be available for usage by others, however. In many cases, these networks are not used on their full capacity and, as far as radio resources are concerned, could be used to provide access to others. Such access can be seen as an essential utility (like water or electricity), but offering it leads to practical, technical and even legal problems.

What Ambient Networks has designed allows private networks to provide public access, in a manner that ensures security, ease of configuration and includes even the possibility of getting compensation for the use of the networks. A typical scenario where this could be applied is a private home with a combined DSL and wireless LAN device that could provide access to others. Another usage scenario is a short-range wireless communication from passengers to the vehicle's high-bandwidth connection to the Internet.

5.4.2 Security for Composition

Composition, the process that enables two networks to automatically work together, is one of the central concepts of the Ambient Networks project. The composition procedure was predominantly a responsibility of the *concepts, architectures and technical coordination* and the *network composition and connectivity* work of AN and is detailed in Chapter 6. For security a bottom-up approach was taken with regards to composition, investigating a number of specific problems related to enabling networks to cooperate with one another.

5.4.2.1 Dynamic Roaming Agreements

Roaming agreements exist to allow a subscriber of one network to use the services normally provided by their home network by means of visited access network with which they do not have a subscription. Today, the most common form of roaming is between GSM operators. The 3GPP standards specify the technical aspects necessary to support roaming, whereas the GSM Association (GSMA) facilitates the creation and implementation of roaming agreements, for example, by providing common templates for commercial agreements, standard test procedures and defining the technical information that needs to be exchanged. Another example of roaming between access networks is WLAN roaming. Although a service still very much in its infancy, some support is provided by the WiFi Alliance through the definition of roaming best practices. Support for intertechnology roaming is not widespread and where offered is provided by proprietary gateway solutions. Specifications have been completed within 3GPP to support roaming between 3GPP networks [35] and WLAN networks [36].

Setting up a roaming agreement today between two operators is a manual process. Typically, individuals from the two operators meet and agree the commercial terms and conditions and sign the necessary paperwork. Then technicians from each operator exchange[1] technical information and configure elements within their own network accordingly to enable roaming services (this may also require coordination with roaming intermediaries). Once roaming services are configured, end-to-end tests are conducted before the roaming services are offered to paying customers. Even with the support of technical standards and industry associations, this process can be both time consuming and costly. With the trend towards increasing numbers of access networks and complexity of access technologies, it is our belief that such a manual process is not scalable and could be prohibitive to business opportunities. We therefore consider a mechanism for automating the establishment of both the commercial agreement between two operators and its technical implementation.[2] AN concentrates on the establishment of a bilateral roaming agreement between two operators.

Credential issuers, operated by industry or governmental organizations, act as trusted third parties to provide a means to establish trust between two operators with no previous security

[1] GSMA also provides technical specifications such as the Transferred Accounts Protocol (TAP), which specifies a common format for billing records.

[2] It should be noted that although the proposed solution focuses on automating the establishment of a bilateral roaming agreement, we do not rule out the use of roaming brokers. Roaming brokers offer a simplified roaming solution by acting as a single point of contact for an operator looking to establish roaming agreements with multiple other operators. However, roaming brokers can also act to restrict the business models available to an operator. For this reason, we have concentrated on providing a solution to provide roaming between two operators without specifying the need for a roaming broker. Nevertheless, we suggest that the same process could be used for an operator to form a roaming agreement with a roaming broker(s).

relationship. As depicted in Figure 5.6, the Federation of Credential Issuers provides a mechanism to bridge circles of trust between operator ANs registered with different industry/governmental organizations. A credential issued by credential issuer provides assertions about an operator AN that are used by another operator AN to decide whether or not to enter into negotiations to form a new roaming agreement. A credential also contains the public key of the operator AN, which is used to establish a security association between the operator ANs, e.g. using an adaptation of HIP or IKEv2. Negotiation of the commercial terms and conditions between the two operator ANs is a function of the composition control FE, which negotiates in accordance with operator-defined policies. Electronic contracts, secured using digital signatures, are exchanged after agreement is reached between the negotiating operator ANs. The parameters agreed are used by the composition control FE to authorize individual FEs, such as AAA, QoS, radio resource management, etc., to exchange and negotiate the information necessary to enable internetworking. In addition to the mechanisms described above to automate the creation of a new roaming service, a protocol is defined to provide nonrepudiation of an individual's usage, based on cryptographic evidence generated by the individual's access device. This is thought to be desirable as the automation of roaming services may result in interoperator fraud becoming more attractive.

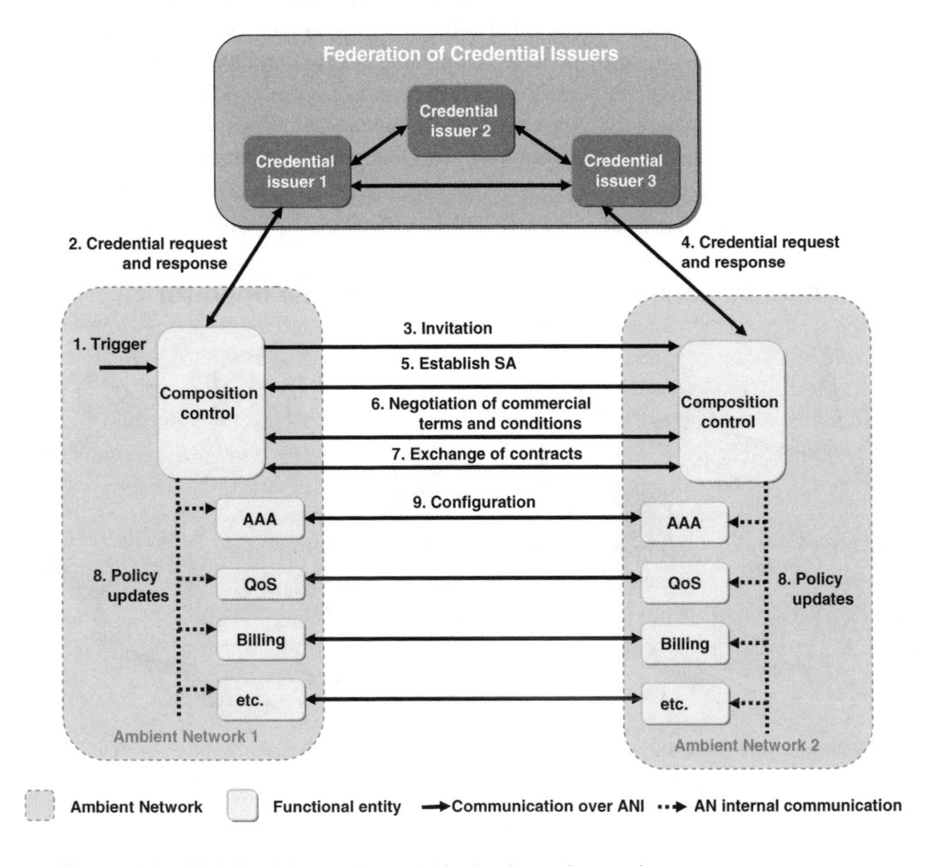

Figure 5.6 High-level interactions within the dynamic roaming agreement process

5.4.2.2 AAA Configuration Protocol for Dynamic Roaming Agreements

One part of the roaming agreement typically covers the interface between the authentication, authorization and accounting (AAA) systems of the operators. Before the AAA systems can talk to each other, there are several aspects that need to be agreed and configured, including addresses, security mechanisms, account identification and the details of exactly what information is to be sent and how. Today, this information is largely configured manually. However, this approach is error prone, scales poorly and makes it difficult to keep the information up to date as existing systems are modified and new systems deployed. Prior work for simplifying the configuration of AAA systems has largely focused on roaming guideline documents and capability mechanisms for AAA protocols. Roaming guideline documents specify how roaming is implemented in some particular situations. Although these documents are useful, they have some limitations. The documents are essentially snapshots of best current practices at some point of time: New versions of the documents have to be produced over time, and in practice, the deployed implementations are always either behind or ahead of the latest document. Some AAA protocols have a capability mechanism that allows the parties to inform each other what protocol features they support. However, these mechanisms are usually quite limited in functionality and assume that basic exchange of AAA messages already works. The AAA configuration protocol proposed in AN specifies a complementary approach for reducing the amount of manual work involved in roaming agreements. The protocol allows the networks to negotiate an agreement covering the technical details involving the interfaces between their authentication, authorization and accounting systems. The basic approach is illustrated in Figure 5.7, where the AAA traffic is shown as solid lines; dashed lines represent negotiation (interoperator) and configuration (intraoperator) exchanges. Instead of embedding a new mechanism to an existing AAA protocol, logically separate negotiator elements are introduced. These negotiator elements are not involved in the actual AAA exchanges;

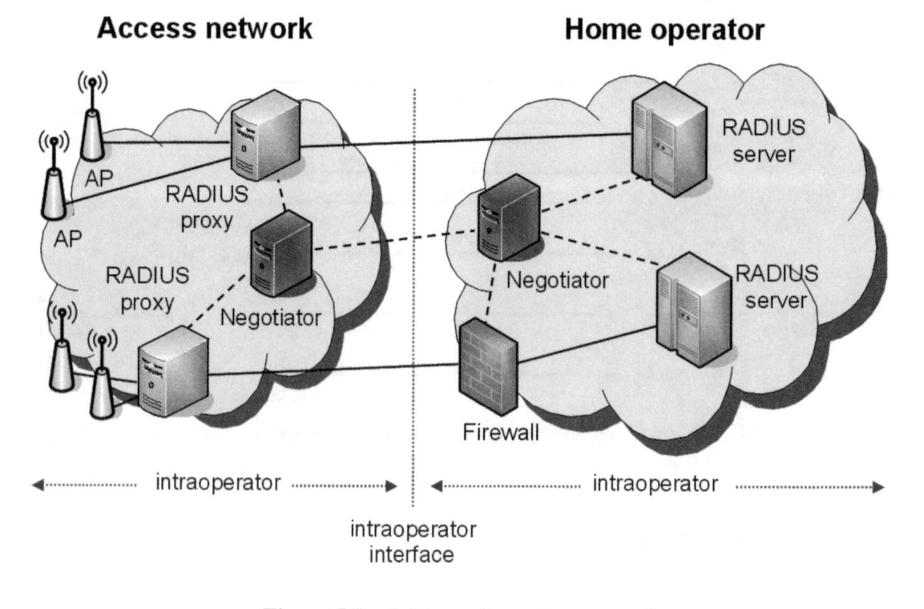

Figure 5.7 AAA configuration protocol

instead, they are responsible for configuring the AAA infrastructure to enable interworking that takes into account the capabilities, preferences and policies of the operators.

The negotiator elements have both interoperator and intraoperator interfaces. The interoperator interface (which can be considered to be a part of the Ambient Network Interface) involves negotiating an agreement containing the necessary technical details. This interface includes two major components: the structure and semantics of the agreement and a negotiation protocol for producing the agreement. These components are to some degree separate. As no protocol can support all possible types of negotiation strategies and policies, different negotiation protocols may be appropriate in different environments. The agreement itself can be independent of the negotiation strategy, but depends on the AAA protocol and the environment where it is used. The work on agreement structure and semantics has so far concentrated on using RADIUS when providing IP-based network access. The intraoperator interfaces (which can be considered as parts of the Ambient Resource Interface) are used for configuring the network elements, such as RADIUS proxies, servers and firewalls according to the agreement.

The AAA configuration protocol does not make setting up a roaming agreement a fully automatic process; the administrators still have to configure the negotiator elements with appropriate policies and set up the connection between them. Nevertheless, it allows more dynamic relationships as less manual work is involved, and the agreement can be renegotiated at any time to react to changes in the AAA infrastructure.

5.4.3 Session Handover and Mobility Mechanisms

5.4.3.1 Security for Bearer Mobility

Bearer mobility is meant to move a data flow from one application endpoint to another, possibly residing on a different physical device. For instance, it should be possible to watch the beginning of a movie on a PDA and the end on a computer.

Bearer mobility should be resilient to well-known attacks on mobility protocols:

- Hijacking and/or impersonation of an endpoint (MitM), in which an attacker might, for example, eavesdrop, tamper or redirect data flows.
- (Distributed) denial-of-service (DDoS, DoS) attacks, in which an attacker tricks parties involved in the protocol to consume a large portion of a limited resource (e.g. computation, bandwidth), for the sake of denying access to one or more network services.

The bearer mobility protocol (BMS) mechanism relies on HIP-like exchanges and assumes that a HIP association exists between the bearer endpoints' owners prior to the endpoint movement. The owner of an endpoint is identified by cryptographic identifiers (i.e. a public–private key pair). The use of cryptographic identifiers allows verifying the authenticity of messages via digital signatures, hence countering MitM attacks. The mechanism shown in Figure 5.8 has been carefully designed not to introduce new (D)DoS attacks opportunities to the HIP protocol (which is already partially protected by its puzzle mechanism). In addition, a partial level of confidentiality with respect to the bearer movement might be enabled via the encryption of bearer names.

There is an implicit assumption that a bearer mobility protocol has shared the required information and acquired approvals of all parties beforehand. As a part of the BMS, there will, however, be a few checks to verify that a host is not tricked.

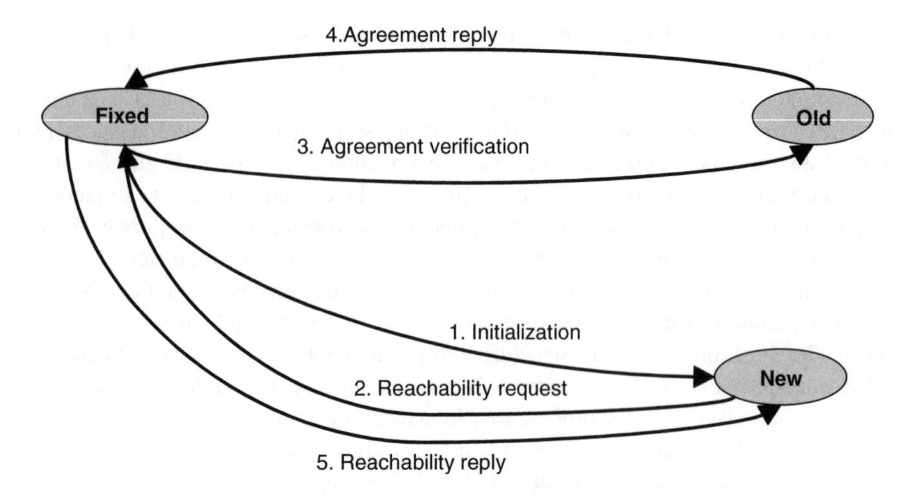

Figure 5.8 Overview of BMS

The initialization phase consists of establishing (secure) connection and authenticating each other. This information must be consistent with the information that the bearer mobility protocol has earlier provided to the hosts. After connection establishment, the new owner will do a simple return routability check, during which the fixed connection endpoint will verify the agreement (agreed during a protocol run of bearer mobility protocol).

5.4.3.2 Middlebox Context Transfer

A middlebox is defined as any intermediary device performing functions other than the normal, standard functions of an IP router on the datagram path between a source host and destination host [52]. Middleboxes could include firewalls, NATs, application layer gateways, etc. The introduction of middlebox devices introduces functionalities that are essential within an access network in order to prevent malicious mobile users to either cause damage or gain unfair usage of shared resources that can severely disrupt normal network operation. Moreover, the security of an access network may be compromised while the user requests network access and at the same time hands off. During a handoff, the interactions between a mobile host, the multimedia servers and the middleboxes must be minimized. One protocol, which will be adopted into the Ambient Network framework which could support this, is what is known as the context transfer protocol. Context transfer aims at minimizing the impact of transport, routing, security and other related services on handover performance [53]. When a mobile node (MN) moves to a new subnet, it must maintain these services that have already been established at the previous subnet. Such services are known as 'context transfer candidate services', and examples include AAA profile [192][193] and IPsec state, header compression, QoS policy, etc. Re-establishing these services at the new access network will require a considerable amount of time for the protocol exchanges and as a result time-sensitive real-time applications will deteriorate. Alternatively, context transfer can forward appropriate state information, for example, from the previous access router to the new access router so that the services can be quickly re-established.

For this research area we have proposed a mechanism of forwarding secure state information between middleboxes belonging to different radio access networks. We have demonstrated how the context transfer protocol could be employed for this purpose to forward certain security information from the old to the new middlebox to support session maintenance during mobility and also at the same time notify the previous middlebox to close unnecessary open ports for improved security and resolve vulnerability. Implementation results have been included for the latter case illustrating how knowledge of the mobile movements could facilitate in closing unnecessary open ports without depending on the long time-outs. This technique could be highly considered in environments like Ambient Networks where security between involved entities is intrinsic to the security architecture. The transfer of secure state information among RANs during the mobility of users could support their security provisioning, minimize security vulnerabilities which they face and also assist in maintaining a mobile user's sessions which may otherwise be dropped.

5.4.4 GANS Security

Generic Ambient Network Signalling (GANS) is the suite of protocols being developed in the Ambient Networks project (see Chapter 7). Like NSIS [60], GANS consists of two abstract layers, GANS transport layer and GANS signalling layer. The terms GTLP and GSLP refer to any protocol implementing these layers.

When designing security mechanisms for the GANS, two important design decisions that have to be made are how exactly the entities involved are authenticated and how authorization decisions (e.g. whether or not to process a request received from an authenticated peer) are made. In many cases, these two aspects are intertwined: Usually the identity (identifier) of the requestor is considered when making an authorization decision, and often the information about how an entity can be authenticated and what resources that entity is allowed to access is stored at the same place.

How exactly the authorization decision is made depends a lot on the GANS Signalling Layer Protocol (GSLP) in question whereby the authorization capabilities of a quality-of-service GSLP are discussed. The investigated approach focuses on a model where the GANS Transport Layer Protocol (GTLP) layer makes only very simple decisions: either it drops a request or delivers it to the GSLP for processing together with whatever authenticated information about the requestor can be found (e.g. identities authenticated by security mechanisms at the GANS layer or below).

In many GANS usage scenarios, it is not realistic to assume that new authentication credentials and user/policy databases would be deployed just for GANS. This is especially the case when the entities are end-user terminals rather than routers or servers in the network. The GANS security work examined how existing security infrastructure can be used to secure communication. As 'existing security infrastructure' heavily depends on the environment, a single solution is unlikely to be appropriate for all usage scenarios. Thus, we examine several options, such as, how

- transport layer security (TLS) authenticated with certificates,
- the 3GPP generic bootstrapping architecture (GBA),

- the Extensible Authentication Protocol (EAP) or
- the Host Identity Protocol

can be used to secure GANS. To disseminate the results of this work, an IETF draft [60] was submitted.

To investigate authorization aspects, the field of quality-of-service signalling was selected. No other signalling applications are currently being developed by the network composition and connectivity work package (WP3). To provide a detailed analysis, the different authorization models (two-party, token-based three-party and generic three-party) and different charging frameworks (New Jersey Turnpike and New Jersey Parkway models) were investigated. Detailed deployment environments, including lying network elements and the usage of EAP, were considered. This work will be provided as input to the ongoing work on QoS signalling and AAA interaction in the IETF (see [55,56]).

5.4.5 Generic Link Layer Security

The GLL, introduced in Chapter 8, should abstract the different radio accesses (RA) which can consist of different radio access technologies (RAT) from the higher layers. To come up with a security solution that cover the new problems of coordinating, enhancing and replacing the different link layer technologies, a new concept, the link layer security controller (LLSC), was introduced. This LLSC will in the first step cryptographically separate the different RA from each other using a key derivation function. This implies that it must not be possible to break into one RA if the key from the other RA is uncovered. Furthermore, the key transfer from one RA to the other should be possible without having online connectivity to a trusted third party. This will reduce the latency needed to get a new and secure RA established.

In Figure 5.9, the LLSC architecture is shown. The LLSC contains an authentication module, which is responsible for the authentication. The RA authentication could be done

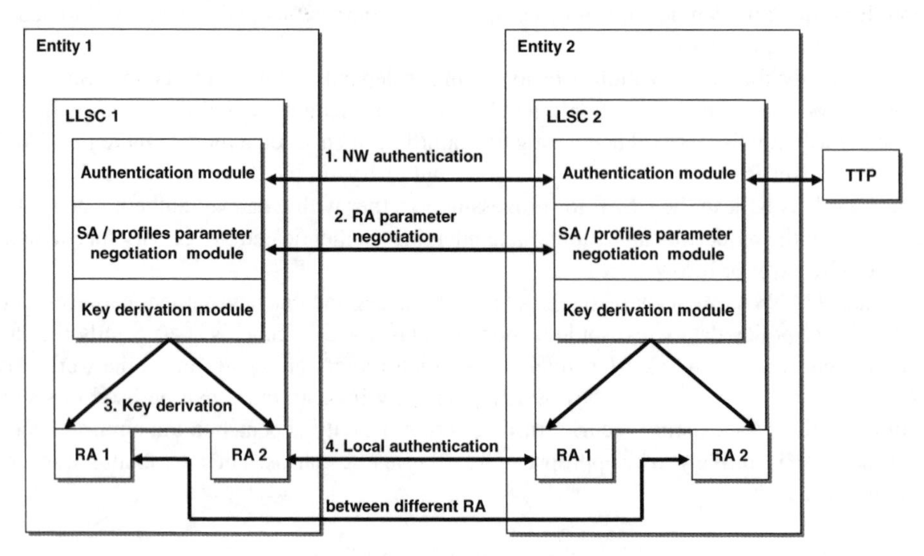

Figure 5.9 LLSC architecture

in conjunction with higher-layer authentication if the endpoints are the same. After the authentication, a negotiation step could be necessary. In this step, the needed parameter (e.g. key derivation function) could be agreed on. This step is separated from the authentication due to the fact that if new RA gets available not the whole authentication should be redone on the new RA. Instead, only the needed parameters could be exchanged and then the next module, the key derivation module, gets in charge. This module derives from the authenticated keys for the different RA and delivers these keys (and the other needed parameters agreed in the previous step) to the appropriate RA. After this delivery, both RA must authenticate to each other using a local authentication procedure. This local authentication procedure could be in the first step a legacy, RAT-dependent one (e.g. WLAN 4way Handshake), and later be replaced by a new one uniform to all RAT.

In Figure 5.10, the sequence of new RA establishment is shown. First the RA1 is established and after the second RA is discovered the fast RA establishment procedure is invoked (if the policy allows it). After this procedure, the RA2 is secure. In this scenario, the signalling traffic to the backend system is handled only by LLSC1 (i.e. the user terminal (UT) and LLSC2 sends all necessary signalling data to LLSC1). Furthermore, in this scenario the threat of node compromising gains importance. If LLSC1 is physically vulnerable, the link between LLSC1 and LLSC2 could be used to inject new keys (i.e. new UT) to the LLSC2 if the node is compromised. Several countermeasures are possible:

- The entities could administratively be protected (i.e. entity is placed on a safe location).
- The used key for this link could be protected via tamper-proof modules.
- The transferred key must be affirmed by a trusted third party.
- The key on the LLSC1–LLSC2 link must be renewed frequently or could be revoked. This would result in a short vulnerable time period only, if the compromise is detected.
- LLSC1 could in the first TTP involvement obtain an assertion which includes the UT identity and is time restricted. Then in the local authentication also the UT identity must be validated (i.e. there is no binding between the transferred key and the UT identity).

The last possible countermeasure seems to be a reasonable choice, especially if the node itself is under the control of the adversary. The assertion could be issued using, for example,

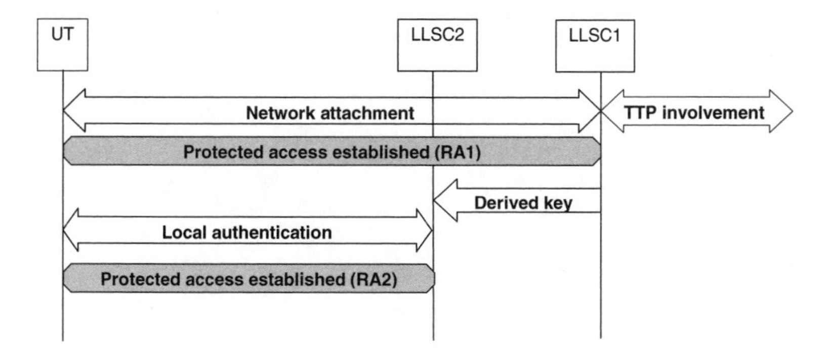

Figure 5.10 Sequence of establishing new RA

extended SAML assertions. LLSC1 would contact the TTP and gets in addition to the key also an assertion, which approves a valid authentication for a specific user. This assertion could then be sent to the LLSC2 in addition to the new derived key. LLSC2 could then verify if the assertion is valid and issued by a TTP. After successful verification, the UT must authenticate against LLSC2 with the identity contained in the assertion and prove the possession of the key.

5.5 Conclusion, Outlook and Further Work

The nature of Ambient Networks required a two-pronged approach to a security architecture, a top-down as well as a bottom-up. The top-down approach resulted in a set of security architecture principles, which include secure identification, security relationships, authorization and the use of security by default. These security architecture principles were used to guide the development of the technical solutions in the bottom-up approach.

The bottom-up approach consisted of the proposal and development of technical solutions to key security problems, which were based on security requirements identified in Ambient Networks. It was concluded that our solutions are effective at meeting the identified security problems; however, other security problems relevant to an Ambient Network may be discovered in the future. It is understood that the solutions provided in this document are not necessarily the only possible choices. Moreover, the solutions discussed in this chapter need to be detailed further and validated by demonstrators. As the specifications and prototyping work of these concepts and solutions are currently under development, further evolvement can be expected.

6

Network Composition

Acknowledgements

This chapter is based on the joint experiences and efforts of the researchers in the first phase of the AN project and particularly the following people listed as contributors and authors (i.e. in alphabetical order): Louise Burness (British Telecommunications Plc.), Rui Campos (INESC Porto), Philip Eardley (British Telecommunications Plc.), Joachim Hillebrand (DoCoMo), Roger Kalden (Ericsson EED), Cornelia Kappler (Siemens AG), Pekka Koskela (VTT Electronics), Gosta Leijonhufvud (Ericsson EAB), Paulo Mendes (DoCoMo), Cornel Pampu (Siemens AG), Carlos Pinho (INESC Porto), Petteri Poyhonen (Nokia), Christian Prehofer (DoCoMo), Gidon Reid (British Telecommunications Plc.), Manuel Ricardo (INESC Porto), Jose Ruela (INESC Porto), Brynjar Viken (Telenor), Thiemo Voigt (Swedish Institute of Computer Science) and Di Zhou (Siemens AG Austria). This chapter has been edited by Martin Johnsson.

6.1 Introduction and Motivation

Ambient Networks are expected to operate in a networking environment, which is characterized by more dynamic topologies and heterogeneous networks cooperating in a responsive environment. New kinds of mobile networks are assumed to gain importance, such as personal area networks (PANs), body area networks (BANs), intervehicle networks and sensor networks, all of which will interwork. The control plane interaction of these networks needs to enable, for example, seamless mobility, end-to-end QoS, integrated security and accounting. A challenge though is the heterogeneity of technologies and solutions employed in the various control planes. For instance, mobility handling is different for a mobile phone, a train network or a BAN. Hence, it needs to be negotiated which specific protocols to use, if such networks want to cooperate. The configuration of control plane interaction of such networks needs to become autonomic to not threaten the objective of realizing on-demand cooperation of heterogeneous networks.

The concept of network composition has origins in many different branches of networking, from the first days of telephony to the most advanced concepts in cellular

---

Ambient Networks: Co-operative Mobile Networking for the Wireless World Norbert Niebert (Ericsson GmbH), Andreas Schieder (Ericsson GmbH), Jens Zander and Robert Hancock

network and Internet research. The most elementary scenario is where networks are actually transparent to each other: as a pair of transparent networks is also transparent, the capability for composition is automatic. Although this might seem a trivial example, transparency is seen as the foundation of the Internet architecture [57]. Of course, networks are never fully transparent, and in the particular case of the Internet, the nontrivial requirement is to understand how the routing policies of different networks can be combined at the points at which those networks interconnect. This interaction of routing functions is provided in today's Internet by BGP [58], which has been extended incrementally in richness and complexity to handle a whole range of policy interactions. Unsurprisingly, an actual formal mathematical foundation for the composition of routing policies is still a research topic [59]; one possible lesson from that work is that a clean break from the complexities of the BGP approach is needed if the concept of network interworking is to be advanced [60].

The seeds of the composition concept can also be discovered in the cellular telecommunications world. The commercial motivation – to allow the owners of assets to cooperate transparently to serve a wider customer base – are the same as in the Internet case, but the approach is quite different. Here, the fundamental goal is to allow direct subscribers of one network to exploit the infrastructure of another. This is achieved by cellular roaming using the MAP SS7 protocol [61], which provides functionality to delegate authentication responsibility and related decisions from one network to another. The roaming concept has deep implications across the whole GSM and UMTS system design, from the format of subscriber identification information to the allocation of functionality to home and visited networks, across nearly the whole protocol stack. Attempts to achieve the same level of interworking between wireless LAN and cellular networks require considerable architectural analysis, even to achieve a limited level of best-effort services [62]. It can be seen, therefore, whichever perspective one starts from, that new approaches to internetworking are needed to make progress.

The concept of network compositions is thus introduced to support self-organized control plane interworking of heterogeneous networks. It enables consistent management over cooperating networks and hides their interconnection details as well as internal structure to the outside. It improves network operation and service efficiency. The details of control plane interworking between composing networks are fixed in a composition agreement. A composition establishment consists of the negotiation and then the realization of a composition agreement. Both negotiation of composition agreement and its realization should be autonomic, i.e. they are usually triggered by internal processes and proceed with minimal user interaction. Policies play an important role in the composition process. The decision whether to compose is policy based, the negotiation of the composition agreement is policy based and the composition agreement itself must meaningfully combine the policies of the composing ANs such that the composed AN has its own policies governing future compositions.

6.2 Composition Procedures

The process of composition is described by means of a composition procedure, presented in this section. The procedure structures the overall composition process in a sequence of five distinct phases. These phases are

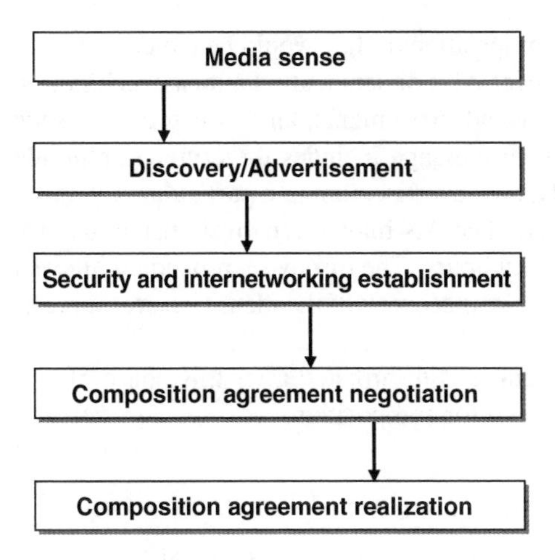

Figure 6.1 Basic procedure blocks for the composition procedure

- media sense;
- discovery/advertisement;
- security and internetworking connectivity establishment;
- composition agreement negotiation;
- composition agreement realization.

The phases are not necessarily passed in a unidirectional and linear way. It is, for example, conceivable that after establishing a security association, further services can be advertised which are available only to certain, trusted Ambient Networks. It is also possible to, for example, update the security association after determining the details of the Composition Agreement (CA) to allow for the required flexibility and efficiency. Figure 6.1 shows an overall and principal flow diagram for the composition procedure.

All five phases are discussed in more detail in the following sections.

6.2.1 Media Sense

The very starting point is to sense a medium that would enable communication with a neighbouring AN. The 'sensing' also includes the case of discovering a link to a remote AN (no physical vicinity). Depending on the particular scenario, different types of sensing are conceivable:

- An operator connects a new access point to its network.
- Two operator-managed networks are connected for the first time.
- A user device is switched on and searches for networks in its vicinity.
- A user initiates a session that requires more resources than allocated.
- A PAN needs to cooperate with a remote AN (virtual composition).

6.2.2 Discovery/Advertisement

Depending on the situation, media sense is followed by either an advertisement or a discovery phase. On OSI layer 2, these messages would typically be broadcasted as beacons. On OSI layer 3 (e.g. for virtual compositions), they would be either sent as targeted composition queries addressing specific peer ANs or sent using multicast techniques.

By making use of active advertisements, an AN can offer resources and services to other ANs. The advertisement message includes the cryptographic identifier used by the AN, which is included to bind the advertisement to a particular AN and may be authenticated and authorized at a later phase. The AN may alternatively listen to advertisements by other ANs or actively discover its neighbours. The discovery procedure allows to select a candidate AN for composition. It allows the discovery of the identifiers, resources, capabilities and network services of other ANs.

The composition procedure is terminated after this phase, if it does not result in an AN being selected as a candidate for composition.

6.2.3 Establishment of Security and Internetworking Connectivity

When the discovery procedure leads to a candidate AN for composition, the two ANs need to establish basic security and internetworking connectivity. An efficient way of doing so is a generalization of the HIP base exchange that includes the generation of a shared session key using the Diffie-Hellmann algorithm. Cryptographic identifiers belonging to the ANs involved in the composition are used to bind the established shared key to the communicating ANs and a cryptographic puzzle is used to protect against denial-of-service attacks.

Where the discovery phase occurs at OSI level 3, it is assumed that existing protocols, e.g. HIP and IKEv2, could be used to establish the necessary security association between the two ANs. However, these protocols may need to be extended to carry additional payloads, e.g. AN-specific credentials, purpose of composition, etc.

The identities of the ANs might be authenticated and/or authorized using a trusted third party. Alternatively, the required trust relationship may be based on a pre-established shared secret or may even be opportunistic, e.g. the ANs make a leap of faith, trusting the unauthenticated identities.

Internetworking connectivity between the two ANs is established at some point during this message exchange or immediately afterwards. We do not prescribe a particular point in order to allow for flexibility and optimizations. However, further studies should lead to a more precise definition of the exact ordering.

6.2.4 Composition Agreement Negotiation

The next step of the composition process is the negotiation of the actual composition agreement (CA). The CA includes the policies to be followed in the composed AN, the identifier of the composed AN, how logical and physical resources are controlled and/or shared between the composing ANs, etc. Where the CA includes commercial aspects, the CA should be digitally signed by both ANs to provide nonrepudiation.

There are two ways for negotiating a CA: centralized and distributed. In a centralized negotiation, the composition control FEs of the two composing ANs negotiate between themselves, each consulting with the other FEs in the same AN. In the decentralized negotiation, each FE negotiates independently about the control functionality it is responsible for, orchestrated by the composition control FE. In this case, the negotiation process must be followed by an internal consolidation phase among FEs in each AN because the partial agreements negotiated by each FE may not be independent from each other.

It is possible that the process of establishing a CA may involve increasing levels of authorization, e.g. negotiation of certain resources and services may only be authorized once the two ANs have agreed the commercial aspects of the CA.

6.2.5 Composition Agreement Realization

The realization phase represents the completion of the composition. During this phase, network elements are configured to reflect the CA. Thereby, each of the composing ANs must also carry out the configuration of its own resources, for example, by updating the resource registry, the Context Information Base (CIB) and policy database, and assigning resources for connectivity and QoS across the ARI.

It should then also be noted that, depending on the outcome and the settlement of the CA, addresses might be reassigned and reorganized. The result of the CA process is either a new AN or an enlarged AN (i.e. one AN is absorbed into the other), or two interworking ANs. The different types of composition as well as their results are further discussed in Section 6.3.

6.2.6 Decomposition Procedure

One or more of the constituent ANs in a composition might decide to invoke the decomposition procedure to terminate the composition (e.g. switch off one AN, relay node leaves coverage).

Decomposition can be seen as 'inverse function' of the composition procedure, which restores the states of the constituent ANs to their original state before these ANs composed. Unlike the composition procedure, decomposition does not require a strict synchronization between participating ANs because once an AN has indicated its willingness to leave a composition it would not be willing to reverse the decomposition process in case the decomposition process fails. The decomposition procedure is thus not considered to be an atomic transaction, which always guarantees synchronization between the ANs involved.

6.2.7 Composition Update Procedure

A composition agreement can be altered by re-initiating the CA negotiation phase. There are various reasons for the initiation of a composition update process conceivable such as changed service needs, mobility, link fading or failures, availability of new links or services.

6.3 Definition of Composition Types

A successful completion of the composition procedure leads to a cooperation between the involved ANs, which can take different forms. The classification of the different types of cooperation resulting from a composition is discussed in this section. Two possible approaches are envisaged. The first one considers the way how resources between the cooperating ANs are shared and the second one takes a network management perspective.

6.3.1 Composition Types from a Resource Management Perspective

To characterize the ways how resources between cooperating ANs can be shared, the following assumptions are made about an AN:

- Each AN implements an ACS.
- Each ACS controls resources $R = \{r_1, r_2, \ldots, r_n\}$, where control implies an (exclusive) right to allocate/deallocate/configure/etc. resources in R.
- The ACS is able to provide information about the status of its resources R to other ANs.
- The ACS has the ability to allocate the resources in R upon request from other ANs, which then can be utilized by these other ANs.
- Resources requested on-demand and deemed 'short-lived' are not considered a part of composition (only the policies which might control such resource reservation).
- Each ACS has a set of policies $P = \{p_1, p_2, \ldots, p_n\}$.
- Resource control, status and information, and resource allocation are defined by the CA.
- Policies P provide detailed information regarding the rules surrounding resource control and allocation procedures.
- The process by which a CA is established between ANs is called composition.
- Prerequisite: A trust relationship must exist between involved ANs to establish the CA.

Figure 6.2–6.4 depict the status before and after composition for each of the composition types foreseen. The terms and definitions listed above and regarding resources and policies are applied. Two alternatives are shown for the case of control sharing, where the first

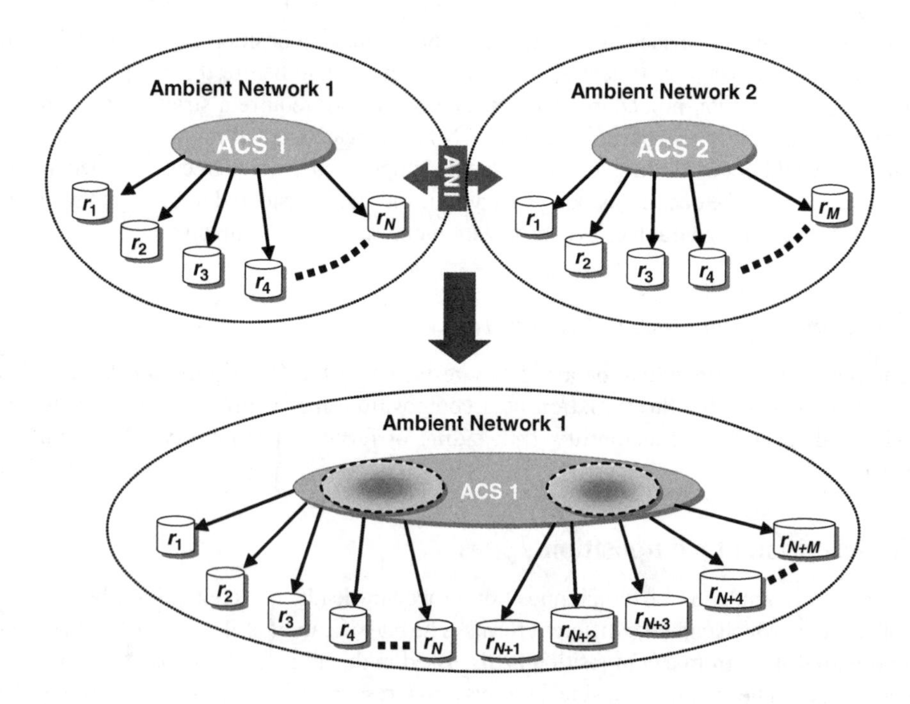

Figure 6.2 Result of network integration

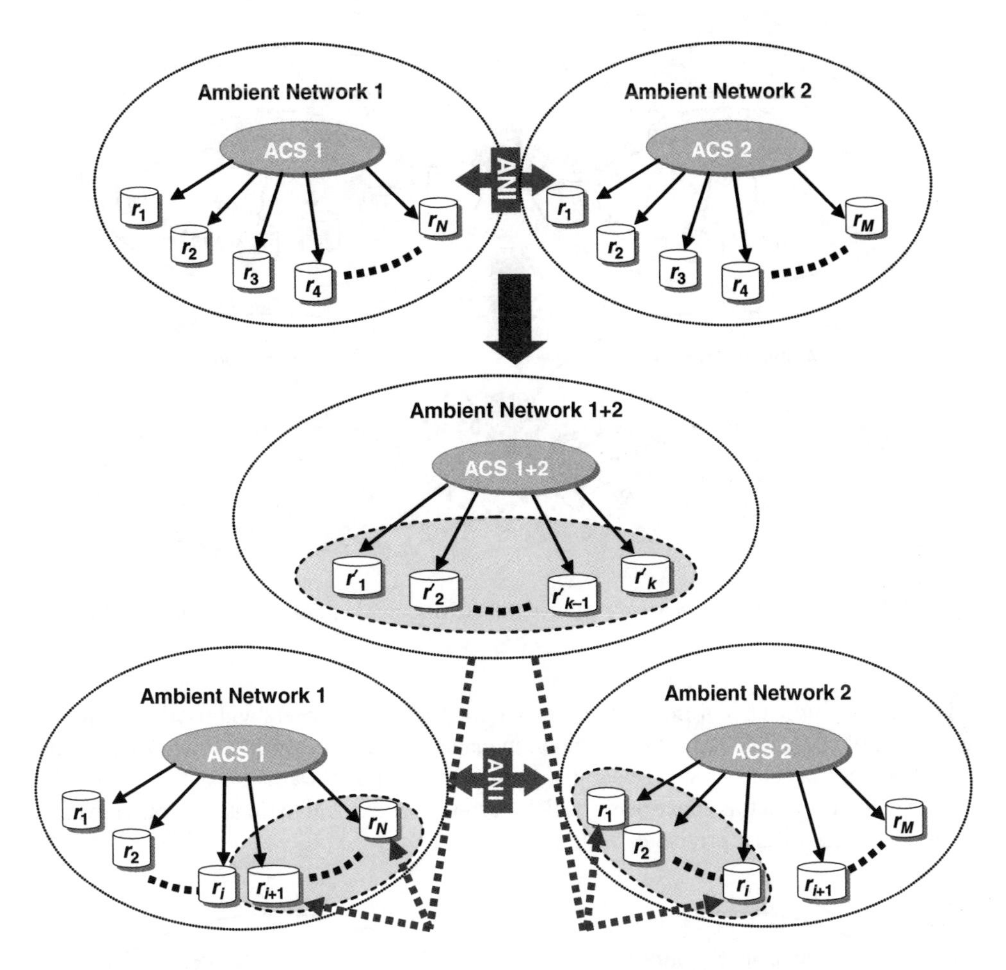

Figure 6.3 Result of control sharing, non-exclusive right to resource control

example in Figure 6.3 depicts a nonexclusive right to control (common) resources, and the following example in Figure 6.5 depicts the case of granting exclusive access rights from one AN to another (also known as control delegation). Control over a shared resource mandates the need for a common/virtual ACS. This common/virtual ACS must at least control the shared resources, but may also control other nonshared resources and may also be needed for other reasons besides resource control. A common/virtual ACS is thus optional in the case of control delegation in Figure 6.5 and network interworking in Figure 6.4.

6.3.2 Composition Effects on Network Management

Network management, which is in more detail discussed in Section 12.3.1, employs a hierarchical P2P overlay to ensure a scalable network architecture for composition in Ambient Networks. This overlay model maintains a separate overlay for every ACS. That means, each AN is associated with an overlay, and the overlay represents the ACS.

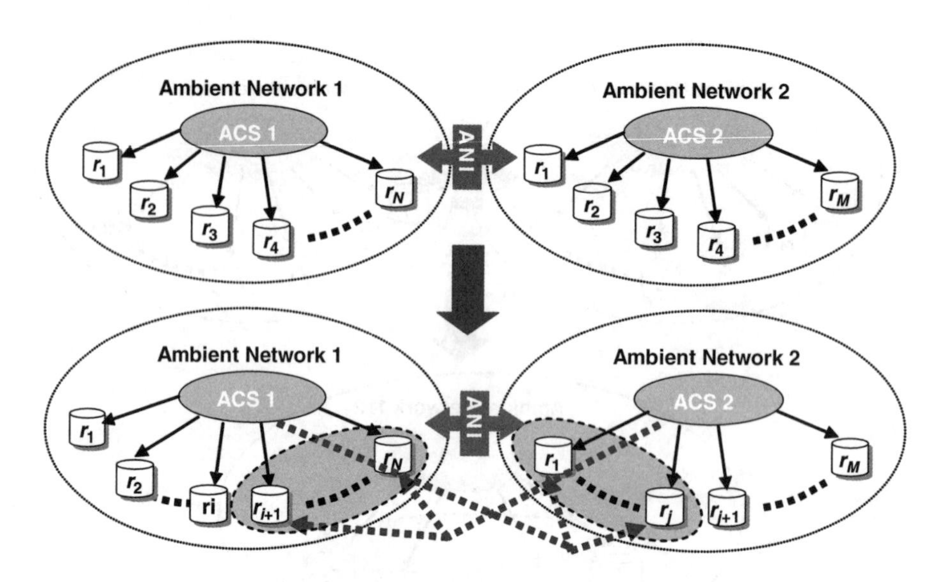

Figure 6.4 Result of control sharing, exclusive right to resource control, also known as control delegation

The basic components of this overlay model are peers, super-peers and overlays. An overlay is a set of peers belonging to a common management domain and forming a virtual network. Each overlay elects a super-peer to represent the overlay towards the outside world. It is important to note that this super-peer is solely responsible for negotiations with other overlays and has no other special privileges within its overlay.

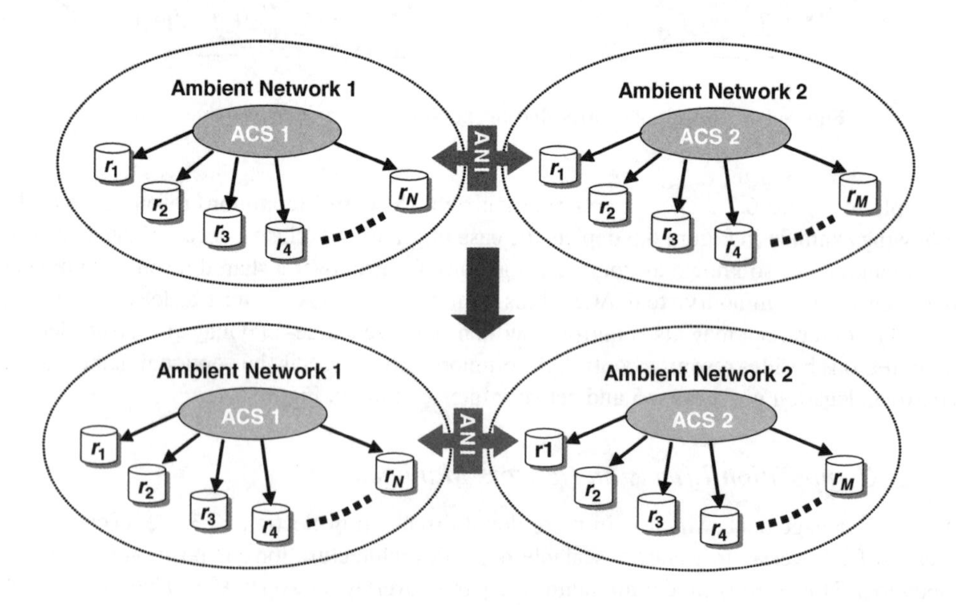

Figure 6.5 Result of network interworking

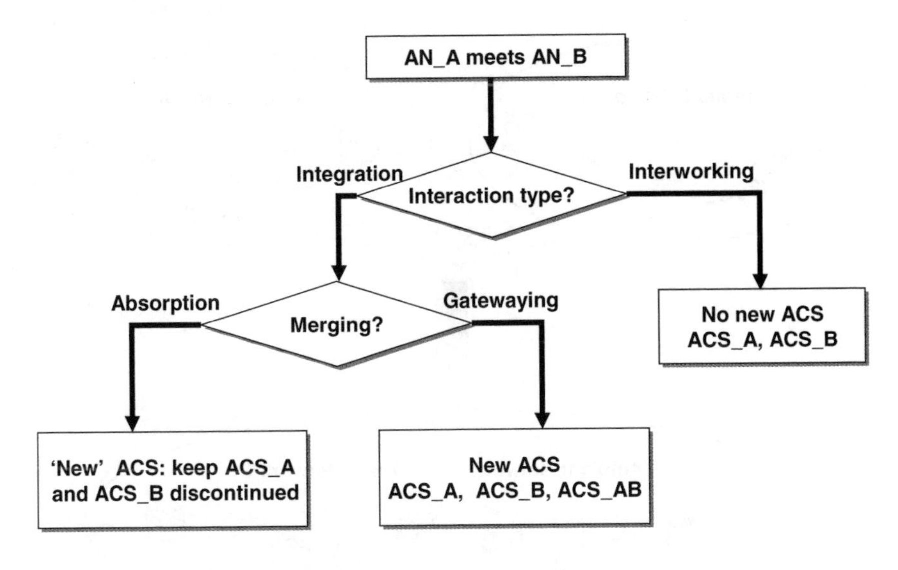

Figure 6.6 Management overlay types

Super-peers may also form overlays at higher hierarchy levels, thereby creating a hierarchical overlay network structure. Another characteristic of the presented hierarchical overlays is that hierarchy levels are not absolute. This means that one cannot assign an absolute hierarchy level index to an overlay. However, the bottommost level overlay is defined for all peers.

In Ambient Networks, network composition processes should work automatically without user interaction (as far as possible) and should be based on the predefined preferences of the users.

Users define their preferences by policy rules. These rules can be either strict rules that express explicit requirements of the user or more permissive rules that reflect only the wishes of the user. Users form networks with users that have similar (or at least mutually acceptable) policies. When an AN is formed, the overall policy reflects the common preferences of the community, which should be continuously maintained according to the current policies of the current members.

The common network policy defines the preferences of the community forming the network. This policy is used when different networks try to interconnect. Based on the policies of the two networks, the composition process may have very different outcomes. Note that the detailed resource sharing is reflected by the composition agreements. The possible management overlay types resulted from the process are represented in Figure 6.6.

6.3.2.1 Absorption (Network Integration) Management Overlay Type

If the policy database of the two networks is close enough to each other (there is no contradiction in the preferences and the differences between policies can be accepted mutually), the networks can join in a way that is called *absorption*. Absorption is the full merging of the

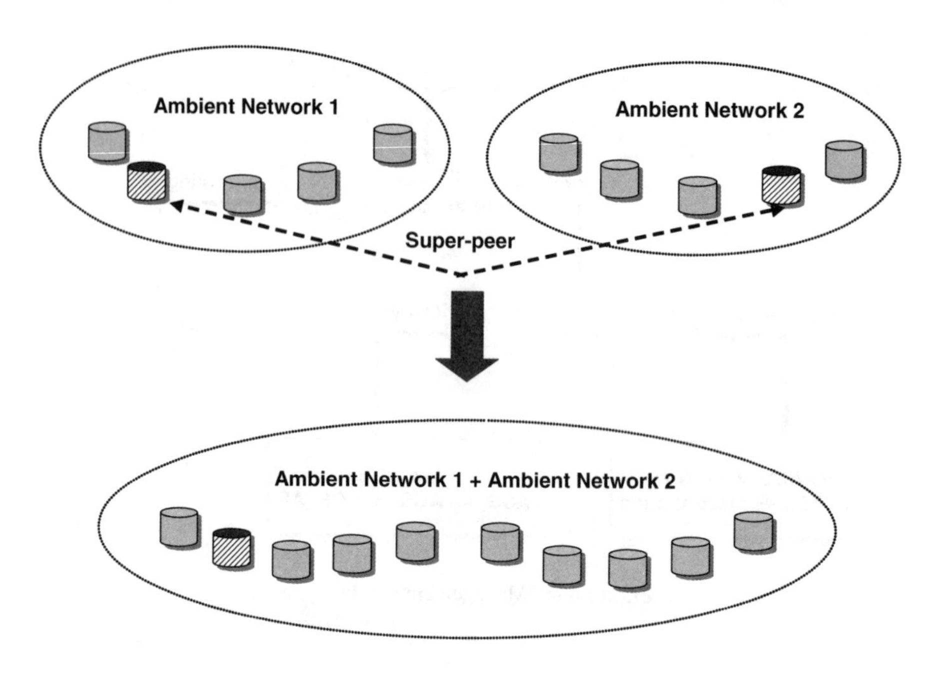

Figure 6.7 Absorption

two networks. They will form a single management domain with one common super-peer, see Figure 6.7.

The super-peer of the new network can be elected through a super-peer election process (i.e. every node can be a candidate for the super-peer role) or one of the previous super-peers can remain the super-peer of the whole network.

6.3.2.2 Gatewaying Management Overlay Type

In case the absorption of the networks is impossible according to the policy rules, there is a trade-off option between full absorption and full separation. This special partly joined way of operation is called *gatewaying*.

It needs to be emphasized that networks connected with gatewaying are not separated networks; they are composed networks but they do not share all resources. This means that they will be seen as a single network from the outside, so the networks have to be represented by one super-peer and one consistent policy database (as illustrated Figure 6.8). Therefore, the super-peers of the joining networks will create an upper overlay level. The members of this network will elect a super-peer that will represent the two networks, interconnected by gatewaying, to the outside world.

In gatewaying, the construction of the common policy database is more difficult because there can be contradictions between the policy rules of the networks. The contradicting policies must be eliminated from the common set of policies. The composed networks and the whole community is described by their common policies and represented by the highest-level

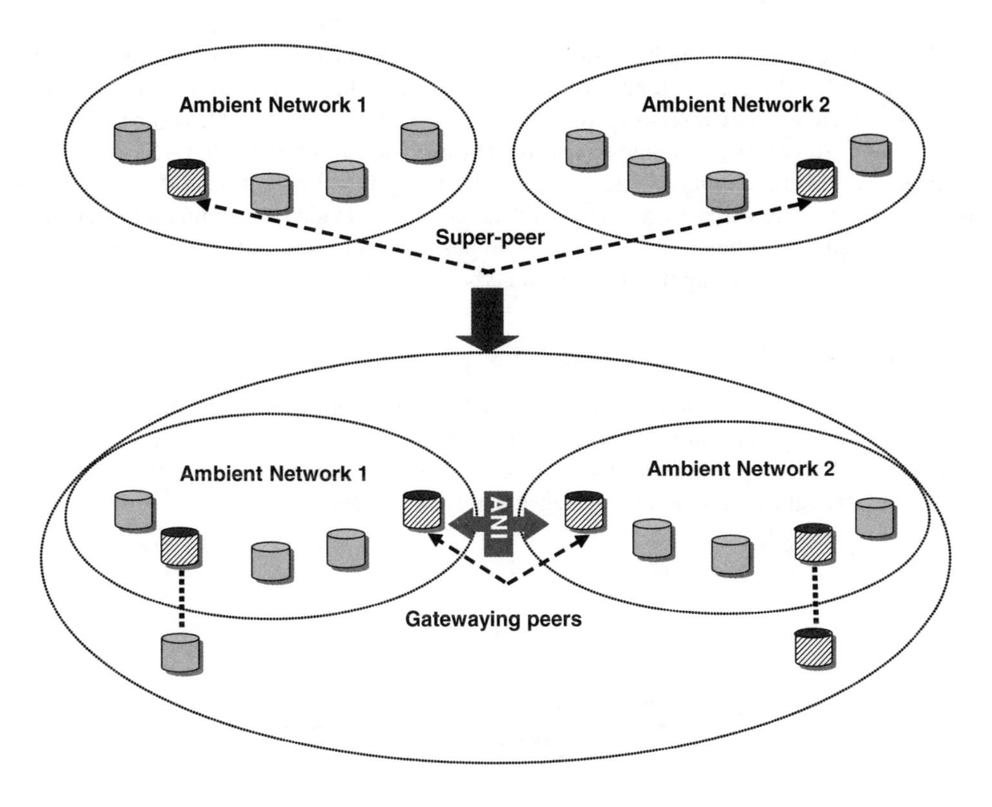

Figure 6.8 Gatewaying

super-peer. The resources not passed to the overlay ACS can be accessed through the overlays and super-peers of the original ANs.

6.3.2.3 Interworking Management Overlay Type

The interworking management overlay type is selected if no new overlay is created. It is a rather independent interworking of the two ANs; therefore, we name it interworking. This also results in the same AN interfaces to the exterior. From the graph representation point of view, this is a nonoperation.

6.3.3 Multilateral CAs

When two ANs compose according to the AN composition architecture, a new common/virtual ACS might be created depending on the negotiated CA. The composition always occurs between two ANs at the time, so it is a bilateral network operation in its base form. But this bilateral operation can be applied multiple times so that multilateral compositions consisting of three or more ANs can be supported. As already mentioned earlier, the composition related signalling takes place in between different ACSs, and for each ACS there exists a related CA. Considering multilateral compositions, there is a CA that is agreed among all constituent ANs. So if the composition does not result in the creation of a common/virtual ACS, then there is no ACS relating to the agreed CA reflecting the established composition. This is the reason why such composition cannot be extended according to the AN composition architecture. So the common/virtual ACS is a prerequisite for multilateral compositions independently on how related resources are shared, as it provides the common/virtual ACS for the agreed CA to be a part of further composition processes and therefore this aspect should be considered during the composition type decision process.

6.3.4 Decision Process and Relations

The composition type decision process involves the analysis of two different factors: (1) common control of agreed resources and (2) need to support multilateral compositions. Figure 6.9 depicts this decision process and the relations between resource management composition types, multilateral CA support and management overlay types.

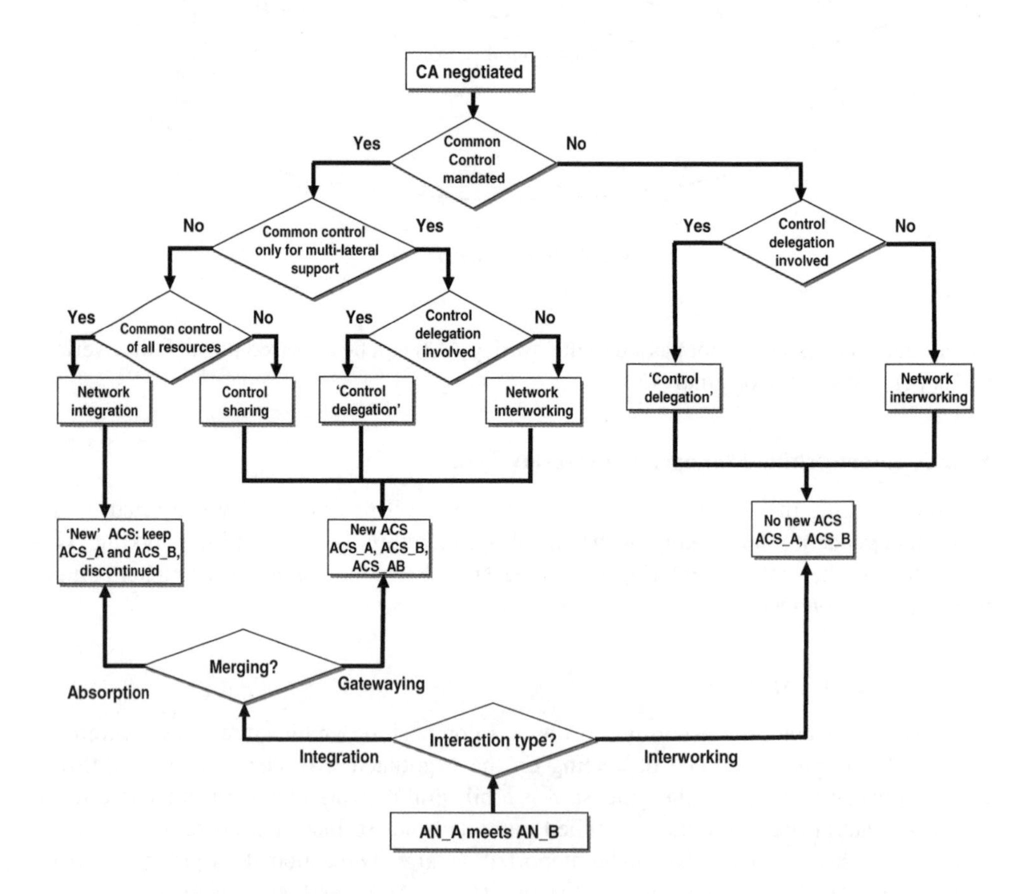

Figure 6.9 Decision tree for selection of composition and overlay types

6.4 Conclusions

It is evident that the composition procedure must be very flexible and efficient, as well as open, extendible and support for security and mobility, as the use cases have such a huge span. It is very hard to predict all possible future use cases which need to be supported by the generic composition framework. All those aspects have been taken into account when defining the composition procedure. The composition types from a (resource) management perspective as well as the AN identifiers provide a framework for the outcome of composition and how it shall be maintained and controlled. Network management is integrated in the overall composition framework to ensure the maintenance of a consistent picture of all networks and their constituents participating in the composition process.

The work is by no means exhaustive, but the focus has been to harmonize the current views on composition as reflected by current status of work throughout the AN project. This has led to a focus on the requirements as well as the composition concept itself.

7

GANS – Generic Ambient Networks Signalling

Acknowledgements

This chapter is based on the joint experiences and efforts of the researchers in the first phase of the AN project and particularly the following people listed as contributors and authors (i.e. in alphabetical order): Nadeem Akhtar, Yaning Wang (University of Surrey, UK), Louise Burness (British Telecommunications Plc.), Rui Campos (INESC Porto), Philip Eardley (British Telecommunications Plc.), Joachim Hillebrand (DoCoMo), Roger Kalden (Ericsson EED), Cornelia Kappler (Siemens AG), Pekka Koskela (VTT Electronics), Gosta Leijonhufvud (Ericsson EAB), Paulo Mendes (DoCoMo), Cornel Pampu (Siemens AG), Carlos Pinho (INESC Porto), Petteri Poyhonen (Nokia), Christian Prehofer (DoCoMo), Gidon Reid (British Telecommunications Plc.), Manuel Ricardo (INESC Porto), Jose Ruela (INESC Porto), Brynjar Viken (Telenor), Thiemo Voigt (Swedish Institute of Computer Science) and Di Zhou (Siemens AG Austria). This chapter has been edited by Cornelia Kappler, Nadeem Akhtar and Paulo Mendes.

7.1 Introduction

In previous sections, network composition was introduced as a central concept in Ambient Networks. When two Ambient Networks compose, the FEs in one AN need to communicate with FEs in the other AN using the ANI in order to realize a composed ACS, see Figure 7.1. More specifically, we need a protocol to realize the phases 'composition agreement negotiation' and 'composition agreement realization' of the composition procedure introduced in Chapter 6. Only some of these communication needs are covered by existing protocols. In this chapter, we introduce GANS, the Generic Ambient Network Signalling protocol, for signalling across the ANI. GANS is an extensible protocol framework for communication between FEs. In other words, GANS complements existing protocols, it does not replace them.

This chapter is organized as follows. In the remainder of this section, we illustrate GANS usage by means of several use cases. The use cases allow us to describe the problem to be solved by GANS in more detail and to present the basic design idea. Furthermore, we consider

Ambient Networks: Co-operative Mobile Networking for the Wireless World Norbert Niebert (Ericsson GmbH), Andreas Schieder (Ericsson GmbH), Jens Zander and Robert Hancock

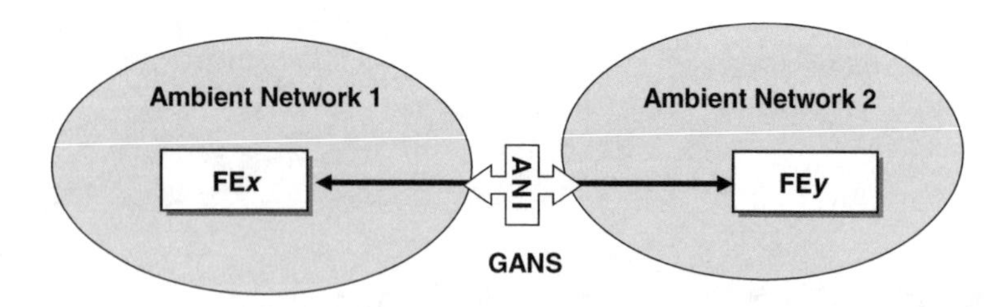

Figure 7.1 GANS signalling architecture

the state of the art. Section 7.3 presents the high-level architecture, followed by description of different components of the protocol suite in Sections 7.4 and 7.5. Finally, Section 7.6 provides outlook and conclusions.

7.1.1 GANS Use Cases

When two Ambient Networks compose, they may need to coordinate the different functional entities that make up the Ambient Control Space. The most basic of the control functions are addressing and routing. In today's IP networks, IP addresses usually are static, autoconfigured or assigned by a DHCP server. New nodes join a network on an individual basis and must adopt the addressing scheme being used in the network. However, when an entire network joins another network, perhaps as a result of a composition for network integration, e.g. merging of two personal area networks (PANs) [63], it is less clear whose addressing scheme should be used. Let us assume that both networks use DHCP servers to allocate private (nonglobal) addresses. In this case, the two networks need to agree on whether one of the servers should be disabled or whether both should continue to run. In the latter situation, the address spaces from which the DHCP servers allocate addresses must be configured to be nonoverlapping. The coordination of DHCP servers can be performed by the servers themselves or by central entities representing the two networks. Either way, coordination of DHCP servers is a typical problem that could be solved by the GANS signalling between the two networks. Besides the actual signalling semantics, the problem that needs to be solved is that, for example, DHCP server A in network A initiating the coordination must locate its corresponding entity DHCP server B in network B and establish a secure signalling relation with it (see Figure 7.2). Clearly, with the coordination of the DHCP servers the addressing problem is not yet solved. For example, duplicate addresses must be detected and new addresses must be assigned. However, this problem must be solved within DHCP scope rather than with the GANS protocol.

A well-known scenario also valid for composition is that of control coordination between networks in the negotiation of service level agreements (SLAs). Here, the QoS-FEs, e.g. bandwidth brokers, negotiate the QoS – bandwidth, priority – that will be provided to user traffic originating from the other network. As above, the bandwidth brokers need to find each other and establish a secure signalling relationship. When SLAs are negotiated in today's networks, the relation between the bandwidth brokers usually is statically configured. However, in future dynamic network scenarios, with entire networks on the move, this is no longer feasible. GANS therefore includes functionality for dynamically locating signalling peers

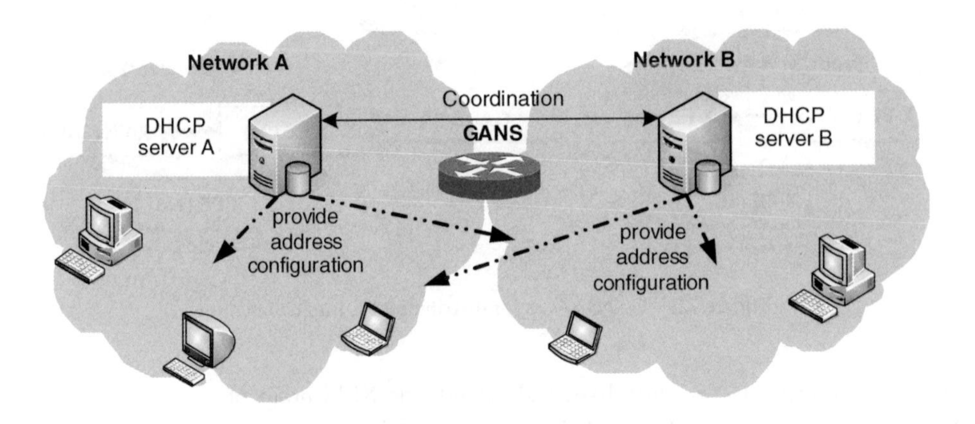

Figure 7.2 DHCP server coordination

and for securing their signalling relationship. Note that we can illustrate the problem with the same Figure 7.2 above, by replacing 'DHCP server' by 'bandwidth broker'.

The last scenario concerns the case of two networks trying to discover information on each other. One could picture a database in each network that provides information on link layer technologies supported, protocols used, etc., similar to the media-independent information service defined in IEEE 802.21 [69]. Before the networks start interworking, they could consult each other's database to find out whether they are compatible. Also, this communication could be performed by means of the GANS protocol.

7.1.2 GANS Scope and Design Approach

The usage scenarios presented above show both common and specific treats. On the one hand, every scenario illustrates a signalling application in a specific field requiring specific handling and processing. On the other hand, all the scenarios require the solution of similar problems. Specifically, the problems common to all three examples include

- *Localization of the signalling peer*: When FE x in AN1 (see Figure 7.1) wishes to communicate with FEy in AN2, it usually does not know FEy's IP address, nor does it care about this address. Therefore, we need an address resolution mechanism in GANS that resolves 'FEy.AN2' into an IP address. When the location of 'FEy.AN2' changes due to mobility or simultaneous other compositions, the address-to-location mapping needs to be updated.
- *Security*: FEx and FEy may need to establish a signalling relation and possibly a security association which enables them to securely transport messages between them.
- *Transporting signalling messages* between signalling peers.

The common problems on the one hand and the specific problems on the other hand lead to a design approach based on a two-layer model: a signalling layer consisting of individual protocol entities handling special needs of each specific application field and a transport layer providing a general message delivery service ensuring efficient and secure signalling data transportation among cooperating networks. This approach is consistent with the architectural approach of the IETF working group NSIS (Next Steps In Signalling) protocol

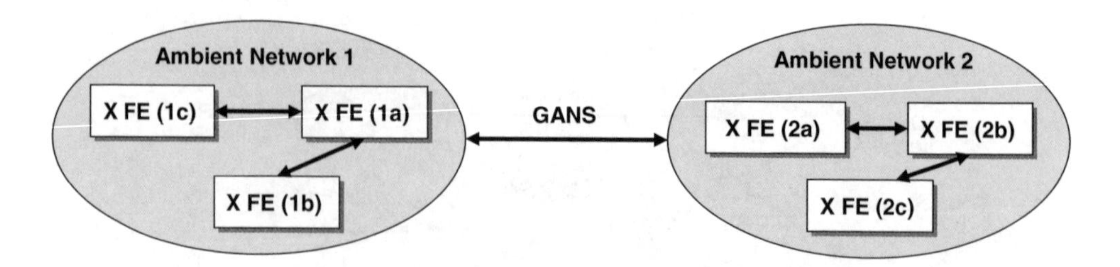

Figure 7.3 GANS usage for distributed FE implementation

suite [68]. In fact, GANS generalizes and extends the NSIS protocol suite. Going beyond NSIS, the lower transport layer offers an address resolution service based on a stateless protocol, DEEP (Destination Endpoint Exploration Protocol). In this chapter, we introduce the design for the lower GANS transport layer, for DEEP, and one specific GANS application for signalling SLAs.

The reader may remember that an 'FE' is not necessarily a single physical node. An 'FE' may in fact be a number of physical nodes that collaborate for achieving a specific function, e.g. QoS. From a GANS perspective, each FE is represented by a single node (the 'contact point'). The contact point is that node in an Ambient Network which (at this point in time) receives GANS messages for a given FE. The contact point may distribute or forward the GANS message to other nodes in the same AN; however, this is not a GANS problem. The contact point may even be a general gateway of the AN, which is responsible for several FEs. This concept is illustrated in Figure 7.3.

7.2 State of the Art

In this section, we discuss protocols with functionality related to GANS and show to what extent they cover or do not cover the functionality desired. Generally, we need a flexible, extensible protocol for exchanging control messages between FEs in the composition process. Recently, a number of protocols have been designed with flexibility and extensible in mind, e.g. the Session Initiation Protocol (SIP), Diameter and the NSIS protocol suite. Furthermore, we need a protocol for resolving abstract names (e.g. FE*x*.AN1) into IP addresses. Here, we must compare our approach with the functionality provided by the Domain Name System (DNS) and the Host Identity Protocol (HIP).

In addition to the general picture, we need to take into account specific GANS applications (i.e. the upper GANS layer). As stated above, GANS complements existing protocols, it does not replace them. One typical GANS application is the negotiation and realization of SLAs as a part of the composition process. We therefore in Section 7.2.5 analyse the state of the art in SLA signalling.

7.2.1 Session Initiation Protocol

SIP [64] is a protocol that supports signalling for creating, modifying and terminating sessions. For example, SIP can be used for agreeing on codecs, ports and QoS before a session is established. SIP is independent of underlying transport protocols and the type of session being established and is designed to run as an end-to-end protocol between user applications at end hosts. SIP is an extendable protocol in the sense that it opaquely carries a session description in Session Description Language (SDL), to which new session types or session characteristics can be added.

From the above description, it is clear that SIP does not cover the functionality we need. A SIP session relates to a specific user session, whereas the control signalling during composition takes place before any user session is established. Furthermore, SIP is an application layer protocol. SIP does however offer a functionality of interest: It resolves location-independent user identifiers, so-called SIP URLs, into the currently valid IP address of the user. This address resolution is integrated into the overall session establishment.

7.2.2 Diameter

The Diameter protocol has been designed to provide an authentication, authorization and accounting (AAA) framework for applications such as network access or IP mobility. The Diameter protocol consists of a 'base' protocol and set of application protocols [65]. The protocol is generic and extensible in the sense that Diameter client protocols can be added on top of the base layer. These clients use the base protocol for signalling specifically related to AAA functions. Thus, Diameter is extensible but its scope is limited to AAA-related signalling. It is not meant for general control signalling as needed here.

7.2.3 Next Steps in Signalling

The NSIS protocol suite [66] is being developed for signalling information pertaining to data flows. The aim is to support different signalling applications that need to install and/or manipulate such state in the network. The NSIS suite has a lower transport layer that provides common functions such as session management, peer discovery, secure channel establishment, etc. On top of the transport layer resides a set of signalling applications for diverse control signalling, such as QoS control, middlebox communications, etc. Examples of NSIS signalling are protocols for reserving QoS [68], configuring Network Address Translation entities and firewalls [67]. A future application may be a protocol for configuring metering entities [71]. Such diverse applications are supported by defining two layers. The lower layer, NTLP (NSIS Transport Layer Protocol), takes care of tasks common to all signalling applications: finding the neighbour signalling peer, establishing a security relation with this peer and transporting messages to upstream and downstream adjacent peers. Although in principle several NTLPs are conceivable, the only NTLP currently defined is called GIST [70]. The upper NSIS layer consists of different signalling applications such as those enumerated above. The signalling application protocols, referred to as NSLPs (NSIS Signalling Layer Protocols), define the signalling semantics that are specific to the requirements of each NSLP. They form a message and pass it to the NTLP for delivery.

NSIS is generic and extensible, but it is geared towards providing signalling support for data flows. Control signalling required for dynamic interworking requires a much more flexible approach, including a wider set of signalling applications as well as a broader operational scope.

The generic and extensible protocol suite we propose in this paper is based on the NSIS protocol suite. However, there are significant differences between the two. Most importantly, unlike NSIS, the scope of our protocol also includes signalling that may have no direct relation to any underlying data flows.

7.2.4 Naming Systems

The previous sections on state-of-the-art protocols dealt with protocols generally related to control signalling, in this section we cover existing protocols for address resolution.

Traditionally, DNS [72,73] has provided global address resolution services to network nodes participating in IP-based communication. DNS resolves names into locators, i.e. IP addresses. Additionally, DNS also has other functions than mapping between names and locators like name-based service discovery, certification and public key distribution, etc. For the static Internet, this has proved to be a feasible and working solution. However, with the tremendous increase in the number of mobile devices and a growing trend towards small, self-contained and often private, mobile wireless networks, the ability to operate with limited and even no infrastructure support is increasingly important. For instance, in a mobile ad hoc network, one cannot assume the presence of conventional DNS servers and other networking infrastructure. In fact, even if such entities were present, the highly dynamic nature of ad hoc networks would render them rather useless due to constant updates. A number of approaches have been discussed in the literature to provide name and service resolution in such scenarios. The Link-Local Multicast Name Resolution (LLMNR) [195] has been designed for name resolution in scenarios where it is not possible to use conventional DNS-based name resolution. As the scope of LLMNR is limited to the local link, it is not meant to be an alternative to DNS. LLMNR provides mechanisms to perform DNS-like operations on the local link when a conventional DNS server is unavailable. Furthermore, it sets aside a part of the DNS namespace for local use; resolution of names in this namespace does not require a conventional DNS server.

In parallel to DNS, another resolution infrastructure for the Internet is currently being developed in the HIP working group of the IETF. HIP aims to separate identity from locator, which so far are both coded by the IP address. The HIP architecture includes an identifier resolution infrastructure consisting of HIP rendezvous servers [51]. As opposed to DNS, HIP resolves cryptographic identifiers rather than names.

The schemes mentioned above are mainly based on the current Internet naming scheme, namely the Domain Name System, which assumes that applications want to reach a locator where the address determines a network attachment point (location) in the network topology. In [78], the authors argue that inherent rigidity of the naming systems used currently have proven to be a big hindrance towards efforts to efficiently enable new services such as mobility, group communication, resource discovery, service location, caching, etc., and they propose an intentional network naming architecture, where applications describe what they are looking for (i.e. their intent), not where to find it.

Based on the research reported above, it is safe to say that future networking environments will have not only diverse naming and addressing architectures but also different schemes to provide name resolution services for different network domains. However, as all these domains eventually have to interwork for the Internet to work, it is absolutely essential to bridge the gap between these diverse architectures and protocols. The presence of potentially diverse address and name space requires a mechanism that allows resolution of names across different networking domains that may have different name resolution systems. DEEP is designed to provide this service as will be clear in the following sections.

7.2.5 Internetwork QoS Control

The control of QoS between networks in heterogeneous and dynamic environments requires internetwork signalling to allow the mapping of the QoS assurances provided by each network. We cluster possible signalling solutions into two groups: one providing flow-based signalling and another group encompassing solutions that are traffic oriented or better to say solutions that provide service level specification (SLS) based signalling. In general, we characterize flow-based signalling approaches as the ones with control messages related to N-tuples defining the paths taken by data flows. For this, devices are needed in the data path to configure network resources for a flow of a group of flows. This type of signalling is implemented by protocols such as RSVP [74], QoS-NSLP [68] or the Simple Interdomain Bandwidth Broker Signalling (SIBBS) [75].The major difference between RSVP, QoS-NSLP and SIBBS is that although the first signals all network nodes in a data path of the signalled flow (data path-coupled) and SIBBS signals network nodes not traversed by flow data (data path-decoupled), NSIS was designed to provide path-coupled signalling but is proposed to being extended to signal also decoupled from the flow data path [196].

SLS-based signalling approaches are characterized as the ones allowing two adjacent networks to establish and maintain a set of bilateral QoS agreements (SLSs) for different types of traffic. This means that signalling entities have no association with data paths. This type of signalling is implemented by mechanisms such as the QoS extensions to BGP (qBGP) [76], developed in the IST Mescal project or the dynamic control for internetwork QoS agreement (INQA) described in this chapter.

The major difference between qBGP and INQA is that although the former is tightly coupled with an internetwork routing protocol (BGP) [77], INQA is independent from internetwork routing schemes. This means that INQA may be used to configure the internetworking not only between big networks, but also between moving networks.

Independently from their functional characteristics, any internetworking QoS control solution must support the use of different intranetwork QoS control technologies and must allow the automatic establishment and maintenance of QoS between moving networks. Moreover, such a solution should allow networks to be aware of the QoS level of communication services available in neighbour networks and should scale with the number of networks, flows and offered communication services.

Flow-based and SLS-based approaches may support different intranetwork QoS control technologies by using a path-decoupled approach, providing a clear separation between internetwork signalling and the signalling used inside a network to configure edge-to-edge data paths. Flow-based path-coupled signalling approaches require all network devices inside all networks to support the same signalling scheme.

Both flow-based and SLS-based approaches may keep an updated state of the QoS established between networks. Nevertheless, although some flow-based approaches are being adapted to react to the movement of hosts, as is the case of QoS-NSLP, in general they have problems handling the movement of networks, as they are only aware of mobility near flow initiators and destinations. Moreover, the movement of hosts requires flow-based approaches to signal the complete new path, as resources are coupled with N-tuples that are different before and after movement. SLS-based approaches are aware of mobility, as routing changes, at any network edge and require only the adjustment of a subset of SLSs due to network movement.

Contrary to flow-based approaches, SLS-based approaches may furnish networks with awareness about the QoS of communication services that their neighbours may provide to them. This is possible in solutions such as INQA, which includes control messages allowing networks to advertise service availability to neighbours. On the contrary, flow-based approaches are based on query or reserve messages that are triggered by a local need for resources, without having any previous knowledge about what QoS neighbour networks can offer. This lack of information about how to reach different networks, via different neighbours, and with certain QoS assurances, even before the creation of data flows also brings disadvantages to multi-homed hosts and networks.

7.3 Protocol Architecture

The GANS protocol suite has been designed keeping in mind the unique requirements of Ambient Networks, especially with regards to network composition and the dynamic nature of the ACS. As mentioned earlier, the architectural principles of NSIS have been followed quite closely for GANS design. GANS adopts the NSIS two-layer approach as shown in Figure 7.4. In fact, the GANS lower layer is a backward-compatible extension of the NSIS lower layer, GIST.

The GANS protocol suite consists of the GANS Transport Layer Protocol (GTLP) and several GANS Signalling Layer Protocols (GSLPs). The service GTLP offers to the GSLPs includes finding the neighbour signalling peer, establishing a security relationship with it and transporting messages, just as NTLP. The difference with NTLP is in the location of the signalling peer. In NSIS, the peer is found by exploiting the fact that it must be the next suitable NSIS hop on the data path towards the destination of the flow. In GANS, the signalling is not always flow related, and the IP address of the signalling peer is not always known either. Rather, it is known that the signalling peer is located in a specific network and that it fulfils a certain role, e.g. 'DHCP server' or 'bandwidth broker'. The GANS signalling application on top of GTLP therefore just addresses the message to a symbolic name, e.g. 'DHCPServer.networkX'. The main challenge for GTLP is then to resolve the role-based identifier into a routable locator, e.g. an IP address; this resolution is transparent to the application. The protocol design focuses on the signalling aspect and assumes that an infrastructure exists for resolving the symbolic name. Very few assumptions have been made about this infrastructure,

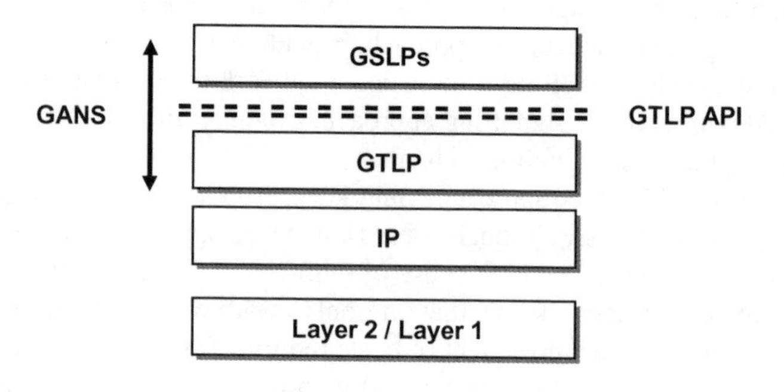

Figure 7.4 Two layer GANS model

e.g. DNS might be an option. The GANS name resolution clearly is an alternative to anycast addressing. The resolved routing address will be stored locally so that subsequent protocol messages could be forwarded directly without using the symbolic address again. The GANS protocol stack structure is illustrated in Figure 7.5.

GTL API is the interface provided by transport layer to GSLPs. It is a simple extension of and backward compatible to GIST API. Signalling applications using the GIST API have to provide the IP address of the destination when sending a message. The GTLP API enables applications to use symbolic names as well to address the destination endpoints of signalling messages.

The message transport function in GTLP is called extended GIST; it is based on GIST and includes additional functionality to meet the requirements of Ambient Network signalling. EGIST is responsible for the transportation of GSLPs signalling messages. Like GIST, it uses services offered by the underlying layers and relies on the routing functionality provided by the underlying network layer.

DEEP is designed for resolving the signalling destinations specified by symbolic names; it implements the symbolic name support in GTLP. The internal interface of DEEP hides local name resolution service details from the rest of GTLP components. Only the GSLP signalling messages addressing destination using symbolic names will involve the use of DEEP name resolution services. The signalling messages using IP addresses are processed without the use of DEEP.

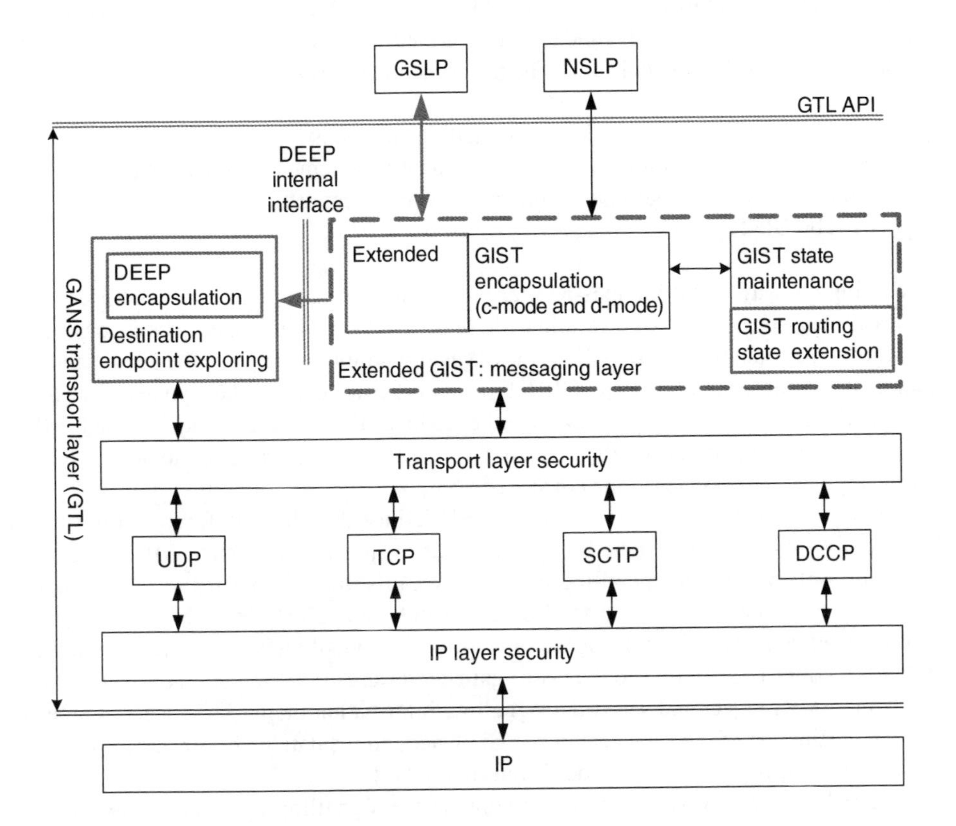

Figure 7.5 GANS protocol stack

7.4 GANS Transport Layer Protocol

The GANS Transport Layer Protocol comprises two main building blocks: DEEP and the extended GIST, which provides the actual transport framework for the GSLP messages received through the GTLP API. The latter is the standard interface provided to the GSLPs and is backward compatible with the GIST API specified in [18], so that applications running on top of GIST can also run over GTLP without any modification. The parameters and primitives defined for the GIST API are maintained. When GTLP receives a message from a GSLP through the GTLP API on the same node, the DEEP module is invoked if a symbolic name is used to identify the destination in order to find out the IP address of the node in the remote Ambient Network to which the message should be sent. Once the name is resolved, EGIST encapsulates the GSLP message and sends it to the IP address returned by DEEP. Depending both on the information passed from GSLP and on local policies, EGIST will use an unreliable (e.g. UDP) or reliable (e.g. TCP) underlying transport protocol to do so. If required by the GSLP, the two nodes also establish a security association.

7.4.1 Extended Generic Internet Signalling Transport Protocol

The EGIST protocol comprises the actual transport framework provided by GTLP to the signalling applications running over the GTLP API. As mentioned before, EGIST is based on the GIST protocol. It comprises all functionality defined by GIST and additional functionality required for the GANS protocol suite. The new functionalities are

- support for symbolic names to identify signalling endpoints;
- interaction with a new protocol (DEEP) implementing a name resolution scheme;
- name binding state storage and maintenance without modifying the routing state table specified by GIST.

Concerning the way signalling messages are routed towards the proper adjacent peer, GIST defines the concept of Message Routing Method (MRM) which defines the algorithm for discovering the route that the signalling messages should follow between the source and destination signalling endpoints at the GIST layer, in order to forward the signalling messages between GIST nodes. Multiple MRMs can be defined, the so-called ‘path-coupled’ MRM being the default one. It defines the activation of the IP router alert option to deliver the signalling application messages to the correct adjacent peer.

In conjunction with the MRM concept, GIST defines the Message Routing Information (MRI) object, which allows signalling applications to specify a particular MRM that GIST should use to forward their signalling messages towards the right destination. For instance, concerning the default MRM, the MRI object includes an MRM identifier (pointing to the ‘path-coupled’ MRM) and details of the flow for which signalling is being exchanged. Based on this MRI object, GIST finds out its adjacent peer towards the destination of the flow identified in the MRI object and sends the signalling information there; the process is repeated until the destination of the flow, or the last GIST node in the data path, is reached. For more information on this process, the reader is referred to [18].

In EGIST, a new MRM is defined to support the signalling applications expected to be designed as a part of the GANS protocol suite. Referred to as the ‘path-decoupled/

single-hop' MRM, it requires modifications with respect to the internal processing of the messages received from the signalling applications, namely regarding the way messages are routed and the way the two operation modes (connection mode or c-mode and datagram mode or d-mode) defined by GIST are used. The new MRM allows applications to use symbolic names for addressing destinations. At the same time, applications can also use IP addresses if such information is available. The support of symbolic names inside EGIST requires the storage of information about the mapping between symbolic names and the corresponding IP addresses. DEEP protocol in its current form is stateless, i.e. it simply passes on the resolved address to EGIST and does not store the bindings locally. However, as signalling applications may need to send messages to the same destination frequently, it is not efficient to invoke DEEP every time a message needs to be sent. Therefore, EGIST maintains such state information. No new data structure is defined for this purpose; instead, EGIST uses the routing state table already specified by GIST and encapsulates the name binding state information inside the MRI object. This is possible due to the flexibility of the routing state table, which stores the MRI object as a whole instead of each parameter separately, hiding the specific details associated with each MRM. In this sense, every new MRI object defined in the context of a new MRM can be stored within the same table.

Simply storing the name binding state is not enough as the mapping may change, especially in the dynamic environments envisaged for Ambient Networks. When there is a relocation of signalling endpoint, which causes a change in the IP address and/or in the port being used by one of the signalling endpoints participating in a signalling session, there is a need for updating the related name binding state. EGIST is responsible for such update, as it is the entity storing the name binding state. For this purpose, the GIST refresh procedure, based on the exchange of Query and Response messages between the signalling endpoints, is reused.

In a nutshell, it can be said that the EGIST protocol is, as the name implies, an extension to the GIST protocol. It does not modify any of the GIST functionalities; rather, EGIST extends the applicability domain of the GIST protocol by providing further features not included in the original protocol. EGIST is backward compatible with GIST, supporting both the applications already defined in the NSIS scope, so-called NSIS Signalling Layer Protocols (NSLPs) and the GANS signalling applications (GSLPs).

7.4.2 Destination Endpoint Exploration Protocol

The Destination Endpoint Exploration Protocol is designed for locating the signalling destinations specified by symbolic names. The main purpose of DEEP is to support a distributed resolution of an abstract name into an IP address. It provides name resolution services to GTLP via a unified internal interface, which hides the implementation details of a local name resolution service. DEEP is not tied to any specific naming scheme; it is designed to support abstract names based on heterogeneous name spaces. DEEP is based on a query–reply messaging model. As mentioned earlier, it is stateless, i.e. no additional protocol or name resolution related state information is stored in any DEEP node. Furthermore, it is transparent to DEEP how local name resolution services are storing name resolution information like caching it locally for further use. Although DEEP does not provide any name resolution caching, the underlying name resolution services (e.g. DNS) may do so resulting in improved name resolution service response times and decreasing the overall DEEP message processing times in DEEP nodes.

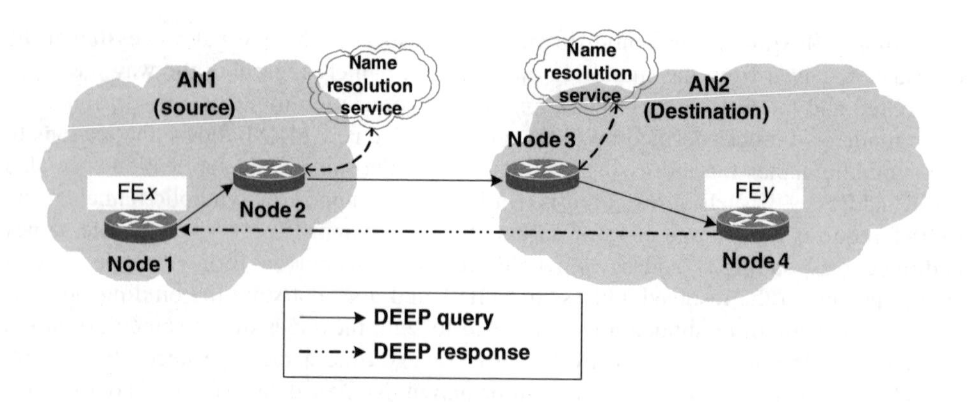

Figure 7.6 An example of DEEP name resolution

DEEP name resolution is typically a gradual process, which resolves an abstract name into a locator (e.g. IP address) in distributed manner involving multiple DEEP nodes and name resolution services that can be independent of each other. Although a variety of mechanisms are conceivable, a sequential approach has been proposed in which the name is resolved in several steps, by sending the same query message from one DEEP resolver to the next one until the IP address of the responsible node is discovered. Each DEEP node is configured to provide name resolution services to other DEEP nodes. DEEP nodes use two external interfaces: a local name resolution interface for accessing local resolvers and a message transport service interface for exchanging messages with other DEEP nodes. Some DEEP nodes may be configured to act as 'forwarders', i.e. based on a local configuration, all received DEEP messages are always sent towards the same DEEP node.

For example, consider a scenario with two Ambient Networks: AN1 and AN2 (see Figure 7.6). A control signalling application (say FE*x*) in AN1 wants to contact its peer application (say FE*y*) in AN2, but it does not know its peer application's contact address. The application passes the message to GTLP where it is handled first by EGIST which requests DEEP to resolve the name FE*y*.AN2. Therefore, a DEEP message with the address FE*y*.AN2 is first sent to a name resolving node local to the originating AN1. This node may know either the IP address of the name resolver at AN2 or the IP address of an intermediary name resolver that is able to resolve a part of the current abstract name towards the complete name resolution by using the sequential procedure mentioned above. Upon knowing the IP address of the next resolver, the current DEEP node sends the 'DEEP query' message there. This process is sequentially repeated until the name resolution process reaches the name resolver at AN2. The name resolver at AN2 knows the IP address of the contact point for FE*y*. At this point, the name is fully resolved and the 'DEEP response' message is sent back to the originating node in AN1.

7.5 QoS Signalling Application

To mitigate the lack of control of inte rnetwork QoS agreements in heterogeneous and dynamic environments, we incorporate in GANS a mechanism for the automatic cooperation between networks. This mechanism aims to provide a clear separation between internetwork

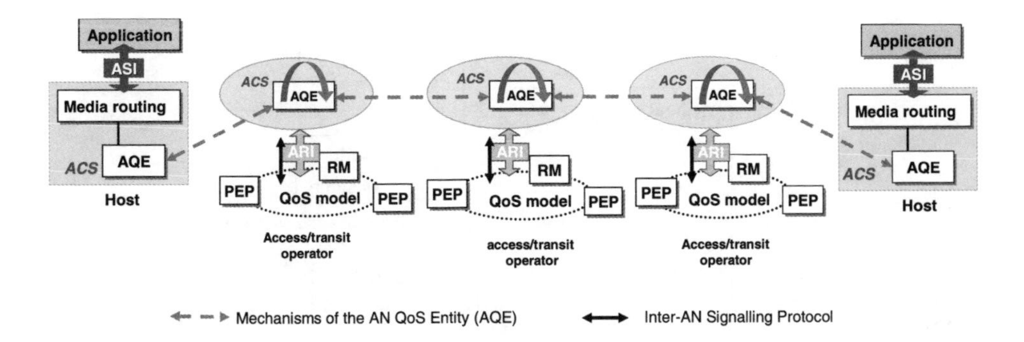

Figure 7.7 Illustration of INQA signalling

and intranetwork QoS signalling, to handle traffic and not flows (to allow the system to scale), to automatic adjust internetwork agreements and to support of services based on elastic and nonelastic traffic.

This mechanism is called INQA, which stands for *dynamic control of internetwork QoS agreements*. INQA is based on modules to advertise (INQA-A), negotiate (INQA-N), monitor (INQA-M) and admit (INQA-AC) QoS agreements, as well as a new inter-AN signalling protocol (INQA-P). In the Ambient Networks framework, the INQA signalling protocol is implemented as a GANS *Signalling Layer Protocol*; in the remainder of this document, INQA-P will always be referred as INQA-GSLP.

Within the INQA framework, each network can have three roles: provider, customer and customer–provider. In the provider role networks advertise local network services, whereas in the customer role networks negotiate network services that were offered in previous advertisements. These negotiated services are to be used by local applications. Networks operate in the customer–provider role when they want to 'resell' services advertised by their neighbours.

Figure 7.7 shows that INQA modules are implemented in a network element called AN QoS entity (AQE). The AQE is one of the components of the QoS-FE, which may include other QoS-related functionalities in the ACS. Three[1] of the INQA modules use INQA-GSLP to signal between adjacent neighbours via the *Ambient Network Interface*. Although INQA-A uses the ANI to advertise SLSs, INQA-N uses it to negotiate previously offered SLSs and INQA-M uses the ANI to negotiate *service level indications* (SLIs), that is, the capability to monitor previously negotiated SLSs. INQA-GSLP is a path-decoupled signalling protocol, in which messages are exchanged between adjacent AQEs.

The green arrows shown in Figure 7.7 illustrate the internal signalling exchanged between different INQA modules in the same AQE, as well as the signalling exchanged between the AQE and network dependent functions via the *Ambient Resource Interface* (ARI). Although INQA-A does not use the ARI, INQA-N uses it to indicate to the local QoS control mechanism the need to realize a set of negotiated QoS agreements. On the contrary, the INQA-M module uses the ARI to collect, from local measurement mechanisms, information about the capability of each of the paths between a pair of edge devices.

[1]The INQA-AC functionality is internal to AQE, having no interaction over any interface.

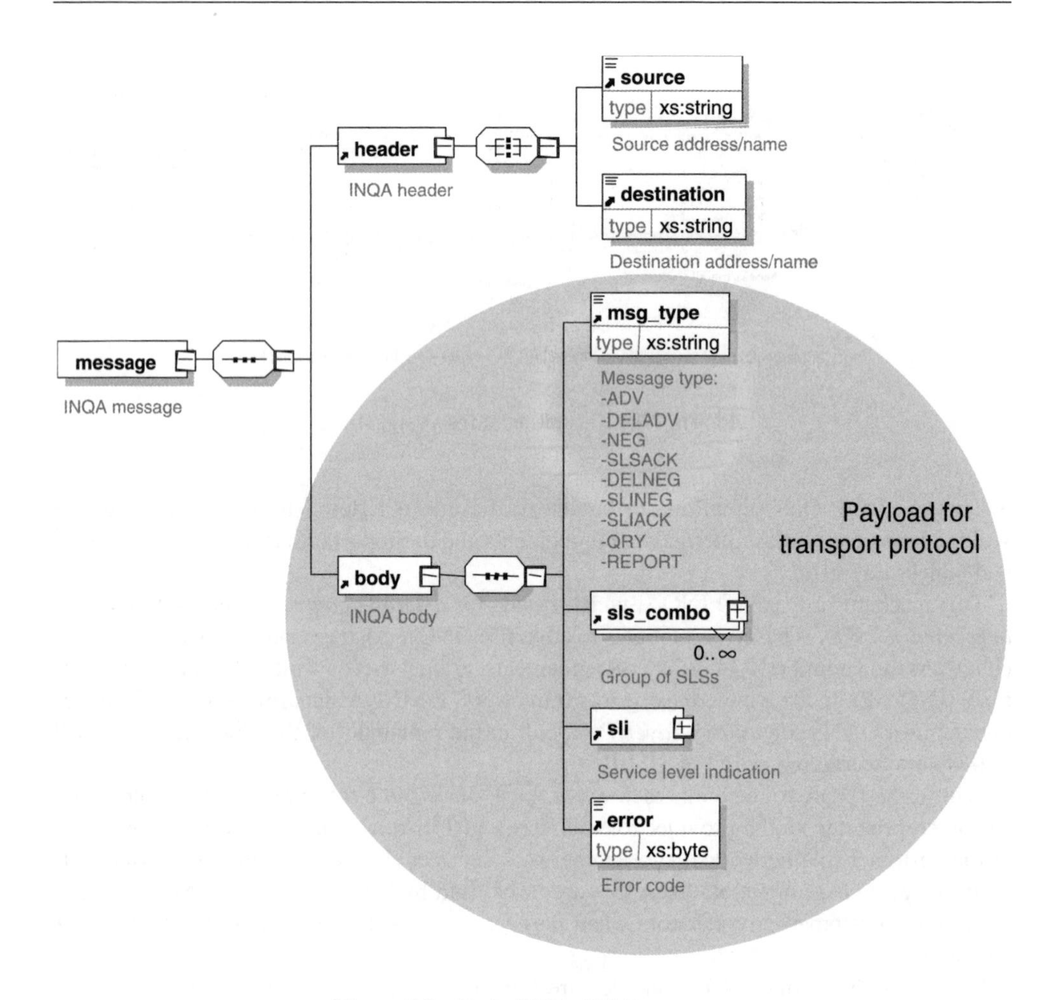

Figure 7.8 Basic INQA-GSLP structure

The INQA protocol is designed to signal bilateral operations between AQEs of ANs. The protocol makes use of an XML-based description to exchange information about SLSs, SLIs and further auxiliary information. Each INQA-GSLP XML message contains a <message> element, which forms the root element of each INQA-GSLP message. The <message> element contains a <body> element and an optional <header> element. Figure 7.8 shows the tree of XML elements for an INQA-GSLP message. The information present in the header may be omitted in normal protocol operation when provided to/from the underlying transport protocol. As shown in Figure 7.8, the INQA-GSLP allows AQEs to combine SLSs to the transmitted, by creating an SLS-Combo. Each SLS-Combo has a common field, created with the parameters common to all SLSs and by a sequence of SLS fields, each of which has the parameters specific to each SLS.

INQA-GSLP is used over the ANI. Figure 7.9 shows the protocol interfaces for INQA-GSLP. There are interfaces to herein defined operational modules (advertisement, negotiation, monitoring), to future modules like an SLS aggregation mechanism and to the transport protocol. For the operation of INQA-GSLP over the ANI, the following mechanisms the assumed:

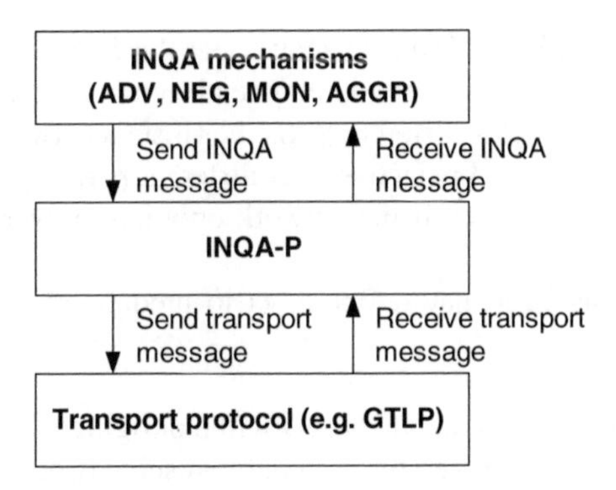

Figure 7.9 INQA-GSLP protocol interfaces

- Discovery of the identifier of the AQE of neighbour ANs.
- Translation from names to locators (IP addresses).

This functionality is provided by the DEEP protocol. However, in order to allow INQA to scale, it is required that the developed naming scheme can also operate without exact symbolic names, e.g. names containing wildcards.

Adjacent AQEs advertise and negotiate QoS agreements by using INQA-GSLP to signal one or more SLSs between them. Different SLSs can be used depending on the role of the network and if the AQE is placed in an end host.

Figure 7.10 illustrates different types of SLSs. From left to right, we can see one SLS to be used by a network to advertise a service of 20 Mb/s with 300 ms of average delay, 1 % of losses and 10^{-5} of availability to reach a specific subnetwork via a specific edge device. This figure also shows an SLS that a network can use to negotiate a service for traffic between exactly two end hosts with hard guarantees, and an SLS to be used to negotiate a service with soft guarantees to reach any end host of a determined subnetwork.

Parameter	*Value*
Ingress ID	ingress
Destination ID	Sub-Net ID
Bandwidth	20 Mb/s
Average E2E delay	300 ms
Average loss rate	1 %
Service availability	10^{-5}

SLS to advertise a network service

Parameter	*Value*
Source ID	Host ID
Destination ID	Host ID
Service type	Hard guarantees Ex: RTP/audio
Peak source rate	16 kb/s
Mean source rate	12 kb/s
Maximum E2E delay	100 ms
Maximum loss rate	1 %
Service availability	10^{-5}
Start time	200508010900
End time	200512310900
Periodicity	Weekly

SLS to negotiate a network service with hard g uarantees

Parameter	*Value*
Source ID	Host ID
Destination ID	Sub-net ID
Service type	Soft guarantees Ex: FTP
Throughput	10 Mb/s
Average E2E delay	300 ms
Average loss Rate	1 %
Service availability	10^{-5}
Start time	200508010900
End time	200512310900
Periodicity	Daily

SLS to negotiate a network service with soft guarantees

Figure 7.10 Example of SLS formats

The SLSs illustrated in Figure 7.10 also shows that the SLSs to be negotiated can have a specific active period. This characteristic may lead to a better utilization of network resources as networks can multiplex different SLSs. For instance, two SLSs valid for one year may be both accepted if they will be active over different periods (e.g. one in the morning and another in the afternoon), even if the network only has network resources enough for one of them.

In any network, the manager may define a set to predefined SLSs to be advertised or negotiated. For instance

- Access networks: It is expected that managers will c onfigured their networks with
  - A list of SLSs to be advertised for each defined subnetwork. These SLSs are similar to the one shown in the figure (on the left).
  - A list of SLSs to be immediately negotiated as soon as some suitable SLS is advertised to them. An example of such negotiated SLS is the one shown in the figure (on the right).
- Transit networks: It is expected that managers will configure their networks to advertise a set of SLSs created based on the SLSs offered by other transit or access networks. Transit network would only advertise the created SLSs after having negotiated the SLSs offered by their neighbours.

Figure 7.11 illustrates the signalling needed to establish bilateral QoS agreements. It is worth to emphasize that the illustrated signalling can be used to set more than one QoS agreement. The figure shows the INQA-GSLP signalling over the ANI and the

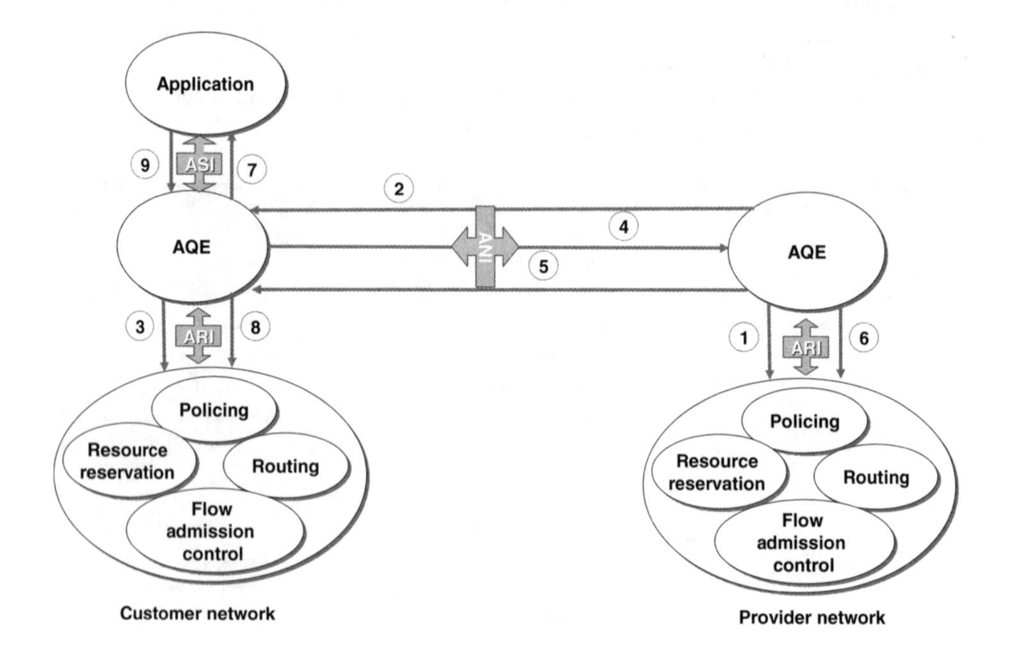

Figure 7.11 Example of INQA bilateral operation

interaction of AQEs with the application and the local QoS mechanism over the ASI and ARI, respectively.

As shown in Figure 7.11, the establishment of bilateral QoS agreements starts when the provider advertises a set of SLSs. This is done in two steps. First, the provider checks what resources are available from the edge device connecting to the customer network (step 1). Second, the provider creates a set of SLSs based on the available resources and advertises them to the customer network (step 2). In order to make the offered services available to local applications, the customer negotiates with the provider a set of SLSs that are supported by the offered ones. To perform this negotiation, the customer starts by checking if its local network can guarantee the quality level expressed in the predefined SLSs (step 3), after which it signals to the provider a request to negotiate them (step 4). The provider verifies if those SLSs can be accepted, by comparing them to the previously advertised SLSs and notifying the customer about the accepted SLSs (step 5). As a result

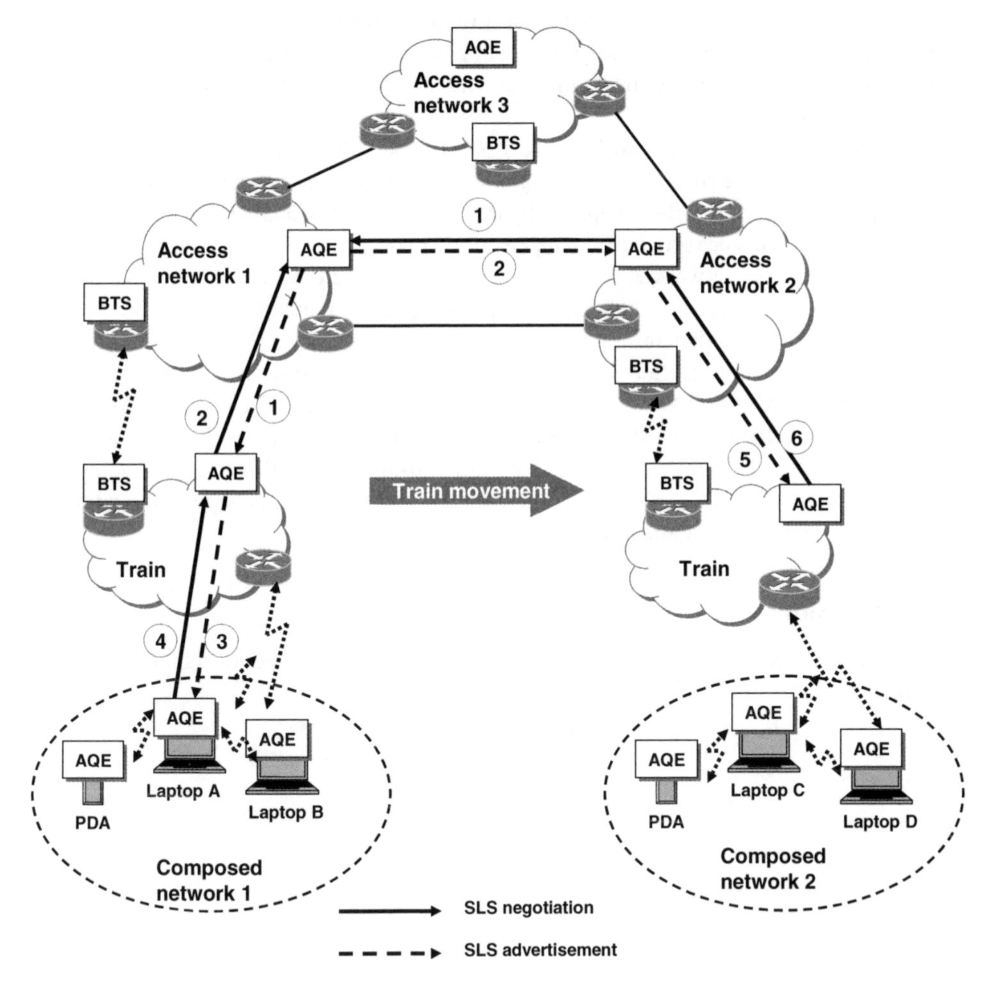

Figure 7.12 INQA operation with Moving Networks

of this operation, the provider notifies the local QoS mechanisms to realize the negotiated SLSs, each time they becomes active (step 6). After receiving the provider confirmation, the customer stores the SLSs accepted by the provider and optionally may notify the application about it (step 7). As a result of this operation, the customer indicates to the local QoS mechanisms to realize the negotiated SLSs, each time they becomes active (step 8). When the application needs to send a flow with some QoS, it requests a suitable service from the customer AQE (step 9). At this point, the customer AQE admits the application request based on a SLS previously negotiated with the provider AQE, without extra inter-network signalling.

To finalize the description of INQA, Figure 7.12 exemplifies the benefits that INQA is expected to bring to scenarios with moving networks. In this situation, access network 1 advertises one SLS to access network 2 and to a train network, allowing the access to one of its subnetworks with a certain quality assurance. Based on these advertisements, both networks negotiate a suitable SLS with the access network 1. Based on the SLS negotiated with the access network 1, the train network advertises a similar service to its local customers. After the reception of this advertisement, one of the local customers negotiates an SLS with the train network, in order to be able to reach in the future a set of devices in the subnetwork advertised by the access network 1.

After moving from access network 1 to access network 2, the train network negotiates with the access network 2 the SLS that it had negotiated before with access network 1. This negotiation is possible only after the reception, by the train, of a set of available services from the access network 2. This advertisement happens immediately after the presence of the train is detected by the access network 2. No extra signalling is needed inside the train and between access networks, as the previously negotiated SLSs are still suitable.

Hence, we can argue that in a scenario with moving networks INQA allows a fast re-establishment of communication sessions, as the moving network is able to find suitable services in new access points, with high probability. Moreover, it allows networks to react to their own movement or to the movement of neighbour networks. The reaction to the movement of networks may not be possible with end-to-end signalling, as that movement may be transparent to end hosts. This means that end hosts will not trigger any end-to-end signalling in order to restore the required state in the new path.

7.6 Conclusions

This chapter focused on the GANS protocol suite that is being designed for signalling between functional elements of different Ambient Networks.

8

Multi-Radio Access

Acknowledgements

This chapter is based on the joint experiences and effort of the researchers in the first phase of the AN project and particularly the following people listed as contributors and authors (i.e. in alphabetical order): Ramón Agüero (University of Cantabria), Eftychia Alexandri (France Telecom R&D), Leonardo Badia (CFR, University of Ferrara), Miguel Berg (Royal Institute of Technology), Fredrik Berggren (Royal Institute of Technology), Marcin Bortnik (France Telecom R&D), Aurelian Bria (Royal Institute of Technology), Johnny Choque (University of Cantabria), Catarina Cedervall (TeliaSonera), Konstantinos Dimou (Panasonic), Giorgio Gallassi (Siemens), Jens Gebert (Alcatel), Stephan Göbbels (RWTH, Aachen Unversity), Ian Herwono (RWTH, Aachen Unversity), Johan Hultell (Royal Institute of Technology), Ralf Jennen (RWTH, Aachen Unversity), Ljupco Jorguseski (TNO ICT), Stephen Kaminski (Alcatel), Peter Karlsson (TeliaSonera), Reza Karimi (Lucent), Ingo Karla (Alcatel), George Koudouridis (TeliaSonera), Franck Lebeugle (France Telecom R&D), Horst Lederer (Siemens), Remco Litjens (TNO ICT), Johan Lundsjö (Ericsson), Per Magnusson (Ericsson), Jan Markendahl (Royal Institute of Technology), Francesco Meago (Siemens), Luis Muñoz (University of Cantabria), Ralf Pabst (RWTH, Aachen Unversity), Rémi Perraud (France Telecom), Dragan Petrovic (Panasonic), Mikael Prytz (Ericsson), Oscar Rietkerk (TNO ICT), Michele Rossi (CFR, University of Ferrara), Joachim Sachs (Ericsson), Eiko Seidel (Panasonic), Rolf Sigle (Alcatel), Ove Strandberg (Nokia), Haitao Tang (Nokia), Riccardo Veronesi (CFR, University of Ferrara), Jens Zander (Royal Institute of Technology). This chapter has been edited by Johan Lundsjö and Peter Karlsson.

8.1 Introduction

One of the key features of the AN concept is the dynamic and instant composition of networks that enables transparent usage of services over different technical and business domains. Probably, the main and commercially most interesting application of this feature is the provisioning of wireless access of moving terminals and networks to the fixed infrastructure using multiple wireless access networks based on different radio technologies. The economical impact of such a technology is significant, as has been outlined in Chapters 2 and 3. Instant access to any

Ambient Networks: Co-operative Mobile Networking for the Wireless World Norbert Niebert (Ericsson GmbH), Andreas Schieder (Ericsson GmbH), Jens Zander and Robert Hancock

network within radio range allows for efficient reuse and complementary deployment of various access networks without sacrificing wide area availability of services providing an 'always best experience' for end users, in an 'always best connection' fashion [79]. It also opens the possibility for users to be served via different service access points (i.e. relays, infrastructure access points) possibly belonging to different radio access networks (RANs) implemented by different radio access technologies (RATs) and operated by different operators. Flexible use of different types of wireless accesses, including selection of a 'best' type of access, by means of efficient management of network resources serves users, network operators and service providers. Furthermore, maintaining service continuity and mobility for users moving between different environments and situations requires support for handover between different RATs, so called *vertical handover*. There is also an emerging need for cooperation among the different RANs and support for rapid establishment of roaming agreements (dynamic roaming) and efficient management of the aggregate resources in the mobile and wireless systems.

Some initial integration steps of heterogeneous radio technologies into one multi-radio access environment have already been taken by both standardization bodies (e.g. 3GPP) and commercial operators. This work has so far been focused on specific access technologies and managed by a single operator. Within the Ambient Networks (AN) project [78], this approached is generalized based on the AN networking vision for cooperation of heterogeneous networks characterized by different access technologies and user/operator domains.

The AN vision puts a number of requirements on the multi-radio access (MRA) architecture controlling and partly embracing the heterogeneous connectivity [80,84,91]. In particular, such an architecture needs

- to enable networks utilizing several access techniques to compose into a new network entity, with fully, or partially, shared network control;
- to allow for the efficient allocation of the radio access resources among concurrent user sessions;
- to provide the ability to utilize more than one radio access for the transmission of certain user data.

In short, the main problem that the Ambient Networks MRA architecture attempts to solve is how to deal with heterogeneity of networks and simplify for end users to reach their services and to realize the best experience. Networks may be heterogeneous in many respects. They may use different radio technologies, and they may differ in size, coverage areas and mobility support. Applications span from personal area networks, via local hotspots in shopping malls, hotels and/or airports, up to countrywide cellular networks. Network ownership and control may differ largely, e.g. between networks owned and controlled by the end user, a local access provider such as a hotel chain or a global operator. Heterogeneity is also a very descriptive term for the many different business models that may be applied for different networks, e.g. monolithic large-scale operators with direct relation to the end user in contrast to, for example, a fragmented business role distribution, where different players specialize in separate parts of the access offering to the end user. In either of these cases, the end user should be able to roam between networks owned and controlled by different operators. Examples of various access networks in the heterogeneous environment are illustrated in Figure 8.1.

An illustration of the set of multi-radio access scenarios explored in this chapter is given in Figure 8.2. Each of these cases takes advantage of MRA, but with some differences in the

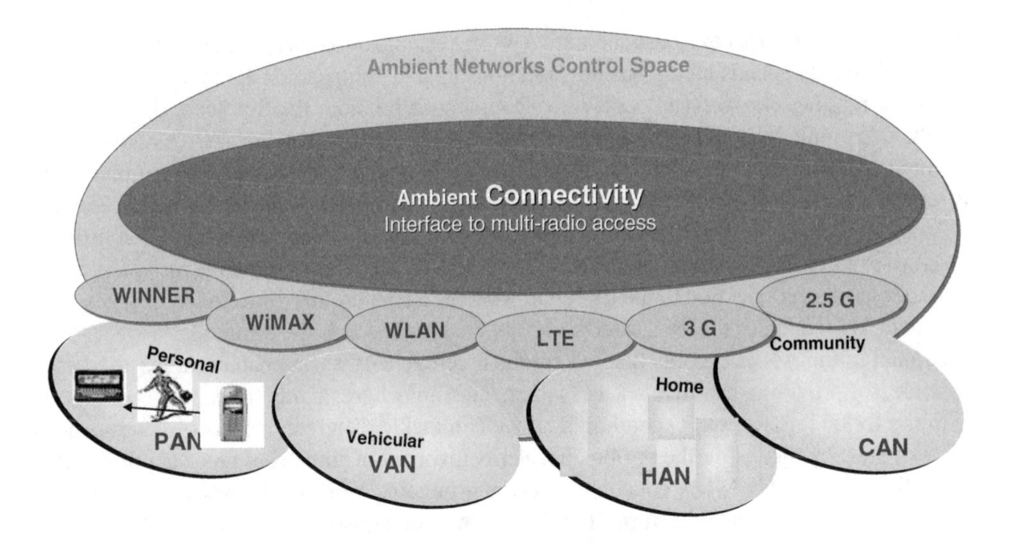

Figure 8.1 The heterogeneous multi-radio access environment

involvement and distribution of the functionality. In the first case (Figure 8.2.1), a multi-radio access capable terminal (AN1) is connected to a single operator (AN2). Note that, in accordance with the AN philosophy, a terminal including ACS functionality is regarded as an AN of its own. The proposed MRA architecture adds generic resource allocation functionality that enables load management over all possible radio accesses (RAs) and instant mapping of data flows to different RAs. The second case (Figure 8.2.2) goes one step further by adding the possibility for information exchange between ANs belonging to different operators (AN2 and AN3),

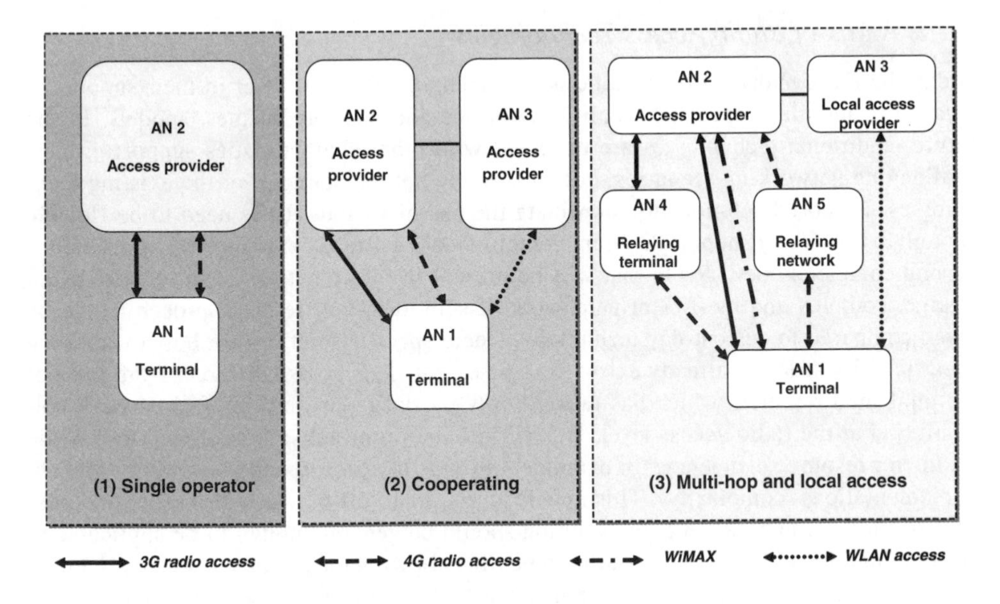

Figure 8.2 Example scenarios

allowing terminals to access these separate ANs in a seamless manner. The level of information to be exchanged depends on the business relationship and composition agreement between the operators, ranging from full competitors to a cooperation relation. The third case (Figure 8.2.3), finally, adds multi-hop and local access provider concepts. An operator (AN2) allows a user terminal (AN1) to establish connectivity in three additional ways: via another terminal acting as relay (AN4), via a set of fixed relay nodes (AN5) or through a local access point (provided by local AN3 access provider). Control signalling and user data may be separated on different RAs. An example could be that a local access provider (AN3), at a certain time instant, provides a direct user data connection to the terminal, whereas a cooperating wide area coverage access provider (AN2) handles the associated control signalling towards the terminal.

Crucial to the AN concept is that is exploits heterogeneity as an enabler, rather than as an obstacle, for providing affordable access to anyone, anywhere, at any time.

In the following, we will expand on how we can tackle different aspects of heterogeneity. We will start by analysing the various problems involved in somewhat more depth. Further, we will relate the AN MRA concept to past and ongoing work in the research community and various standardization efforts. Finally, we will demonstrate how this can be achieved by means of schemes and architecture support for multi-radio access selection (MRAS) and multi-radio resource management (MRRM) [80,91] in combination with generic link layer (GLL) [85,86] components. These concepts are built on previous research, e.g. [87–89,90], which has been generalized and extended with functionality to address the requirements posed by the dynamic Ambient Networking. MRRM is responsible for joint management of radio resources and load sharing between the different RANs and GLL provides a toolbox of configurable link layer functions that allows cooperation between different RATs [85,86].

8.2 Multi-Radio Access – Problems and State of the Art

8.2.1 A Mix of Radio Access Technologies

Today there are many different radio access technologies that differ in their support of data rates, mobility, coverage, quality of service and possible business models. In the future, additional technologies are expected with other characteristics supporting new challenging networking scenarios, but most likely not replacing all of the existing technologies. Proposed solutions to coordinate the use of various RATs need to be flexible enough to efficiently support different combinations of already existing, evolved versions of, and completely new, RATs that will be present in future networks. In order to avoid a large, complex and inefficient patchwork of solutions, a different approach has to be taken compared to current-day technology, where specific solutions for how to combine specific pairs of RATs already exists to some extent. This is also different from most of the published research, which has tackled only partial issues towards full network collaboration at the radio access level. In addition, the proposed solutions need to provide means for resource efficiency, for example, in terms of spectrum and hardware/software implementations (complexity). This may lead to a trade-off between generality and possibilities for optimization, i.e. the solution should be general enough to be applicable to any combination of RATs, while at the same time allowing the use of sufficiently detailed information in, for example, resource management algorithms to be able to deliver high performance.

The fundamental multi-radio access problem is that of selecting which access technology to use for transmission of a data flow. We will use the term access selection for the decision-making part of this function. Access selection may be based on one or many different parameters, such as the access technologies' general capability to satisfy the QoS requirements of the application, current link quality status depending on radio propagation conditions and interference, network load, end-user device capabilities and possibly price for the access, end-user and operator preferences and policies. Access selection decisions may, depending on the optimization criteria and also business model related aspects, be taken by the end user itself (or as an automated process in the terminal device), by one or more entities in the network, or in a combined fashion. Access selection principles and algorithms will be further elaborated in the following sections.

Two functions that are intertwined with access selection, as well as with each other, are *load sharing* and *admission control*. Using multi-radio admission control mechanisms with the access selection procedure allows for efficient load management between different RATs. Such admission control mechanisms will not be limited to admission or rejection of a new connection establishment within a particular RAT, but it will also allow redirecting incoming connection requests to the most appropriate RAT taking into account the load situation. In this way, a better utilization of overall network resources can be achieved, possibly resulting in decreased demand for infrastructure investments, i.e. fewer base stations and other network nodes may be required to satisfy the overall capacity demand. The cost saving potential of multi-radio load sharing capabilities is further discussed in the following sections.

The possibility to move user sessions between different RATs imposes demands on efficient inter-RAT handover procedures. Requirements for fast and low data loss (or even lossless) handover arise from an end-user perception perspective. For some types of applications like voice, video and in particular real-time gaming, there is typically a requirement that the end user should be delivered a continuous service while moving between different RATs, without noticeable interruptions. In addition, from a resource efficiency perspective, retransmission of already successfully received segments of not yet successfully received complete packets should be avoided. A way to tackle these problems, besides from optimization in signalling procedures, is to introduce context transfer mechanisms on link layer level. Thereby, context such as ARQ state information and security parameters can be transferred from the old to the new link layer entity in order to minimize latency and data loss. Such mechanisms will be further discussed in the following sections.

With many networks around, efficiency in the required scanning procedures to be performed by terminal devices becomes a challenging problem. Already in today's cellular standards, substantial effort has been put on the design of energy-efficient signalling mechanisms, where the time and effort required for terminals to scan for new candidate networks and base stations. Such mechanisms include (possibly dynamic) deterministic patterns for broadcast information, paging schemes and signalling of neighbour cell lists. Some of these mechanisms have been standardised by 3GPP for the specific combination of WCDMA and GSM. In the scenarios envisioned for Ambient Networks, further extensions for efficient access discovery mechanisms including network advertisements are required that scale with the number of technologies and networks. Carefully designed access advertisement procedures will also be required to announce capabilities to provide certain services, which may differ between networks and access technologies.

8.2.2 Multiple Operators

Getting access while moving, i.e. the extensions of a connectivity service in a location that is different from the home location where the service was registered, is usually referred to as roaming. In the AN context, we are extending this definition to also include moving between technologies and multiple networks controlled by the same operator as discussed in the previous section, and now also to include moving between different network operators. In such multi-operator/multi-RAT scenarios, where user terminals/networks are equipped with multimode terminals and access providers either cooperate or compete to (i) improve the customer offering or (ii) increase the aggregated capacity of their systems by, for example, load sharing.

Critical issues in such a scheme are resource control and context management, i.e. how resources and user information should be treated when multiple business entities are involved in providing the access service. The Ambient Networks architecture allows for a decentralized solution, where networks are joined to form a 'composed' network. A more straightforward solution, on the contrary, is again a centralized one, involving intermediary actors or virtual operators (MVNOs) that deal with resource management and internetworking issues on behalf of the user. From a resource management perspective, we are with this solution back to a single-operator case.

Most challenging and intriguing from a research perspective are more 'competitive' scenarios. Some of the key features of such cases and how they compare with the traditional, 'cooperative', scenarios are summarized in Table 8.1. The extreme competitive case is probably the 'user-centric' access case, where most of the resource control stays with the user or rather with a user agent acting on behalf of the user. This agent may physically reside in the user terminal or in the network in the shape of an 'access broker'. Very little work has been done in this area, although a number of new technical, business and regulatory problems need to be solved in order to establish the feasibility of such an operating mode. One significant new complication is that in competitive access the user agent has to continuously evaluate different access offerings from a technical as well as from a business perspective to obtain the best 'value for money'. The user agent is not just a resource manager in the traditional sense (evaluating technical performance criteria (signal strengths, bit rates, QoS,

Table 8.1 Features of traditional versus competitive access

	Traditional/cooperative	Competitive
Advertising	Conventional advertising: newspapers, TV, brand names	Network advertising
Access selection	Central/network control	Distributed (user agent) selection
Authentication	Known customer	Anonymous access
Payment	Customer account, billing	E-cash, clearing house
End-user relation to network provider	Subscription	Weak brokers, MVNO

etc.), but also a trade agent, entering agreements and performing electronic financial transactions. Today, such choices are typically done beforehand, e.g. when a user subscribes to the services of a certain operator or when a operator make agreements with roaming partners.

In competitive environments, the problems of *network discovery* and *advertising* become of significant interest. A mechanism where the presence of one network is advertised via another network seems to be a viable possibility only in single-operator or cooperating operator scenarios. How can the user terminals discover the presence of networks operated by fierce competitors in their surroundings, and how should different networks advertise their presence, their services and their price model, and how can multimode terminals detect this information? Even the discovery of network candidates is not a trivial problem as there may be many frequency bands to search in and many options to evaluate. Networks actively advertising their presence on several modes and directory-based solutions (accessed over a wide area network) seem to be promising solutions currently under investigation. After detecting the presence of several candidate networks, the content of the advertisements (service offer) and own measurements are evaluated by the user agent in order to make an access selection. A key problem is that a decision has to be made in a distributed way on input data (service offer, signal quality, etc.) that may change in the near future. A special characteristic of a competitive environment is also the problem that the networks (like in all types of advertising) may provide false information in order to attract customers. In the next step, a business agreement between the user agent and the selected network has to be reached, and finally a financial transaction has to be completed. An important issue is to keep the cost of such transactions low, as the number of transactions may be high. The ultimate challenge is to provide access anonymously, without pre-arranged subscriptions and with full protection of privacy.

The solutions should enable new business models by including means for cooperation as well as competition between wireless network operators. New players, and even end users themselves, should be able to participate in this cooperation and competition in order to achieve affordability.

8.2.3 Relaying and Multi-Hop Access

Adding the possibility of relaying or transmission over multiple (sequential) wireless links to the multi-RAT and multi-operator aspects introduces not only new possibilities, but also new challenging problems. Multi-hopping techniques have the potential to enable extended coverage, extended range of high data rates and deployment at lower cost.

Challenges include even more importance of network advertising and discovery mechanisms as not only the base stations, but also user terminal devices are potential access elements. Access selection over multi-hop paths turns into a routing problem and should permit both per-hop decisions and path access selection. There are potential benefits in realizing the routing/access selection decisions at a low level in the protocol stack, making use of radio information. At the same time, it is necessary to handle cases with fairly loose collaboration between the involved nodes.

8.2.4 Related Work

During the past three or four years, the academic and industrial research communities have been increasingly concentrating on the integration of distinct wireless access networks, e.g.

within (inter)national joint research projects as well as in several standardization fora. As an exhaustive discussion of all related work would be infeasible within reasonable space, considering the large body of literature currently available, this section intends to provide an up-to-date review of the existing literature at a fairly high level.

For handover between different radio access technologies, IP mobility schemes are commonly investigated. Such handover schemes and extensions, e.g. to transfer context between IP access nodes, are developed and standardized in IETF and IRTF [81–83]. Several research projects have investigated IP mobility schemes for intersystem handovers (e.g. EU FP5 Projects Brain, Mind, Drive, OverDrive, MonaSidre, etc.). IP mobility schemes are also the basis for mobility management within the Ambient Networks multi-radio access architecture, which interacts with the IP mobility schemes and complements it with a radio mobility management layer. The IETF has developed guidelines for the efficient design of link layer protocols and subnetworks in order to achieve a good performance for Internet services in particular when based on the Transmission Control Protocol. These guidelines have to be considered within the multi-radio access architecture.

A radio abstraction layer with some similarities to the generic link layer but narrower in scope has been defined in previous EU IST projects and in [104]. The standardization fora 3GPP, 3GPP2 and IEEE are proposing the integration of 3G and 2G cellular systems as well as the integration of WLAN and 3G systems. The integration of broadcast technologies and support for multi-user services has been investigated in previous EU IST projects (e.g. Drive, OverDrive and others).

The working group IEEE 802.21 for media-independent handover services is developing standards to enable handover and interoperability between heterogeneous networks. The target networks include personal, local and wide area networks according to IEEE 802 specifications and mobile networks defined in 3GPP and 3GPP2.

The key architectural elements of the media-independent handover solution being developed are

- media-independent handover (MIH) function;
- event indication (also known as triggers);
- information service.

The MIH function resides just below the IP layer and is uniform across bearer types. The MIH function can use inputs from layer 2 such as trigger events, hints, access to information about different networks and can help in the handover decision-making process. Link layer events can help anticipate user (mobile terminal) movement and help prepare the mobile terminal and network in advance. An abstraction of such event notifications from link layer with other relevant parameter information can help reduce handover latencies. The standard shall provide an information service that provides detailed information about various networks that can assist in network discovery and selection.

A large body of research concentrates on the common (or joint) resource management of multiple different access technologies, with a significant focus on radio access selection schemes that incorporate user-, network- and/or service-specific information. Some of these studies are of a rather conceptual nature, whereas others concentrate on performance gains using simulations or experiments. Another common resource management mechanism is spectrum sharing or dynamic spectrum allocation among multiple networks, in order to enhance aggregate system capacities. The evolution of a multi-radio access network architecture, considering operational and deployment cost aspects, is another area that has attracted considerable interest. Aside from the optimization of common resource management of independently deployed access networks, the joint planning of diverse access networks is also studied.

Aside from the 'cellular' focus of perhaps most multi-radio access studies, a significant amount of research is targeted towards multi-hop or hybrid networking (e.g. EU projects MIND and others [105]), where intentionally deployed (fixed) nodes or user-associated (mobile/nomadic) terminals act as relay nodes to enhance overall service coverage and quality. In contrast to most prior work, the Ambient Networks does not so much concentrate on fundamental multi-hop networking issues, but rather investigates how multi-hop networking can be integrated into the Ambient Networks multi-radio access architecture and assesses the benefits it may bring, with a relatively strong focus on scenarios where different radio access technologies are used for multi-hop networking.

The related work listed above differs from the Ambient Networks approach to multi-radio access in several ways. A significant part of the prior work concentrates on (only) two specific radio access technologies that are tightly integrated and may even allow joint radio resource management. The proposed tailor-made approaches cannot be easily extended to other radio access technologies. Another part of prior research focuses on loosely integrated radio access technologies, which can comprise a number of radio access technologies. Due to the loose network integration in a radio agnostic way, these cases allow only for a limited degree of joint resource management – usually restricted to availability-based access selection and not considering system load or link quality variability on small time scales.

8.3 The AN Multi-Radio Access Architecture

8.3.1 Architectural Constraints and Interactions

In this section, we will discuss how these ideas map on the overall Ambient Networks concept and how they relate to other key functions.

The problem of how to deal with heterogeneity is not only central to multi-radio access concepts, but really the fundament of the Ambient Networks idea. One of the key concepts of Ambient Networks is that of network composition, allowing for networks to share resources and functionality with each other, to various degrees (ranging from agreements at different levels up to full composition into a single logical network). Multi-radio access features like network advertisements, network discovery and resource aware access selection bring value to the network composition concept, allowing for efficient use of scarce radio resources within and between networks and access technologies. These features are also there to minimize battery and processing power of terminal devices in their efforts to detect and attach to any of the possibly vast number of available networks in their ambience.

It can be anticipated that the potential of the proposed multi-radio access concepts is best exploited in an 'as complete as possible' networking concept. That is, multi-radio functions should be able to efficiently interact with, and benefit from, other networking control functions. An example of this is the interdependence between radio resource based access selection, mobility management and context management of AN. The decision to select a certain network and access technology when moving around can be based on

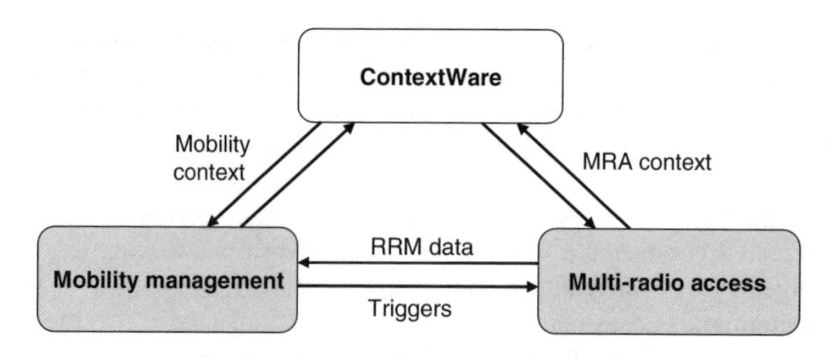

Figure 8.3 Interactions between AN functions

radio resource information jointly with information related to, for example, other types of network resources as well as operator and end user related policies. Here, context management, 'AN ContextWare', comes into play (Chapter 11). Triggers for the access selection decision may be provided with help from the triggering framework discussed in Chapter 9. Interactions between the MRA FEs and ContextWare and mobility management are shown in Figure 8.3.

Finally, when the decision has been taken to move a session to a new network and/or access technology, the handover should be executed with the best suitable schemes or tools, again described in Chapter 9. A clean, consistent and general architecture framework has a great value in itself. Adding a specific solution for coupling each pair of access technology and network type that appears in the market will most probably be highly inefficient in the end. Multi-radio access functions being an integral part of a broad networking control concept such as Ambient Networks strive for avoiding that type of patchwork of solutions.

8.3.2 *Functional Overview*

A structuring towards MRA architecture into a high-level functional layer architecture model is shown in Figure 8.4. It illustrates the two main interrelated functional groups MRRM and GLL in a layered model, including user plane data flow (solid lines) and MRA (MRRM and GLL) signalling (dashed lines) through the layers. Arrows indicate control interfaces between different functional blocks, carrying information exchange and control commands, for example, for configuration or for measurement data retrieval. Note that only one communicating peer (network or terminal) is depicted. For the single-hop case, the model can simply be mirrored at the other end. The GLL is introduced on top of, and is partly replacing, the RA-specific parts of the link layer. The toolbox of link layer functions within the GLL provides a unified interface towards upper layers in the user plane and provides adaptation towards the underlying remaining RAT-specific link layers.

The MRRM functions are built upon or mapped onto the network intrinsic radio resource management (RRM) functions, which belong to the underlying RA and are therefore not within the explicit scope of the AN MRA. Two possibilities are illustrated for how MRRM signalling is conveyed between communicating entities, either over IP or directly mapped onto the GLL. Finally, Figure 8.4 illustrates information exchange between MRRM, GLL and other

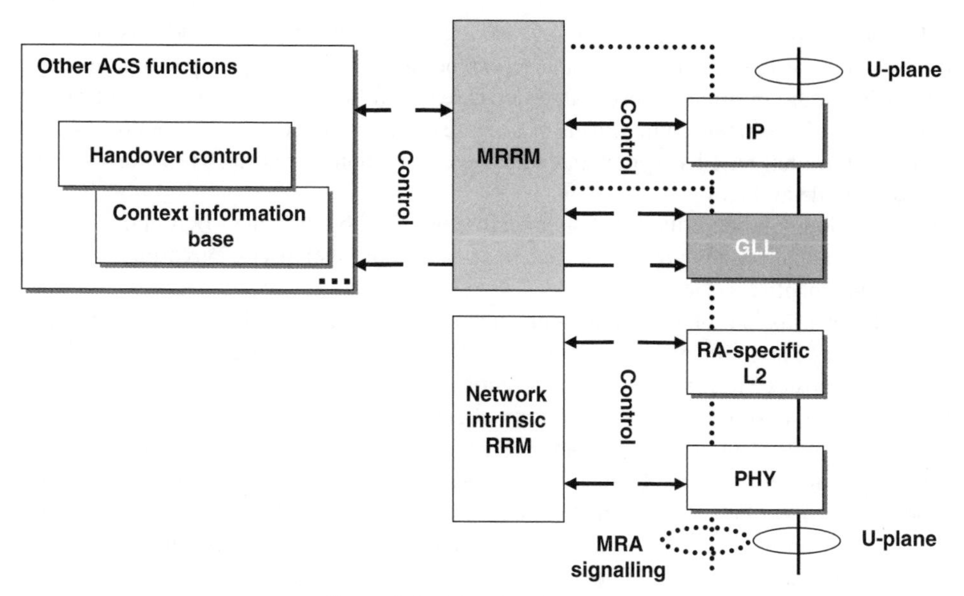

Figure 8.4 Functional overview

ACS functions, here exemplified by handover control for execution of appropriate handover procedures (e.g. based on MIP, HIP, GTP) [81–83] and a context information base (CIB) for retrieval of context information, e.g. policies such as user preferences (see Chapter 11).

In general, the GLL encompasses functions that are located close to the user plane of a data flow. One example is that GLL provides and reuses context information that is transferred between GLL entities at RA reselection for seamless access selection. Another example is the RA selection for which a hierarchical distribution of functionality between MRRM and GLL is proposed, where the GLL dynamically handles the mapping of data flows to any of the RAs selected by MRRM.

8.3.3 Architecture Model

In this section, we will present a proposal for general-purpose target MRA architecture that has been based on feasibility evaluations of potential multi-radio access functionality, where efficient support for access selection has been identified as the most critical issue when it comes to architectural guidelines. More background and details on access selection performance will be given in Section 8.4. In short, the basis for the architectural analysis has been a qualitative weighing of the observed advantages (capacity, throughput, etc.) against the sensitivity to the conditions for these advantages (signalling delays, location of functionality) and against the probable costs of implementing the required schemes (transmission requirements, changes to legacy standards). The outcome was that the gains associated with fast access selection, although clearly visible, were judged not sufficiently extensive when compared to the costs. This does not preclude fast access selection entirely in the architecture, but the focus when defining functional entities, (logical) nodes and necessary control protocols is on supporting slow access selection.

The proposed target MRA architecture also focuses on support for network-based decision and execution of RA selection to be able to exploit the associated capacity and performance gains when radio information over larger areas (e.g. cell load) is collected and utilized. The architecture also supports other forms of access selection and control distribution schemes, including terminal-based selection and adaptive distribution of control. These topics will be discussed in detail below.

This section will detail the MRA architecture proposal by defining MRA entities, logical nodes, a principal mapping of functions to nodes and additional discussions around the nodes and its functionality.

The basic principle of the architecture for a single-hop, single-operator case is shown in Figure 8.5. The main Ambient Networks MRA entities are shown as small solid-line or dashed-line boxes. The arrows between the entities indicate their relationship with each other. The user plane is shown using solid lines and control is shown using dashed lines. Hence, solid-line entities lie in the data path and participate in user data processing, and solid arrows indicate passing of user data. Dashed-line entities are for control and dashed (dotted) arrows indicate exchange of control information. One of the entities is shown in bold face to indicate that it is in control of access selection. The dashed arrows indicate the (technology-specific) access links that traverse the radio interface; the corresponding boxes with dashed borders indicate the link layer termination points for the access links. There is also a dash-dotted arrow indicating context transfer functionality, see below.

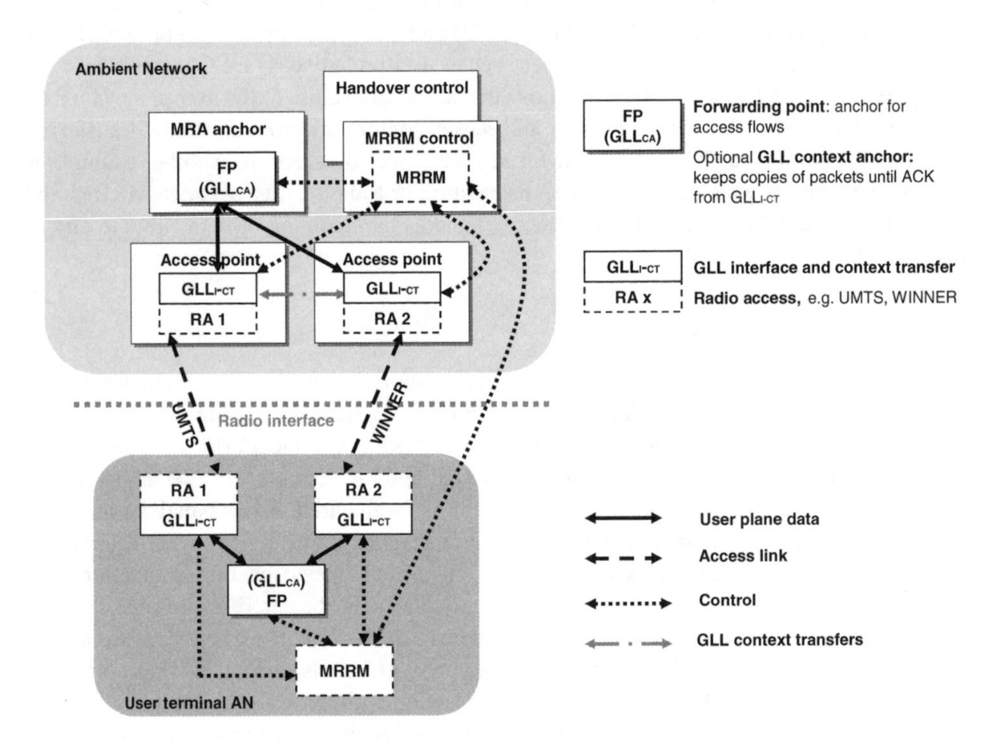

Figure 8.5 Multi-radio access architecture model

The main Ambient Networks MRA entities with functionalities are

- *MRRM*: MRRM is the key control MRA entity. It performs access selection decisions based on a number of different input parameters, e.g. the performance and quality of access flows, the resource costs and current resource availability, the operator and user policies, and the service requirements and user terminal behaviour, and the extent to which these can be handled efficiently by different access flows. The MRRM access selection function is a part of other MRRM functions, e.g. admission control and load management. MRRM maintains three different sets of access links for its operation: the detected set (DS), the candidate set (CS) and the active set (AS). Section 8.4 discusses in detail these sets, the operation of MRRM and the required input parameters.
- *GLLI-CT (generic link layer – interface and context transfer)*: The GLLI-CT provides a generic interface and support functionality for transmission of user and control data over an access link. It embeds access-specific transmission methods and protocols; hence, there are different entities for different types of accesses. The GLLI-CT monitors the performance of the access link and the QoS that is perceived by an access flow. It also observes the resource costs and availability (including load) of the access links. Based on certain rules and thresholds (event filtering and classifications), it reports link events (triggers) to MRRM, e.g. when the performance of the access link changes by a certain amount, when a new access link is detected or lost, when a QoS requirement of a flow cannot be met any more and when the resource costs for access link pass a threshold or resources become scarce. GLLI-CT receives configuration information from MRRM. It can also receive measurement queries to report on access link and resource status. In case that a flow is handed over between different GLLI-CT entities, GLLI-CT can support context transfer and provide the corresponding link-specific context.
- *FP (forwarding point)*: A routing decision point where MRA access selection handovers are executed. In addition, the FP may be combined with GLL context anchor functionality GLLCA to support another form of link layer context transfers. Here the context anchor keeps copies of data packets until the GLLI-CT entities signal that the packets have been successfully transmitted over the access links. The GLLCA may also exercise flow control to minimize queued data at the access link endpoints.

One important remark here is that GLL is not used over the radio interface between the two ANs. Instead, the GLLI-CT and GLLCA entities interact only within the same network, either the AN (or composed ANs) that is providing access for the user terminal AN or the user terminal AN.

Note that the control relations shown in Figure 8.5 are not complete. The GLLI-CT and GLLCA entities may exchange control information for the context transfer and flow control functionalities. In addition, to reduce cluttering the Figure too much there is only one control relation shown between the two ANs. However, as the MRRM entity in the top AN is shown to be in control of access selection, it will enforce (via handover control) the bearer to access flow mapping in both of the FPs. For the user terminal AN, this may be channelled via its MRRM entity (as shown in the figure) or it may go directly to the FP.

The MRA functional entities can be implemented in different ways to suit different AN MRA networking scenarios. They may also play different roles, which may need to be dynamically negotiated. To be able to define the required external interfaces, it is necessary to consider physical implementations. To this end, a number of logical nodes have been defined. A logical node is something that can be implemented as a physical node on its own. However, several logical nodes can also be combined into a single physical node. Figure 8.5 shows the case with a single user terminal AN connected to a single network AN where the following nodes are identified:

- *MRA anchor*: A user plane node where the multi-radio access selection is executed, i.e. it includes the FP unit.
- *MRRM control*: A control node that can decide on access selection and derive wide area radio information, such as cell load, by collecting MRRM control information from other nodes.
- *RAP/RBG (radio access point/radio bearer gateway)*: A logical node where the access link layer is terminated. It is defined for a single type of radio access, i.e. it is RAT specific. The node includes the GLLI-CT unit for embedding access-specific transmission methods and protocols and for supporting link layer context transfers.

Note that there may be other RAT-specific nodes in between the RAP/RBG and the radio interface, for example, when it is co-located with a single radio bearer gateway having several radio access points beneath it.

8.4 Access Selection

8.4.1 Principles and Algorithms

As discussed earlier, MRAS refers to the MRRM access selection function which decides which (radio) access flows(s) (among the available ones) should be used for the end-to-end bearer in a multi-radio access scenario. An access flow can contain a single access link, in the case of single-hop communication (Figure 8.6) or multiple access links in the case of multi-hop or multicast/broadcast communication. The radio access flows are the elements managed by MRA functionality. Management of the access flows is achieved by means of access sets that are established and maintained by the RA coordination functions.

- *Detected set* is the set of all access flows that have been detected by MRRM for an AN through, for example, scanning or reception of RA advertisements.
- *Candidate set* is the set of access flows that are candidates to be assigned by MRRM access discovery function to a given active bearer; it is always a bearer-specific subset of the DS.
- *Active set* is the set of access flows, assigned by the access selection MRRM function, to an active bearer at a given time; it is always a subset of the CS.

In case when a GLL entity controls two or more tightly integrated radio accesses, we have an additional access set:

- *GLL active set (GLL AS)* is the set of access flows assigned to a given GLL entity by MRRM to serve a given data flow at a given time; it is always a subset of the AS. The GLL AS is used for fast access selection when multiple single access nodes are connected via GLL to a common multi-access anchor node or for access flow forwarding in multi-hop situations.

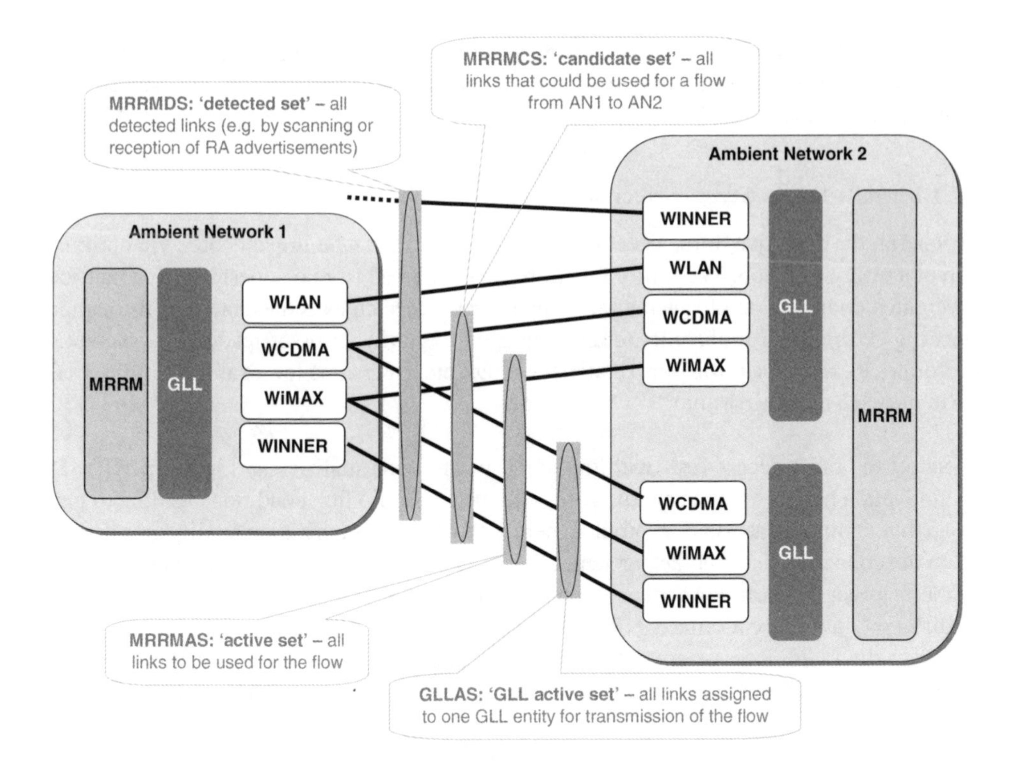

Figure 8.6 The concept of access sets

As the fundamental MRRM function, access selection uses knowledge about available access flows for a particular terminal to assign one or more of them to each active AN bearer. As mentioned previously, AN flows could be recursively defined so that a flow can contain subflows. Such a 'flow within a flow' may use a different set of endpoint locators, which are not visible at the higher level. Although access selection encompasses fixed accesses, the focus here is on radio accesses. The term radio access selection is often used interchangeably with access selection.

Access selection algorithms may consider many parameters when determining the best bearer to access flow mapping. They also need to continuously react to any changes in conditions, e.g. deteriorations in radio signal quality, and reallocate resources accordingly. For the purpose of access selection, MRRM interacts with other FEs which corresponds to other AN control functions such as handover control, context management, security control, etc. The following list illustrates some examples where

1. MRRM sends requests for measurements over access flows or activating an access flow towards GLL. On the contrary, GLL sends the measurement information and trigger information (if some thresholds are exceeded) towards MRAS function.
2. MRRM interacts with handover control and context management functions when an access flow is changed or a new access flow is added to the active set for handover and context transfer purposes.

3. MRRM interacts with other AN control functions in order to select the access flow. This access flow selection might be restricted by policies and/or constraints imposed by these AN control functions.

8.4.1.1 Multi-Radio Access Selection Objectives

In general, the objective of the access selection algorithm can be implemented via optimization of a utility function. The utility function can be derived from one performance metric or a weighted combination of several performance metrics such as achievable user throughput, blocking or dropping probability, communication costs (in terms of resource consumption and/or price), resource utilization (load balancing), etc. Here are a few example objectives for the *access selection* algorithm:

1. Select the radio access flow with the highest radio link quality (e.g. highest SINR). The motivation here is to select a radio path with the best radio link conditions leading to high-quality communications. The problem here is that the communication quality also depends on the congestion level of the particular RA. Additionally, this comparison should consider the fact that different RAs (depending on their functionalities and efficiency) might have different quality for a same SINR.
2. Select the radio access flow with the lowest congestion level. The motivation here is to evenly distribute the load in the available RAs (i.e. load balancing). However, because the radio link quality is not considered in the selection, the user might end up with bad radio link and consequently low communication quality.
3. Combination of selection criteria 1 and 2 estimates the communication quality based on the radio link characteristics and the congestion levels. This objective combines the advantages of 1 and 2 above. However, the drawback here is that this approach requires more complex processing and also it does not include cost-related information.
4. Predefined priority list (e.g. preferred RATs and/or operators) that is used in selecting the radio access flow.

A special objective is the so-called *fast radio access flow selection*, i.e. fast switching between the radio accesses done at GLL level and independently from the MRRM functions. The objective is to increase the spectral efficiency (i.e. Mbit/Hz) of the transmission and also user throughput. This fast selection requires tightly integrated radio accesses and instantaneous radio link characteristics as input (e.g. instantaneous SINR) to the GLL functions.

8.4.2 Multi-Radio Access Selection Input Parameters

In general, for any decision process there is information necessary upon which the decision should be derived. For the access selection algorithm, the input used in the decision could contain one or more parameters that characterize the candidate access flows, user's terminal and the desired service. These parameters can be divided into the following two broad categories:

- *Static parameters*: The values of these parameters are changed on a time scale that is much longer than the usual lifetime of a flow (or session) and is not dependent on current radio and load conditions. These parameters are access point (AP) capacity, service QoS

requirement, RAT preference, financial costs (Euro/min or Euro/Mbyte), terminal capabilities and level of integration among RAs.
- *Dynamic parameters*: In this category, the parameters are dynamic because their values vary on a time scale (e.g. hours, minutes, seconds or even milliseconds) that is comparable to the usual session lifetime and depends on the current load conditions, user's speed and location, etc. These parameters are AP load and congestion level, instantaneous or averaged radio link characteristics (signal strength, interference level, SINR, etc.), amount of resources needed for satisfactory communication quality, financial costs and terminal velocity.

Note here that these parameters could also be used by the access discovery algorithm in defining the candidate set.

The input information for the access selection algorithm should be signalled between the MRRM entities either via backhaul fixed network (on the network side) or transmitted by the network or the terminals over the air. For signalling over the radio interface, the network and/or terminals could use either broadcast or dedicated type of radio channel. This is necessary in order to make this information available to the MRRM entities throughout the MRA system for their access selection decisions.

8.4.3 Multi-Operator Multi-Radio Access Selection

The access selection algorithm in the case of multi-operator networks is strongly influenced by the level of cooperation and/or competition between the network operators. In order to present the effects on the access selection, we can define the following three categories.

Fully cooperative operators. In this case, the different RAs from different operators fully share the control over their respective resources. This level has been agreed upon during the composition process and also configured in the MRRM functions. Effectively, this situation is merging into a single-operator case. All relevant information that is required for the access selection decision is available, e.g. operator-sensitive information such as current congestion level, resource consumption per user, pricing information, etc. is freely shared between the MRRM entities and could be used in the access selection decisions. The physical location of the access selection algorithm is flexible but it is not envisaged that it can be implemented in the multi-radio access point from one of the operators. The reason for this is that fast access selection is unlikely to exist in multi-operator network due to the physical separation of the different access points (from different operators), which introduces longer signalling delays. Fast access selection is the main motivation to locate the access selection procedure close to the multi-radio access points.

Partially cooperative operators. In this case, the different operators only partially share the control over their respective resources. The information available to the MRRM entities for the access selection decision is limited because some of the operator-sensitive information is filtered out (i.e. not exposed to the other operators). This information could be the congestion level, pricing, resource consumption, etc. However, a certain amount of trust is established and also some compensation schemes are agreed upon. For example, if a user is transferred from operator A to operator B, then operator B shifts other user(s) back to operator A in order to have fair sharing of traffic and revenues. The access selection algorithm is expected to be located in the MRA anchor point that might even be outside the domains of the different operators.

Noncooperative (fully competing) operators. In this case, there is no shared control of the operators' resources but rather fierce competition to serve as many users as possible. The access selection algorithm here has rather restricted information available to make the access selection. Regarding the physical location of the access selection algorithm, it is expected that it can be located either at an MRA anchor node such as 'access broker' [99] or in the terminal. The 'access broker' has trust relationship with all operators that are involved in the communication (as they do not have trust relationships between each other). For the case of executing access selection at the terminal, some approaches are foreseen with congestion pricing broadcasting and personal advertisements that can help the terminal in selecting the desired RA. However, an important consideration at this kind of access selection is the system stability, i.e. the risk of having a (large) group of terminals frequently change between the operators due to rather variable pricing information.

Note here that in reality any degree of cooperation is possible, i.e. the cooperation level could be interpolated between these three categories.

8.4.4 Performance Evaluation

To validate the proposed concepts, a number of feasibility studies have been performed. The majority of these studies have focused on access selection through MRTD. The considered performance measures are spectral efficiency, system throughput, user throughput, user satisfaction index, delay and amount of signalling overhead. The evaluation was performed in different scenarios which span over a space defined by two dimensions: number of radio access technologies available and number of network operators (e.g. single-/multi-RAT under the control of single/multiple operator(s)). Under these scenarios, the studies considered varying assumptions regarding

- The level of coupling among RATs: tight coupling versus loose coupling.
- The utilization of load balancing in access networks consisting of multiple RATs.
- The area coverage of the available RAs: full coverage versus partial coverage and the location of the cells and radio access points.
- The influence of signalling delays and aged information in the process of RA selection.
- The level of cooperation between the network operators: full or partial cooperation versus competition.

Throughout the studies, the criteria used for deciding the access selection are various, including radio link characteristics, cell load and capacity, RAT preferences, terminal capabilities, terminal velocity, service type and required QoS, etc. Further, depending on the used criteria for access selection and the state (offline, registered, active) of terminals and sessions, the MRAS decision can take place in the terminal, in the network or be distributed between the terminal and the network.

8.4.4.1 Performance of MRTD Realization Alternatives

Multi-radio transmission diversity can be realized either at MAC PDU level or at IP packet level (MRTD@MAC and MRTD@IP, respectively). The former allows, at least in theory, higher performance gains as it can react on a very short time scale, but it might be rather complex and costly to implement as it requires tight integration between RAs. However, it can also be utilized as an upper bound on the achievable gain in a practical system; if the gain for this idealized system is very small or nonexistent, the additional complexity induced by MRTD cannot be justified in that setting, whereas a large gain means that there is a big potential to explore in a practical system.

Several of the evaluation studies assume very tight integration of available RAs combined with MRTD@MAC. Performance also depends on the scheduling algorithm and whether switched or parallel MRTD is employed. In particular, [93] shows that the benefit from using parallel instead of switched MRTD is mostly visible when there are a large number of RAs per flow. With switched MRTD, this would result in several unused RAs. This situation is likely to happen in low-load conditions. In other cases, the gain vanishes and the simpler switched MRTD is almost as good.

The importance of scheduling is shown in [94,95], where it is seen that the gain of max-rate over round-robin scheduling increases with the number of users. In fact, MRTD with round-robin scheduling is inferior to independent RAs (no MRTD) with max-rate scheduling when the number of users is large. An explanation for this is that the max-rate scheduler benefits from multi-user diversity within each RA whereas the round-robin scheduler does not. In contrast, for a small number of users, multi-RA diversity is more important and MRTD performs better than independent RAs in this case. In the mentioned studies, the gain of MRTD in terms of throughput and spectral efficiency varies between 15 and 60 %. In some cases, this might not be enough to justify MRTD@MAC, as the resulting complexity is high and it implies large changes to existing RATs.

A looser type of integration by means of MRTD@IP has also been considered for use between UMTS and WLAN [96]. MRTD@IP level is relatively easy to implement compared to MRTD@MAC level, as it requires a few changes in lower layers of existing RATs. The time scale for MRTD@IP is longer than that for MRTD@MAC meaning that smaller gains are reported.

Here, an evolved RNC, with extra functionalities necessary for joint operation of WLAN and UTMS, performs the access selection on the downlink. This selection is based on both WLAN and UMTS measurements, whereas RA transmission is performed in the WLAN AP or the UMTS node B. Such a selection requires the use of flow control in order to adapt to the instantaneous throughput of each RAT as too many pending packets for a RAT that is no longer selected could give performance problems (throughputs between UMTS R99 and WLAN are by nature quite different).

The MRTD gains are measured in terms of user throughput as a function of different traffic loads and for different file sizes and UL/DL traffic mixture. Simulations have been conducted, using realistic models of UMTS and WLAN, showing improvements in average user throughput, independent of traffic load. It is shown that for bigger file transfers, gains are higher, as MRTD provides a better traffic splitting between the two available accesses. For low-size files and low traffic load, use of (slower) UMTS link decreases average user throughput (as opposed to using only WLAN). The solution can be to introduce a more intelligent traffic splitting algorithm in the GLL (taking into account, for example, the number of IP packets waiting in the GLL buffer). Also, some gain was achieved in decreasing WLAN cell load for medium traffic load via MRTD.

Figure 8.7 shows the impact of the uplink traffic on the average user throughput. The four series of data relate to two different uplink traffic load values with and without MRTD.

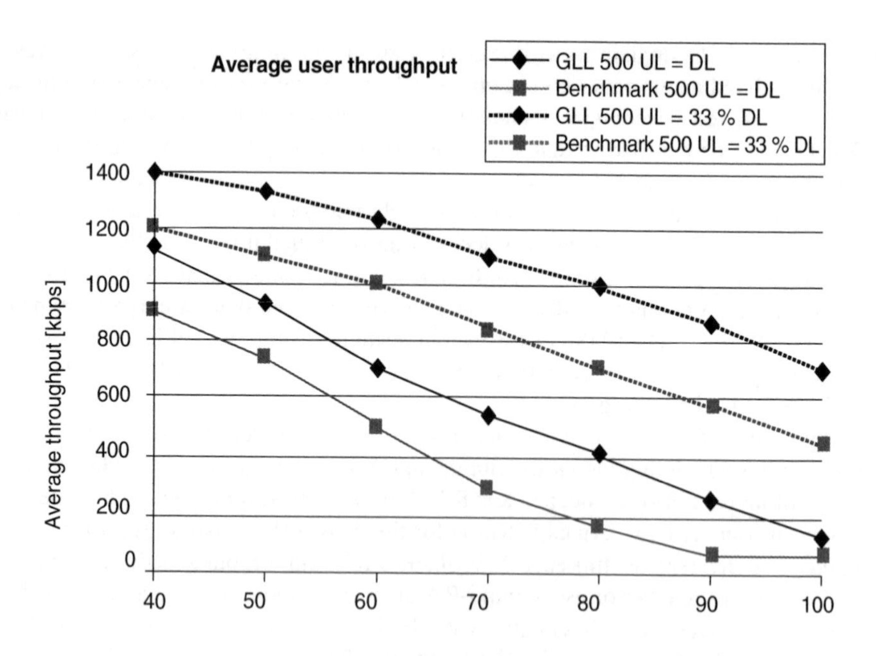

Figure 8.7 The impact of the uplink traffic on the average user throughput. GLL – with MRTD, benchmark w/o MRTD

In two of the series, the uplink traffic load is equal to the downlink traffic load. The two other series represent a case where the uplink traffic load is only one third of the downlink traffic load.

The difference in DL average traffic per user is about 500 kbps between high and low UL traffic cases. This is due to the more efficient use of the medium in WLAN cells, when the uplink traffic is lower and the number of retransmissions is much lower. MRTD gain is also slightly higher in the low UL traffic case. This can be explained by more efficient use of UMTS capacity, when less DL traffic is sent via UMTS.

8.4.4.2 Load-Based Selection Decision

In [97], load is taken as input for the MRAS algorithm. Available RAs are a 3GPP LTE like access and WLAN. The results show both capacity and QoS improvements (e.g. average bit rate) when considering load. Capacity gains of 15–20 % are obtained under high-load conditions, whereas virtually no gain is noticed for low loads. The largest gain also appears in scenarios where the geographical distribution of traffic does not correspond to the deployed cell capacity in each RA. Similar trends are observed regarding the 'trunking gain' when combining the WLAN (802.11a) and UMTS resources in the radio access selection procedure [98]. This 'trunking gain' results in reduced blocking probability and increased average downlink data rates per radio access point.

Further, one study [95] assumed RAs providing only partial coverage through randomly deployed WLAN APs with best-effort traffic and found that system (as well as user) throughput gains of up to 70 % are possible with MRAS (no scheduling). In suburban environments, loose integration and terminal-based access selection performs as well as tight integration with network-based selection. In hotspots on the contrary, the terminal-based scheme gives only about 10 % gain, whereas network-based schemes (requiring node coordination) can increase performance by 20–50 %. Fast reselection can give an additional improvement of up to 10 % when combined with the network-based scheme, but also require tighter integration and it is likely that the additional gains achievable cannot compensate for the resulting increase in signalling overhead [92].

8.4.4.3 Impact of Signalling Delays and of Aged Information

Signalling delays and realistic capabilities for information exchange result in that the MRRM and GLL entity have to base their decisions on potentially non-up-to-date, aged, system information, which is thus potentially no longer accurate. Results show that aged information can considerably deteriorate the system performance compared to the theoretical reference case with always up-to-date and accurate system information.

In the case of very fast channel variations, like MRTD at MAC level with time scales of the order of milliseconds, these signalling delays show a huge impact on the system performance and may even result in that the GLL system performance becomes inferior to a system using the best RAT without GLL-MRTD. In contrast, slow MRAS, as it is performed by the MRRM entity, operates typically in the time scale of seconds. Although the signalling delays have an appreciable impact on the theoretical system performance without any signalling delays, its system performance is still considerably better compared to separated systems without MRAS.

Figure 8.8 illustrates the impact of aged system information on the system capacity [93,94]. Distributed MRRM entities control two RATs (GSM and UMTS) within one AN, where the information exchange is carried out according to the real capabilities of 3GPP, i.e. attached to intersystem handovers only. The real-time traffic was set to obtain a certain QoS measured

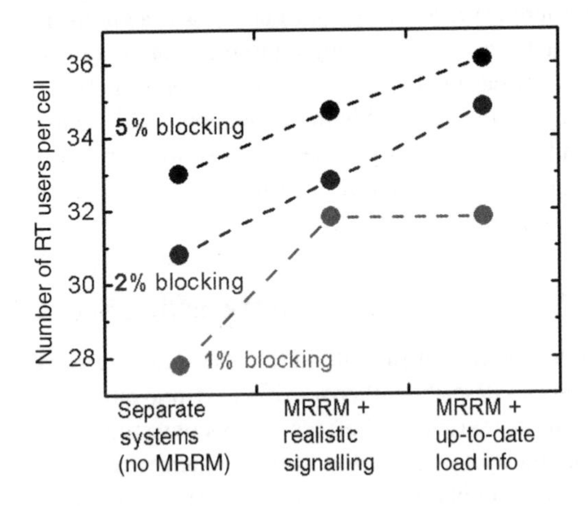

Figure 8.8 Effect of realistic signalling with aged information on the system capacity

here in terms of blocking probability. The nonrecent and nonaccurate system information lead to that the carried load is decreased by up to 6 %, compared to the reference case with always up-to-date and accurate system information. On the contrary, and despite the degradation by aged information, the MRRM performance is still by 5–14 % better than in the case of separated systems without any MRRM functionality. In the shown example, the MRRM decisions are based on cell load information, which are between 12 and 25 s old and differ by on average 8 % from the accurate value.

The signalling delay should not be larger than the typical time scale of channel variation, otherwise the MRTD and MRAS decisions are based on potentially arbitrary system information. Thus, MRTD works best when the time scale of channel variations is not shorter than the time scale of signalling delays and the age of available system information. This condition can in particular be fulfilled well for slow MRAS in the MRRM entity. As MRTD@IP uses only slow MRAS, it seems to be less sensitive to the problem of information ageing, but on the contrary, its reporting delays are larger as the decision is probably not taken in the measuring node. In particular for fast access selection, like MRTD at MAC level, the fast channel fluctuations require fast and frequent information exchange which leads to a large signalling overhead. For all cases, there is a trade-off between the MRTD/MRAS system performance and its therefore required signalling overhead.

The system performance can be tuned by additional mechanisms, dedicated to exchange system information, which lead to more recent and more accurate information. To reduce their additional signalling overhead, it is suggested to use selective triggers based on the age of present system information and based on the variations of the channel properties: competitive versus cooperative operators.

Distributed MRRM in the form of MRAS performed at the user terminal has been considered in [92] and it looked at a competitive case, where two wide area CDMA operators advertise an instantaneous access price to the users, proportional to the amount of transmission power required from the base station. This approach provides a form of congestion pricing mechanism, which dynamically adjusts the price as load in the network varies. It was shown that letting the users decide on their own which network to use could give similar performance (Figure 8.9) as in the case when operators fully cooperate and dictate the 'best' user allocation. The reference (no cooperation) curve is drawn on the assumption that each user can access only the base stations belonging to the operators it subscribed to. The results suggest that although user-based access selection may lead to efficient resource utilization, there is a trade-off between the achievable improvements and system stability. Unpredictable and unstable load conditions can easily arise due to short-term churning of users from one network to another (i.e. ping-pong) due to continuous adjustments of the advertised prices. Similar results were also presented in [95].

8.5 Challenging Multi-Radio Access Networking Scenarios

In this section, we highlight three different multi-radio access networking scenarios, deemed to be challenging as they all describe rather specific and innovative multi-radio access deployment. In this respect, the scenarios can be seen as nonconventional. A common denominator for the scenarios is that they all target low-cost deployment as a primary objective. Using the identified scenarios and their characteristics as starting point, focus has been put on deriving requirements on MRRM functionality to efficiently support the scenarios, and how legacy

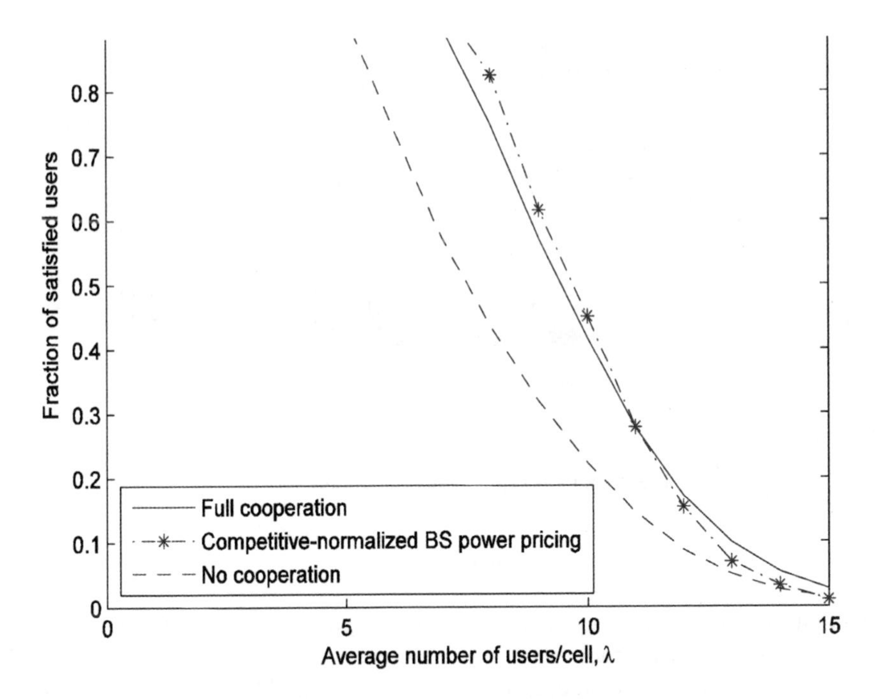

Figure 8.9 Example of capacity gain for fully cooperative, partially cooperative and noncooperative operators

RRM functionalities can be evolved and embraced by MRRM and the overall multi-radio access architecture.

8.5.1 Scenario 1: Low-Cost and Smart Coverage Extension for Rapid and Temporary Deployment

Scenario 1 focuses on situations where temporary, low-cost infrastructure has to be deployed rapidly due to increased demand during a constrained time period (e.g. during large sport events). In this and similar scenarios, the new infrastructure serves as a complement to the already existing one (wired or wireless) and it will typically enhance the capacity over an extensive geographical area.

Within this scenario, a wired network operator may extend its services to the mobile paradigm, providing service ubiquity. This is depicted in Figure 8.10, where user B utilizes user A as a forwarding node in order to reach the temporary local access point (LAP) deployed by the legacy wired operator. The LAP might, for example, be placed on a tourism site where third party content providers could adapt services tailored for users present within that particular area, therefore, creating new business opportunities.

Furthermore, nontraditional operators, such as railway companies may decide to exploit their existing communication infrastructure to offer wireless access. By doing this, they may benefit from the situation, e.g. by increasing their revenue, without the need of any upfront infrastructure investments.

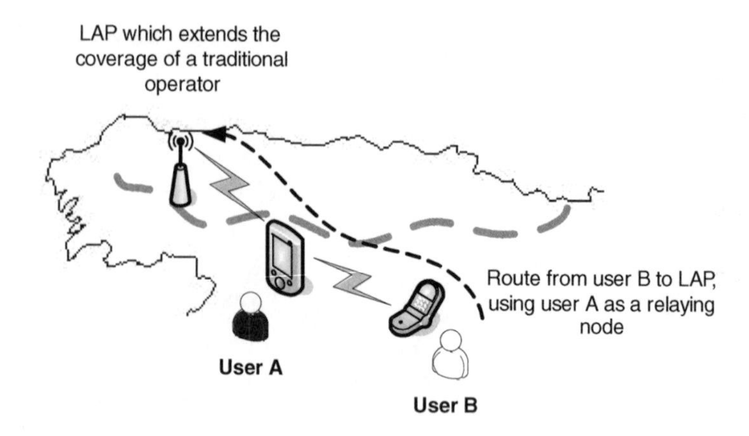

Figure 8.10 Application scenario for scenario 1

Two deployment models are foreseen within scenario 1:

- Traditional communication operators deploying low-cost infrastructure (LAPs), inherently associated with short payback time, to extend their services and/or coverage area. This can be extended to Greenfield operators, which would need to establish agreements with broader ones.
- Nontraditional operators, e.g. companies in the utility sector, railway operators, etc. They see the opportunity of obtaining benefits at low cost and therefore use their existing infrastructure to deploy LAPs for offering low-cost access to their customers.

As can be seen, the anticipated deployment models are characterized by operator initiated and supported deployment of temporary networks (including rapid deployment) with access points, fixed and mobile relays and user terminals using multi-hop strategies with single or multi-radio access technologies to form flexible access networks (including multi-operator case). Furthermore, the 'forwarding node' concept introduces a new business dimension, as end users might try to obtain benefits from relaying traffic coming from other users.

This scenario poses some clear challenges on the network advertising and discovery procedures. Here the concept of proxy advertisements is of interest. Different entities should be able to advertise themselves as access points on behalf of one or several wide area network providers (or other access points) and vice versa.

In addition, it highlights the necessity of routing procedures for multi-hop topologies, such as route discovery methods, distributed admission control considering multiple hops as well as maintenance procedures, so that the communication continues when the relaying node is not available anymore. In case a number of accesses are available, the main issue is to select the most appropriate access, including wide area networks, LAPs as well as other users' terminal devices acting as relay nodes. Various parameters need to be considered (type of service, user preferences and policies, agreements, etc.) in the access selection process. Before a handover is executed, an admission control procedure, potentially involving all participating relaying nodes as well as the LAP, needs to be invoked.

8.5.2 Scenario 2: Public Access to Privately Operated Local Access Networks

Scenario 2 is characterized by privately deployed, owned and managed LAPs allowing public access in order to augment capacity and coverage provided by traditional cellular wide area operators [100]. By 'privately' we typically mean private persons and small businesses, here denoted as local network operators (LNOs). A similar concept was presented in [101], where it was argued that the there will be a primacy of private, unlicensed systems with full transparency between public and private communication systems in future communication systems. A special case of the scenario 2, where the local access providers act as network franchisees for a wide area operator, has furthermore been described in [102].

The main idea behind scenario 2 is to increase capacity and coverage in areas where traditional wide area operators are unable or unwilling to satisfy traffic demands for some services. Major causes for these coverage/capacity 'gaps' are too high costs and limited deployment flexibility when using conventional technology and business models. Scenario 2 addresses the problem by utilizing privately operated LAPs for public access and the main application areas are deployment adapted to local traffic demand; low cost, local area coverage for very high bit rates, possibly varying QoS; and deployment in homes, offices and public hot zones.

A LAP deployment example is shown in Figure 8.11, where the large ellipse shows the coverage from the wide area access point (WAAP) and the smaller ellipses show the coverage zones of the local access providers. The LAPs are typically connected to the Internet through the deployers' private broadband connection (xDSL, cable modem, fixed wireless, etc.) but could also be connected directly to a wide area operator's access network as LAP A in the figure. As the individual LAPs are unreliable, it is difficult to give strict quality-of-service (QoS) guarantees. Instead, we assume that a wide area operator can provide support

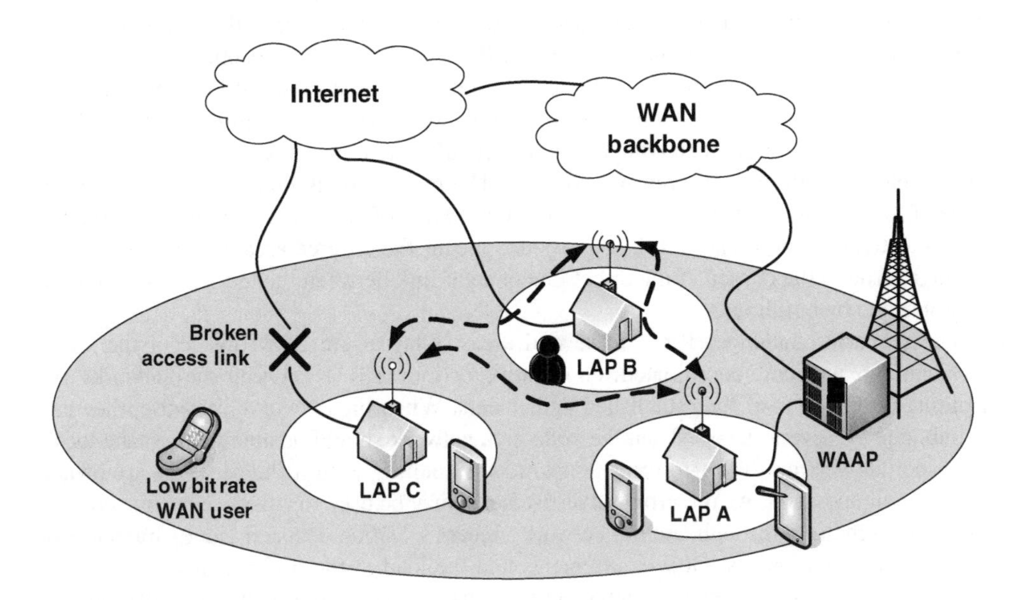

Figure 8.11 LAP deployment example for scenario concept 2

for real-time services and mobility (e.g. paging) if needed. For increased reliability, several LAPs in an area can form a wireless multi-hop (partial mesh) network in order to mitigate failure or congestion situations in the individual broadband connections. This is shown as the LAP-to-LAP links in the figure.

In this scenario, it is critical that local network operators can advertise their existence and functional capabilities (RATs) so that roaming users can find, evaluate and possibly utilize the LAP. The presence can be advertised to the users either by the specific network itself or by adjacent networks (proxy advertisements). In the latter case, revenue sharing will probably be necessary.

To enable cooperative resource management between networks belonging to different administrative domains, the LNOs will also advertise their existence and capabilities towards other LNOs in the vicinity. Two or more cooperating networks can share information such as free capacity, used RATs and backbone transmission resources. The AN MRRM and GLL FEs enabling common power and spectrum control can be used to reduce both intra- and intercell interference. The AN cooperation should also support more evolved forms of load sharing and admission control. For example, in an area containing several LAPs, the LAPs can form a meshed network in order to share each other's backbone transmission resources as shown in Figure 8.11.

Access selection parameters of particular importance include end-user QoS requirements, end-user cost–capacity performance, multi-operator network capacity, revenue of specific operator, fair share of traffic and revenues among operators. Typically, the end-user terminal will have to take an active part in the access selection.

8.5.3 Scenario 3: Wide Area, High Bit Rate Networks with Fixed Relays

Scenario 3 proposes an architecture consisting of a relay-based local or metropolitan network and a wide are network. A nonconventional way to increase capacity and coverage at low cost while providing appropriate QoS to any kind of mobile and wireless communication services is the long-term (preplanned) deployment of fixed relay stations within the local/metropolitan network [103]. Wide area networks can still provide a basic connection and they may make use of the relay concept too. The relay-based and operator-owned local/metropolitan network will provide both outdoor and indoor broadband access to terminals with medium velocity of movement and can compose with a cellular radio network to support a high terminal velocity with medium transmission rate. The relay stations should be installed at the edge of the coverage area (as can be seen in Figure 8.12) of an access point in order to relay the data stream in either layer 1, 2 or 3. By this means the covered area is extended or even 'brought around the corner' if the direct line of sight link between the user terminal and AP is obstructed by buildings.

The scenario challenges RRM in several areas: relay aware RRM inside of the local/metropolitan network; coordinated, cooperative or joint RRM between the networks depending on the level of the established agreements. When the networks compose, they can collaborate on several levels. Then the wide area network should be able to assist the local/metropolitan network by performing cooperative signalling even if the networks are owned by different operators. Again, proxy advertisements may be used to advertise one network by means of signalling through another network. Access selection decision input parameters of importance include users' QoS requirements and the load in the different networks. In the multi-operator case, similar parameters as in scenario 2 come into play, e.g. cost–capacity performance and fair share of revenue between the operators.

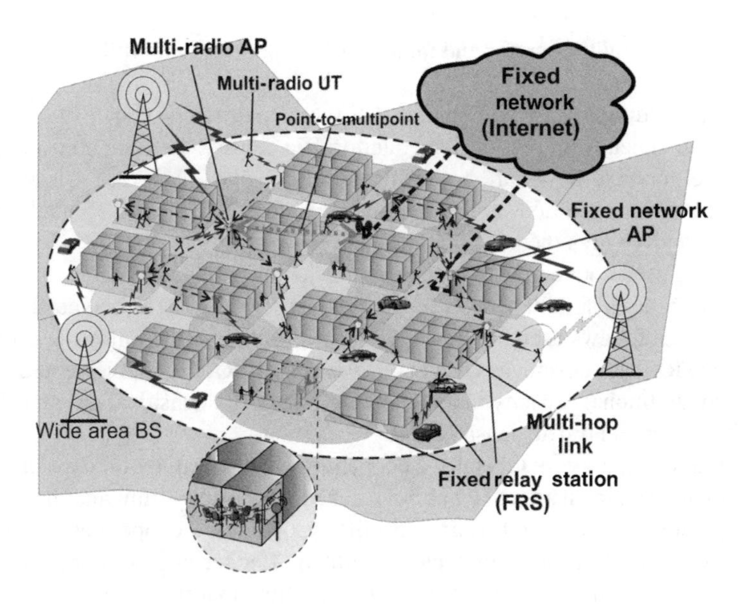

Figure 8.12 Application scenario for scenario 3

8.6 Deployment Cost Savings

The multi-radio access architecture within the Ambient Networks concept opens up opportunities for new business roles within the mobile and wireless domain. The new business roles take advantage of the control plane functionality that enables cost-efficient connectivity over heterogeneous radio access technologies. With access to 'any network', the users will experience a higher quality of service level with increased availability and reliability. Network and service providers can offer this at a lower cost if they utilize their different RATs in an efficient way enabled by common radio resource functionality and/or by cooperating with other providers with the purpose to buy and sell radio capacity from each other.

The economic feasibility of the AN concept has been evaluated by means of deployment cost savings from a MRA perspective through qualitative and quantitative analysis of different scenarios. AN MRA technology enables very efficient and tailored use of all available radio capacity across business boundaries. The benefit is that less infrastructure capacity is needed to support the same level of traffic at a given QoS level. Less capacity means lowered total costs from an overall perspective. This type of benefit is similar in nature to the infrastructure and network sharing seen in today's mobile networks, but taken to a much more detailed and refined level through the dynamic and flexible AN cooperation technology. The AN multi-radio cooperation within and between operators allows, for example, a user entering a particular area to have his ongoing data communication be handed over from one radio access in one operator's network to a different radio access in another operator's network. The second operator could be a small local provider that in turn utilizes a third provider's access. Another capacity saving strategy enabled by AN is the ability to dimension for lower amounts of 'spare' capacity and rely on another access, even across business boundaries, for example, during peak hours.

In the specific evaluation, operators use up to three different RATs in each of three scenarios described: (1) no AN MRA supporting multi-RAT, (2) multi-RAT cooperation with AN functionality and (3) multi-operator and multi-RAT cooperation with full AN functionality:

1. Each RAT is used separately without any AN functionality. The choice of which RAT to use in an area depends on the local user and traffic density. There is no overlap in RAT coverage (i.e. multiple RATs are not available in any given place), as it is assumed that there is no RAT cooperation. A user can use different RATs when located in different areas, but in a given area there is only one choice.
2. The RATs are used cooperatively with AN functionality. In a given area, multiple RATs can exist and be used simultaneously to handle the user traffic. The choice of which RAT to use in an area follows user and traffic density according to the following principle: the most suitable RAT for coverage is used everywhere; this RAT is supplemented with a second RAT for additional capacity in medium user and traffic density areas, and also a third RAT in high user and traffic density areas.
3. Multi-operator cooperation: Operator 2 cooperates with operator 1 for capacity whenever operator 1 has a site in an area that has not reached the capacity limit and that can accommodate operator 2's traffic in this area. Similarly, operator 3 cooperates with operator 1 for capacity if the site is still not capacity limited after serving both operator 1 and 2's traffic and can accommodate operator 3's traffic. Note that the decision to cooperate is made per area reflecting the flexible cooperation schemes made possible by AN.

Figures 8.13 and 8.14 show the cost per user per Gbyte/month for a suburban scenario. Figure 8.13 shows the comparison between scenarios (a) and (b). In this case, there is no

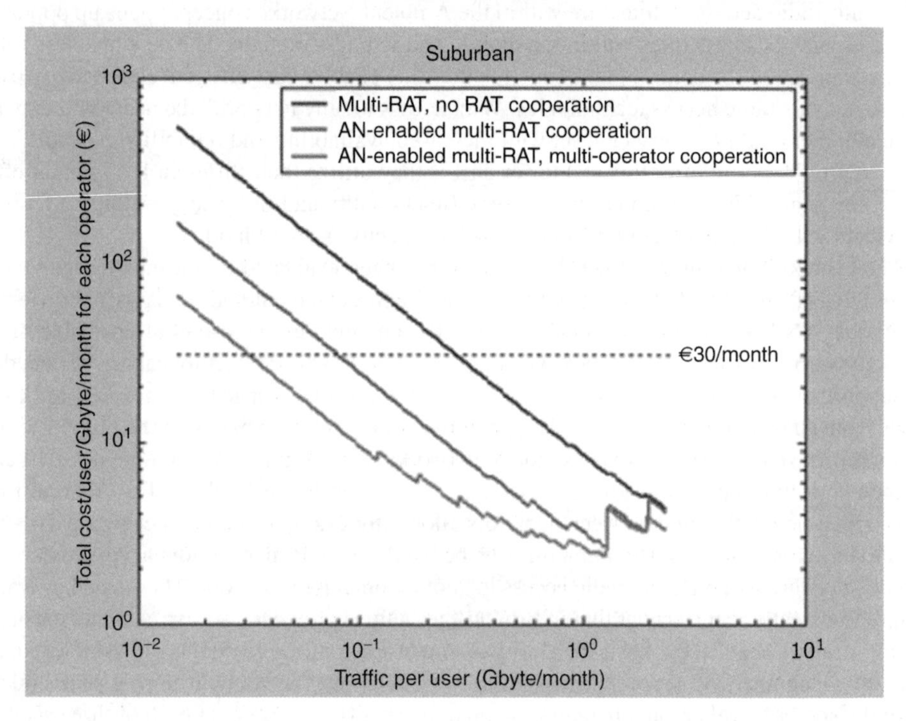

Figure 8.13 Suburban scenario – cost per user per Gbyte/month for the three operators with and without AN-enabled multi-RAT cooperation

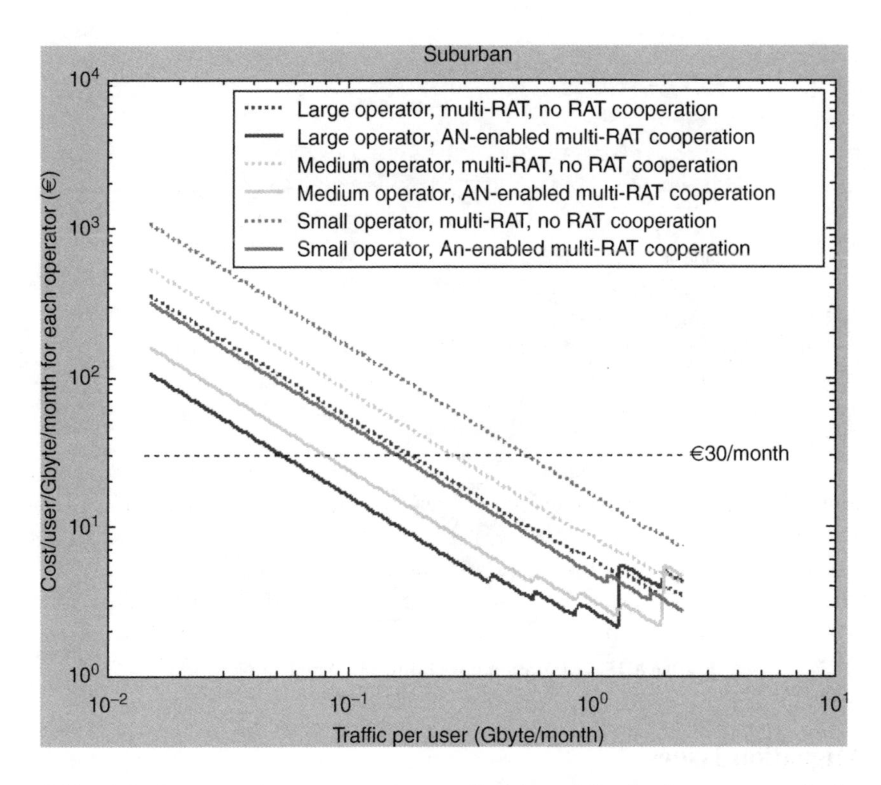

Figure 8.14 Suburban scenario – cost per user per Gbyte/month for the three operators is shown for the separate multi-RAT deployment scenario, the AN-enabled cooperative multi-RAT scenario and finally for the AN-enabled multi-operator multi-RAT cooperative scenario

multi-operator cooperation, so the performance measure, the cost per user per Gbyte/month, is given for each individual operator. figure 8.14 shows a comparison for all scenarios including multi-operator cooperation. Here the measure is the cost per user per Gbyte/month taken over all operators' users. This market view of cost circumvents the need to consider compensation and cost distribution between operators, but is still indicative of the cost benefit of AN technology as such.

For comparison, the Figures also show the 30 €/month line – this is a rough figure of the amount of money a user is willing to spend per month on communications (corresponding to the popular rule 'a dollar per day').

Figure 8.13 shows clear benefits in terms of deployment cost savings with the AN-enabled multi-RAT cooperative scenario for all three operators. Note that the number for this scenario includes the cost from AN MRA servers and AN functionality in the individual RATs. The cost savings also translate into a shift of the lowest traffic per user when the cost curve intercepts the 30 €/month line. figure 8.14 shows that there are additional benefits in terms of cost savings with multi-operator cooperation. These savings mainly follow from the ability to pool traffic from multiple operators to 'fill up' the capacity of the used RATs. At low traffic densities, the gain is almost as large as the gain from the intraoperator AN-enabled multi-RAT cooperation.

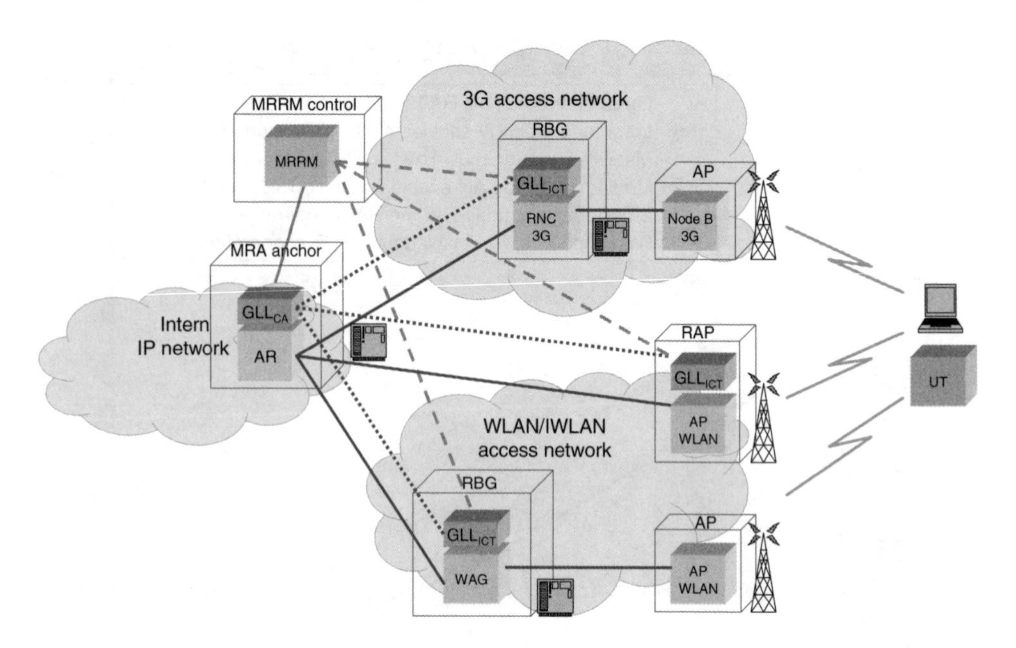

Figure 8.15 Migration possibilities in Ambient Networks

8.7 Migration Issues

At an early stage of the AN project, it was already recognized that business motivations are important aspects of migration. Thus, the MRA solutions have addressed concerns that migration to AN should be business motivated and not only a new technology push. The migration to complete multi-access architecture with AN functionality will not be as hard as, for example, migration from GSM to UMTS, where all access nodes, core network nodes, user equipment and service components had to be launched before commercial launch. As oppose to migration from GSM to UMTS, the migration to a native AN multi-access solution should be seen as a stepwise process of changing and/or enhancing legacy control functionality towards the Ambient Control Space (ACS). The ongoing standardization efforts of architecture and handling multi-access in 3GPP SAE and IEEE 802.21 must be able to fit in the ACS. The emerging access solutions like 3GPP LTE and IEEE 802.16 as well those not yet brought into standardization such as WINNER type of systems should also be able integrate with the ACS through the ARI. The migration towards a native ambient MRA must follow general progress in standardization in order to avoid being too specific in the architecture description. The control and user plane paths and interfaces for migration related to nodes in 3GPP and WLAN standards are given in Figure 8.15.

8.8 Conclusion, Outlook and Further Work

The Ambient Networks multi-radio access architecture is one important key towards management of heterogeneous networks. The Ambient Resource Interface has been developed as an abstraction of the resources provided by the underlying multi-operator connectivity.

The functionally related to connectivity layers in the Ambient Control Space MRRM and GLL are transparent to any particular data transfer technology, maximizing the applicability of the results and opening the range of migration possibilities. The MRA design is, in this sense, flexible enough to adapt to new requirements that may appear. Among these, multi-hop and multi-operator scenarios have received special consideration. In this chapter, we have described the MRRM and GLL functions of the MRA architecture and evaluated various approaches of radio access selection algorithms in a number of feasibility studies.

The obtained results on fast access selection over tight-integrated RAs show clear gains for the multi-RAT, single-operator case under idealistic settings. However, in realistic network settings these improvements are significantly reduced as the signalling delays increase. In general, the improvements obtained by fast selection were not of the orders of magnitude that would justify the complexity and the costs it implies on the architecture. Slow access selection using cell and network load information also shows capacity and user throughput gains, although these are lower. The feasibility analysis has also highlighted the importance of using up-to-date information, in order to select the most appropriate access; a trade-off has to be done between the overhead required and the benefits that may be obtained from this extra signalling. This chapter has also studied the impact of cooperative and competitive scenarios; in these studies, the results are better when the operators agree to cooperate, especially in high dense (e.g. hotpot areas) environments.

Using the results of these studies, a target MRA architecture has also been proposed characterized by slow access selection that would be based on load information. Additionally, the proposed architecture supports distribution of control so as to allow for the cooperation among network operators.

The concluded target multi-radio access architecture presented in this section is still defined only at a rather high level. It provides a first step towards a final architecture proposal by identifying the (based on feasibility evaluations) most promising functionality and where it is provided (logical nodes).

In the second phase of the AN project, the architecture proposal will be refined further. Interfaces and protocols will be defined and elaborated. Performance and scalability will be evaluated through comprehensive quantitative analysis and simulation. A prototype will be developed to further prove and validate proposed functionality.

One example mechanism for which thorough protocols/interfaces investigations are needed is the MRRM interaction with peers and other functional entities in the access selection process. Feasibility of abstracting RAT-specific input to the access selection algorithm, and the corresponding effect on performance, needs to be evaluated for various MRRM role distributions between actors (interoperator, cooperative/competitive operators). Access selection algorithms need to be defined in order to validate the principal behaviour of MRRM. Another example is the proposed GLL context transfer mechanism, which needs to be further elaborated in terms of signalling procedures and performance evaluations.

9

Ambient Networks Mobility Management

Acknowledgements

This chapter is based on the joint experiences and effort of the researchers in the first phase of the AN project and particularly the following people listed as contributors and authors (i.e. in alphabetical order): Ramon Aguero (University of Cantabria), Leonardo Badia (Consorzio Ferrara Richerce), Ambroise Boni (France Telecom), Servane Bonjour (France Telecom), Roksana Borelli (National ICT Australia), Johnny Choque (University of Cantabria), Jochen Eisl (Siemens), Michael Eyrich (Technical University Berlin), Michael Georgiades (University of Surrey), Inge Grønbæk (Telenor), Eleanor Hepworth (Siemens/RMR), Jan Höller (Ericsson), Daniel Hollos (Technical University Berlin), Yuri Ismailov (Ericsson), Tony Jokikyyny (Ericsson), Marko Jurvansuu (VTT Finland), Vesa Kyllönen (VTT Finland), Julien Laganier (Docomo Eurolabs), Mikael Latvala (Nokia), Mikko Majanen (VTT Finland), Jukka Mäkelä (VTT Finland), Francesco Meago (Siemens), Jan Melén (Ericsson), Daniel Migault (France Telecom), Marco Miozzo (Consorzio Ferrara Richerce), Takatoshi Okagawa (DoCoMo Eurolabs), Jussi Paakkari (VTT Finland), Eranga Perera (National ICT Australia), Anand Prasad (Docomo Eurolabs), Kaushalya Premadasa (National ICT Australia), Jarno Rajahalme (Nokia), Teemu Rinta-aho (Ericsson), Michele Rossi (Consorzio Ferrara Richerce), Masahiro Sawada (DoCoMo Eurolabs), Kristian Slavov (Ericsson), Abigail Surtees (Siemens/RMR), Stein Svaet (Telenor), Jari Tenhunen (VTT Finland), Bruno Tharon (France Telecom), Tiziana Toniatti (Siemens), Ralf Tönjes (Ericsson), Ville Typpo (VTT Finland), Shintaro Uno (Motorola Japan), Jukka Ylitalo (Ericsson), Adam Wolisz (TU Berlin), and Michele Zorzi (Consorzio Ferrara Richerce). This chapter has been edited by Jochen Eisl, Jukka Mäkelä, Ramon Aguero Calvo and Shintaro Uno.

9.1 Background and Motivation

Mobility is one of the most liberating features having a tremendous impact on how communication is evolving into the future. At the beginning of this section, we give a retrospective view of mobility management including a brief state-of-the-art analysis. Since the early days

Ambient Networks: Co-operative Mobile Networking for the Wireless World Norbert Niebert (Ericsson GmbH), Andreas Schieder (Ericsson GmbH), Jens Zander and Robert Hancock

of mobile communication, new mobility solutions and enhancements have been proposed and implemented, but nowadays mobility like described by the Ambient Networks concept is still a vision. A set of demands stemming from this vision shows why more research was (and still is) necessary for mobile communication. In the following sections, we explain how Ambient Networks addresses the challenges.

Mobility has always been a characteristic of wireless communication systems. In the early days of radio networks, network topologies were simple and the radio range of individual terminals and sets was very large. Global networks could be designed with virtually full connectivity, and there was basically no need for mobility management. It was not until the advent of cellular systems with user terminals roaming over many small cells, the relevance for mobility management became evident. In such systems, mobility has mainly been about supporting real-time communications to mobile devices connected to a fixed infrastructure (e.g. cellular telephony). With the rapid transformation of the communications needs (Internet, media, etc.), new challenges in mobility management need to be met. As will be seen in this chapter, there is a new level of mobility support required by the networks supporting emerging ambient and ubiquitous communication needs. On the one hand, the cellular systems of today, 2G and 3G, have mainly been designed as dedicated and vertical systems, optimized to accommodate a certain set of features including their specialized mobility support. They do provide access to the Internet, but transparent mobility, e.g. handover between different access technologies, is still far from a reality. The Internet architecture, on the other hand, was originally designed to be a homogeneous interconnection architecture of networks of stationary computers. The focus was on providing end-to-end transparency for any type of applications to run on top of the Internet. It was never within the scope to provide an architecture, which had mobility built into it from the start. Rather, there is a continuous effort of retrofitting mobility solutions to the architecture, thus resulting in a patchwork of specialized solutions at different layers of the system. At the same time, emerging wireless technologies enable a multitude of new networked devices, not only being used and carried about by humans but also being attached to machines or objects of daily life. For such devices, connectivity to a stationary infrastructure is less important than ability to communicate with and via anything in the surrounding environment in a highly dynamic fashion. Communication has to be continuously adapted to changing user needs and application context. There is evidence that research is necessary to break with traditional mobility concepts and to start with a new approach to address the above-mentioned problems. However, this new approach needs to take into account a set of state-of-the-art mobility macisinisms, which should be re-used where appropriate.

Subsequently, we have formulated the basic needs for taking this new approach, where the need for mobility for a diverse range of endpoints – user devices, groups of applications and groups of users – has been identified (in the AN scenarios [190]).

Analysis of the mobility requirements needed for Ambient Networks has highlighted a number of key aspects that have to be considered:

- Currently deployed mobility solutions tend to be designed within dedicated systems and, therefore, may be defined to accommodate a limited set of features or optimized for a particular environment (essentially those supported by current cellular networks).
- On the contrary, a large number of protocols have been defined to support mobility across the Internet, most of which have been retrofitted into an architecture that was never originally intended to support this type of operation. This has resulted in a patchwork of solutions with unclear interoperability and deployment guidelines.

- Emerging wireless technologies have enabled a multitude of new mobile devices, placing further requirements on the network to support seamless mobility across heterogeneous link technologies.
- In addition, there is an increasing requirement to better support mobility of groups of nodes both in ad hoc configurations and in transportation, e.g. vehicular environments. As services and applications are generalized from traditional client–server or terminal to network cases to include peer-to-peer and collaborative applications as well, there is an increasing need to optimize the network to handle mobility patterns of this sort.

Mobility exposes the users to heterogeneity and dynamics on different levels, e.g. access technologies, networking and trust domains, device capabilities and user contexts. This implies the requirement for coexistence of versatile mechanisms in different networks and even requires harmonization of mobility among different layers in the communication stack. A large number of mobility management solutions have been developed in recent years for various purposes. The challenge of the project is to integrate appropriate existing mechanisms, enhance some of them with new functionality and provide new solutions, so that the requirements for future mobility in Ambient Networks are met. To be coherent with the concept of composing networks, these mobility mechanisms should be deployed like a set of tools and a controlling entity is needed to select the most appropriate tools for a specific mobility event.

State-of-the-art handover is mainly driven by changed radio layer conditions. However, future mobility solutions may consider various types of triggering events, such as application-related events or triggers generated in different Ambient Networks. A framework is needed to enable processing of events from various sources, classify them and allow prioritization among conflicting events.

9.2 The Framework for Mobility Management

9.2.1 The Ambient Networks Approach

The goal of the Ambient Networks mobility solution is to provide a framework within which existing mobility solutions can be deployed and interoperate, whilst ensuring that new mobility solutions can be added as and when they become available. This section introduces the architecture within which the mobility solutions are defined. The goal of the architecture is to illustrate how the mobility functions are placed with respect to the overall Ambient Networks architecture and to support a modular approach to mobility management that can be applied to any type of mobility endpoint.

The various types of triggering events that have been identified require a functional entity within Ambient Networks that is responsible for the collection of those events, classification and filtering of the same, and finally delivering these events to ACS functions (consumers). The number of consumers and triggering events handled by the system is configurable; this offers a very flexible and dynamic framework. A large set of possible events has been identified, which are classified into a set of trigger groups according to some defined set of classification criteria. We have identified trigger policies as a set of classification/filtering rules and criteria to enable the triggering engine to be configured with context-related information. The concept of distributing triggering functionality (e.g. classification, filtering, etc.) among different Ambient Networks increases the possible set of features provided by this framework.

An implementation of the triggering framework has been carried out and has proved the added value of the concept.

A group of nodes may show common mobility pattern and as a consequence these nodes may benefit from internal routing optimizations while the group is staying together. Mechanisms need to be introduced, which allow the recognition of such common movement, establishing and maintaining relationships among the nodes and the announcement of special functionality provided for the whole group, such as connectivity to other networks via gateways. Additionally, functions are needed for the mobility of the group to enable delegation of mobility handling to specific nodes and to optimize handover for a larger number of nodes, which usually can lead to scalability problems.

Finally, the functions needed to satisfy the described mobility challenges are not deployed in an isolated way but have to be integrated into the Ambient Networks architecture getting full advantage of the dynamic plug-and-play concept introduced by the composition feature.

9.2.2 Framework Principles

Mobility is considered to be a fundamental part of ANs, especially when considering novel challenges coming from the concept of network composition along with network heterogeneity and the need for integrated mobility, which are not yet handled by existing solutions.

Moreover, in heterogeneous environments, scalability and manageability required (easy to use/deploy, many networks everywhere) in addition to 3G migration and integration of legacy networks. Under these circumstances mobility framework principles have been defined for the Ambient Networks architecture.

In order to scope the mobility problem space, and to allow for the definition of interoperable mobility solutions, a number of architectural delimiters are required with a stable set of interfaces about which we can make certain assumptions. The following section discusses the overall architectural model and its impacts on the mobility solutions for Ambient Networks.

The mobility control space (MCS) is defined as a subcomponent of the Ambient Control Space (ACS) and is shown Figure 9.1. This diagram illustrates the relationships between the FEs defined as a part of the MCS and other FEs defined within the ACS. There are two sets

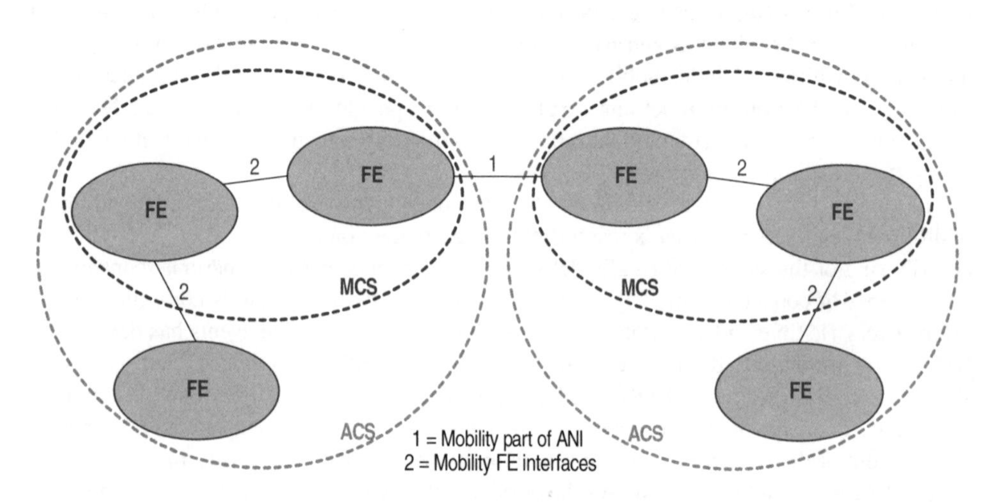

Figure 9.1 The MCS and the ACS

of interfaces that have been considered within the architecture definition work, the mobility part of the ANI and the mobility functional entity (FE) interfaces that support the exchange of primitives between functional entities within the ACS.

The rest of this section begins by outlining a number of key simplifying assumptions derived from the overall ANs architecture work and discusses how these affect the mobility architecture. A mobility architecture is then introduced with the purpose of providing the framework within which different mobility approaches can be discussed. The design considerations that were considered during the development of the architecture are also highlighted. Finally, the overall functionality that the AN mobility solution must support is introduced, along with the identification of the key functional entities that the mobility control space will implement.

9.2.2.1 Assumptions

Within the Ambient Networks project, a number of abstractions have been defined that describe how objects (or actors) in the network communicate with each other.

In terms of mobility support, the following key architectural decisions are noted.

The ASI exposes a nonchanging identifier to the applications above it.

A nonchanging identifier is exposed to applications and can be bound to application sessions for the purposes of communication. For example, an identifier such as a host identity tag (HIT) or even a home address defined by Mobile IP could be used. The key property of this identifier is that it hides any changes to the underlying connectivity (in terms of what locators are used by a communication exchange) from applications and application sessions. *Note*: Although this seems to eliminate the need for session mobility, this does not prevent applications from deciding to change identifier if they wish (although this would appear as a new session to lower layers) and does not address the legacy case where applications handover from a legacy network, potentially requiring a change in identifier. In these cases, session layer mobility may be needed to support applications changing identifier bindings, and this is expected to be application specific.

The ARI exposes one or more locators to the functions above it.

The locator is exposed across the ARI and represents the point of attachment to a network. It is used to deliver user data to the device and as such represents the location of a device in the network, for example, the IP address of the end host system. The locator exposed across the ARI may have certain properties. For example, the locator may change due to a mobility event (as the device is now communicating via a different network point of attachment) or it may be nonchanging (which is true if the underlying connectivity hides mobility events, as could be the case in a cellular network where the radio access network can hide local mobility events from the IP layer above). In addition, locators may have different scopes over which they are valid, from locators that are usable only within a small localized segment of the network to those that are valid across an entire network. The properties of the locator will influence the mobility mechanisms employed by the ACS. It is assumed that the locator is always available for use (i.e. idle mode operation is not considered within this deliverable).

Flows are supported by forwarding state in the network that may change due to mobility events.

Flows represent the basic connectivity provided by the underlying network technology and model the path through the network taken by user data. In order to route traffic correctly across the network, it is assumed that forwarding state information, such as explicit host routes and tunnel encapsulation information, is stored within various nodes within the connectivity. After a mobility event, this forwarding state may need to be updated.

The mapping between identifier and locator may be altered by a mobility event.

When the locator exposed across the ARI changes, one or more mappings that map identifiers onto locators may need to be updated. This in turn may result in update of rendezvous and name resolution state held within the network.

Changes to flows beneath a bearer do not affect the relationship between the bearer endpoints.

Flows connect single 'bearer hops' together, for example, one flow connects one bearer intermediary to another, and another flow is used to get from the bearer intermediary to the bearer endpoint. A flow endpoint is a network point of attachment. Flow mobility (a change of network point of attachment) may occur between two bearer intermediaries or endpoints, but it does not affect the relationship between the bearer endpoints (as the bearer endpoints and intermediaries remain the same). Any re-routing of the bearer through new intermediaries is triggered by a notification of a topology change followed by the establishment of new flows between the appropriate bearer endpoints and intermediaries (as opposed to requiring a flow mobility event).

9.2.2.2 Mobility Reference Architecture

The Ambient Networks architecture has identified a number of interfaces that delimit the boundaries of the ACS. This section illustrates the Ambient Network interfaces in the context of a host end system and highlights the key aspects that must be considered from the mobility perspective.

The MCS (illustrated in Figure 9.2) has control plane interactions via which it manages the connectivity layer itself through the ARI and interacts with applications above the ASI. These interactions include exchange and configuration of mobility-related state, such as triggers. The MCS also exchanges control messages with peer MCS entities in other devices/networks, which are carried as user plane traffic across the ANI.

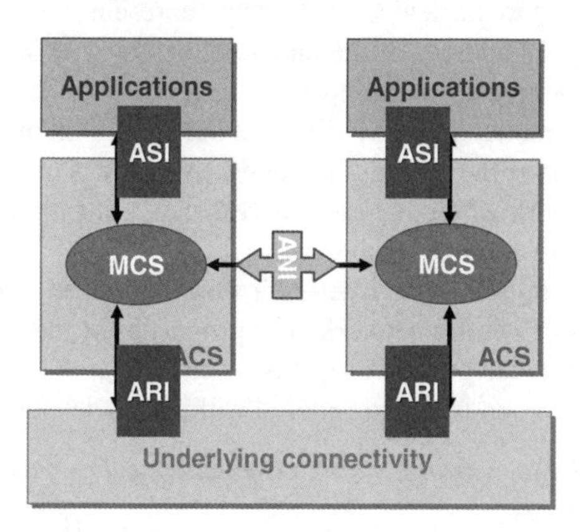

Figure 9.2 The MCS and the AN interfaces

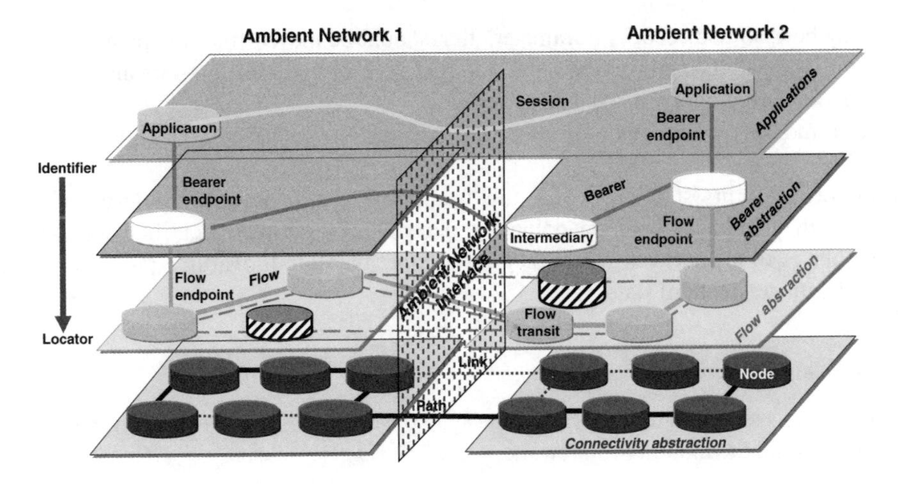

Figure 9.3 Identifiers and locators

Figure 9.3 illustrates the location of identifiers and locators within the Ambient Network abstractions.

Applications (and ACS FEs) bind to identifiers for the purposes of communicating with other end host systems (and other ACS FEs in different ANs). This identifies the source or destination for the bearer endpoint. The choice of which identifier to associate with is up to application policy, and multiple sessions from different applications can potentially use the same identifier.

In addition, traffic for transmission over the underlying connectivity is exchanged using technology-dependent locators. These locators are used by the underlying connectivity to deliver packets between flow endpoints.

Figure 9.3 shows the placement of identifiers and locators with respect to the bearer and flow abstractions from applications through to the connectivity infrastructure. Within the end host system, many-to-many mappings between identifiers and locators can be established to support the transfer of application session data. The precise mappings that are established are dependent on the mobility solution and may involve multiple levels of indirection (e.g. an identifier may be mapped onto an intermediate identifier/locator before being mapped onto an appropriate locator).

In addition to these mappings within the end host system, there is a basic set of information that has to be installed and maintained in the network to support end system to end system communication. These can be categorized into

- Forwarding state that determines the next hop to which the forwarding entity should send the data of the ongoing communication exchange to reach the current location of the host end system.
- Rendezvous and name resolution state that supports the resolution of an identifier onto a locator that represents the current location of the host end system.

There is also quite often information associated with an application session that is installed at various points in the network (at the flow, bearer and even application layer). For example, there may be QoS and security parameters that should be moved from the path taken by an old flow to the path taken by a new flow. Alternatively, communication state and application state may be moved between host end systems.

After a mobility event, some or all of the information described above may have to be updated.

Therefore, mobility solutions within this framework can be described in terms of the assumptions they make about the mapping and dependencies between identifiers and locators, and the places in the network where they install and maintain forwarding (and also rendezvous and name resolution) state.

9.2.2.3 Design Considerations

The following section highlights the key design considerations that were taken into account when developing the mobility framework.

Modularity

One key design consideration for the mobility architecture is the issue of modularity. This issue arises because there is a wide range of mobility solutions already specified for today's networks, with no universal mobility scheme commonly available. This is partly because different network environments place different requirements on the performance of the mobility solution and partly because operators may choose to enhance mobility operation for their networks by offering additional mobility functionality (such as state transfer). Therefore, it is important for the AN mobility architecture to allow integration of multiple mobility solutions, including legacy protocols and networks through to new (and even yet to be thought of) mobility schemes.

Modularity support is not restricted to just mobility functions, but is possibly also a feature of other ACS functionality. Therefore, a common registry (outside MCS) which can be used by any ACS function to determine the current set-up of the ACS is required.

In addition, these mobility solutions place different requirements on what identifiers are used, how applications should bind to these identifiers and how identifiers are mapped onto locators. This creates a number of different 'communication endpoints' that the mobility architecture must support. Support for different types of communication endpoints enables flexibility within the mobility architecture as to how identifier/locator relationships are managed, ensuring that the mobility solutions are applicable to a wide range of mobility scenarios.

Therefore, ANs do not put any bounds on the solutions to be used, nor establish any compulsory solution that has to be provided. On the contrary, it defines a flexible, modular approach, by means of which, whenever two ANs meet and want to communicate, they need to agree on the mobility procedures that will be used afterwards.

However, in order to ensure interoperability between different ANs, and to build a scalable and feasible solution, a minimum set of requirements (or functions) that any AN should implement must be identified. In terms of the general AN architecture, the composition procedure has been identified as the core aspect that any AN should be able to support and is necessary to identify what mobility functionality is supported by different networks. As discussed, composition is used whenever two ANs want to integrate their control planes (to share resources, to create a bigger 'composed AN', etc.).

The level of interoperability between ANs that require mobility functionality is guaranteed through the identification of a basic set of mandatory mobility functionality along with other mobility functionality that is conditional or optional. The mandatory functionality is captured by the basic mobility management (BMM) mechanisms, whereas all other conditional and optional functionality is correspondingly provided by advanced mobility management (AMM) mechanisms. It is noted that definitions of 'mandatory' and 'optional' depend on the implementation. This division allows slim implementations, as well as dedicated realizations targeting different optimizations and functional and deployment requirements of specific ANs. It is worth noting that BMM is generally not defined as containing a particular functional entity or function thereof. A FE, or a part of it, will generally have mechanisms that can belong to either AMM or BMM.

Multi-homing

In future ubiquitous networks, many services and many access technologies are expected to be internetworked and harmonized for better services provision and delivery. To do so, a mobile node must have different access technologies and different network interfaces for each access technology.

Multi-homing of a mobile node or a mobile router is an important and nontrivial issue. When a mobile node has multiple interfaces, it can use heterogeneous network interfaces providing ubiquitous access. However, this creates the need to extend existing solutions and maybe to provide new approaches with capabilities of taking advantage of multiple simultaneously available accesses.

The area of multi-homing support is relatively new, despite that there are a few proposals in IETF trying to provide solutions for such communication scenarios. This chapter provides mainly brief overview of existing and emerging solutions with the emphasis on their importance and need to be integrated in the context of the notion of AN. Currently, it is very difficult to provide exact design of multi-homing support in the AN with relation to the overall AN architecture design; however, during next phase of the project, it is planned to provide more details of how multi-homing support will be an integral part of the AN architecture.

9.2.2.4 Challenges

The main challenge is the ability to support a wide range of endpoints that could be mobile, from application sessions right the way through to moving networks.

The scope of the mobility problem can be phrased in terms of a number of features the solution has to consider. These include

- Bootstrapping and locator management – This is concerned with detecting network points of attachment, selecting the most appropriate access, configuring a set of locators to be made available across the ARI and initializing state information in the underlying connectivity. The formation of a moving network (routing group) is considered a special case of this bootstrapping functionality, where requirements for detecting and overlaying a network on the underlying connectivity require additional functionality.
- Triggering, handover and routing group management – collecting and processing triggers to initiate suitable mobility events and performing the handover operations. The handover operations performed may be dependent on the type of endpoint that is being moved and what state in the network needs to be updated or moved. For example, in the case where an entire network is moving, the mobility management mechanisms can be optimized to reduce the signalling overhead for this operation.

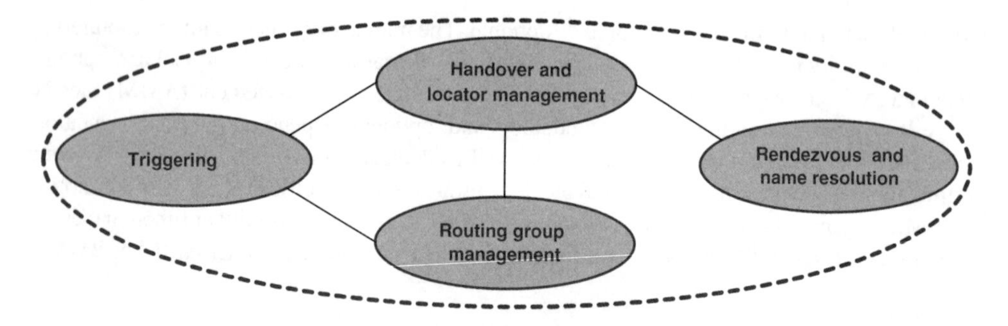

Figure 9.4 Mobility functional entities

- Rendezvous and name resolution – maintaining the state information that provides the lookup services required to resolve an identifier (for a user, service or end host system) onto the location of the host end system.

From this list, a number of functional entities are identified to address the more complex aspects of the mobility solution, focusing on those areas where existing mechanisms are not powerful enough.

9.2.2.5 Functional Model

The functional entities represent the capabilities that are required in an AN in order to support mobility events. The identified functional entities are illustrated in Figure 9.4.

There are three functional entities that are defined within the scope of this deliverable – triggering (TRG), handover and locator management (HOL) and routing group management (RGM). In addition, the diagram also shows additional functionality that is not an FE in its own right, but highlights the need for mobility specific behaviour associated with rendezvous and name resolution. This functionality may, in future work, be integrated into the node identity FE defined as a part of the overall AN architecture. For convenience, this functionality is referred to as rendezvous and name resolution in subsequent text.

9.3 Functional Entities

The following sections outline the four mobility-related functions in more detail.

9.3.1 Functional Entity 'Triggering'

The functionality provided by the triggering FE is depicted in Figure 9.5. The TRG FE provides the following functions:

- Collecting and identifying various events from different sources, which may trigger mobility management actions such as

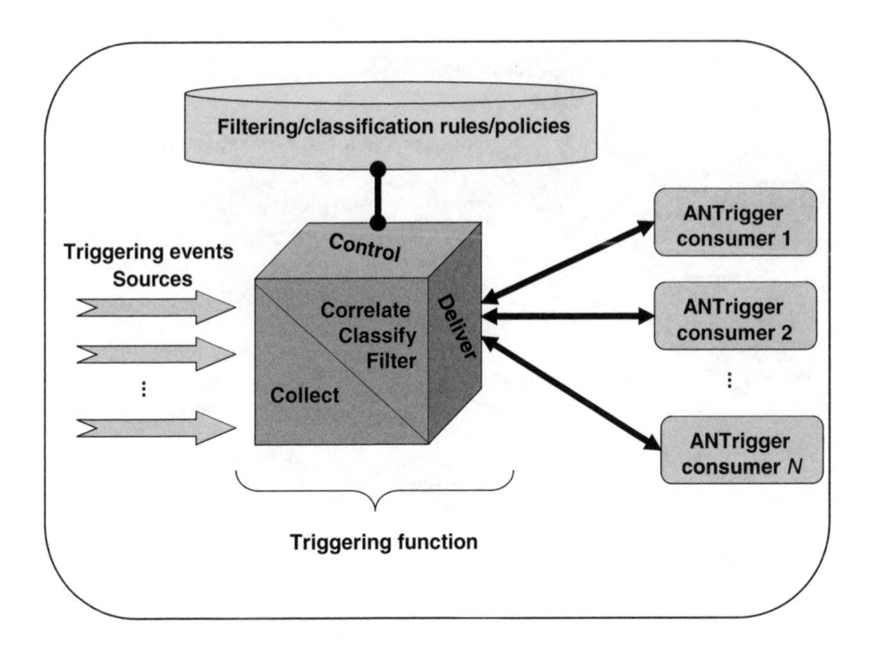

Figure 9.5 General framework for triggering

- handover decision (within handover and locator management FE);
- routing group formation (within routing group management FE);
- triggering events could also be used for triggering other actions within other FE (i.e. for any ACS function).

- Filtering and classifying mobility triggering events according to
  - criteria (six criteria have been identified so far: event source, type, frequency, persistence, time constraint and related mobility dimension);
  - triggering policies (set of filtering and classification rules).
- Performing triggering event correlation for reducing the number of events.
- Making those triggering events available for handover and locator management FE, routing group management FE or even others by temporary storing them with additional information (timestamp, priority flags, etc.). Those functions are seen as ANTrigger consumers.

9.3.1.1 Functional Entity 'Handover and Locator Management'

The handover and locator management functional entity (HOL_FE) aggregates all the procedures needed to perform various types of handovers to support mobility in the ANs HOL_FE is illustrated in Figure 9.6. These include handovers between communication access points within a single radio network, between different access technologies, mobility between different IP address spaces, multiple service provider domains (i.e. network layer mobility) or application-level handovers between different terminals.

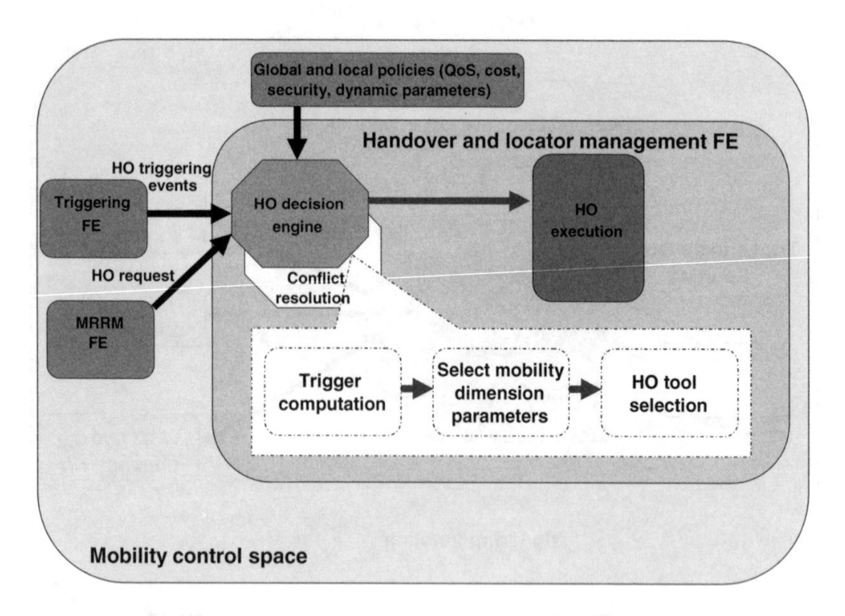

Figure 9.6 Handover and locator management FE overview

Handover decision is a functionality distributed between MRRM (multi-radio resource management) FE and HOL_FE, the latter having the task of setting up and triggering the appropriate protocols within the AN nodes.

Functioning of the HOL_FE is based on the selection of appropriate HO tools for a given HO need. This may result in the selection and configuration of the appropriate communication protocols, which may be compiled in a dynamically configurable communication stack, or it may result in the selection and execution of specific functions within a given communication protocol.

Protocols that may be impacted by AN for mobility-related matters include SIP, HIP and MIP. However, design of new communication protocols inheriting HO toolbox concepts is also possible.

9.3.1.2 Functional Entity 'Routing Group Management'

The routing group concept has been adopted in the framework of the MCS to describe a number of nodes which are moving together. In particular, this FE is concerned with the formation, maintenance and management of such routing groups (RGs), it also deals with ensuring the connectivity to external networks, by selecting and maintaining the most appropriate gateway (depending on different user-defined policies). One of the most important aspects about the RG concept is that it enables several optimizations, in particular when the mobility management of the whole network is considered, rather than treating each node individually.

Figure 9.7 shows high-level functions belonging to the FE and the information flows inside this FE and towards other FEs within the MCS. The routing group management FE registers interest in and receives triggers from the triggering FE (there are some illustrative examples shown in the figure). Some different algorithms may be used to create the RG and select the

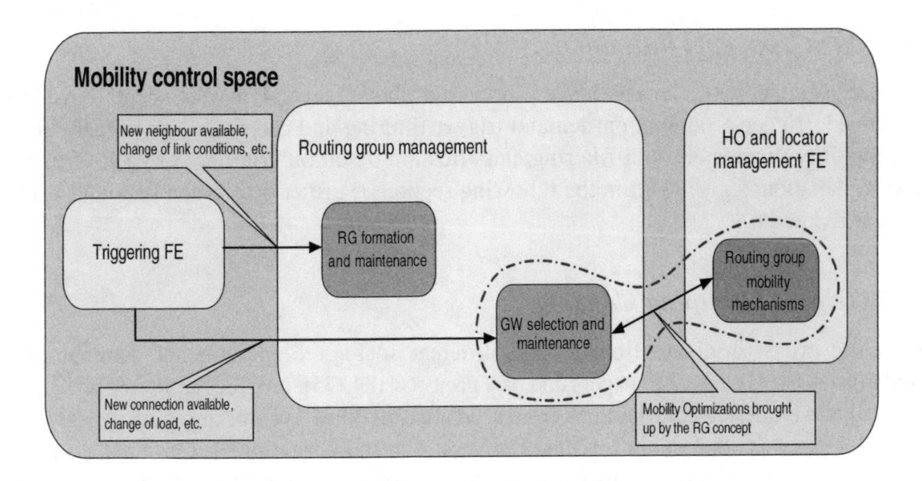

Figure 9.7 Routing group management FE within the MCS

GW, and the corresponding structure is maintained during the lifetime of the RG. As mentioned before, having such a comprehensive knowledge about the network topology may be used by special optimization processes (residing in the handover and locator management FE), which will take advantage of such information so as to bring up improvements, especially on an inter-RG level (i.e. communications inside the RG or to external entities), although intra-RG communications may also be optimized, as specified later in this chapter.

9.3.1.3 Rendezvous and Name Resolution Functions

Rendezvous and name resolution is responsible for maintaining information in the network to allow communicating endpoints to contact each other at their current locations. This involves providing a mechanism to allow an endpoint to resolve some identifier onto the current topological location of a mobile object – another endpoint. Traditionally, this functionality has been provided by DNS, where domain names are resolved onto IP addresses of the end host system. In general terms, it is expected that information mapping the end host system identifier onto a (set of) locator(s) will be stored at one or more locations in the network. After a mobility event, one or more of these locators may change, and this locator change must be reflected in the rendezvous and resolution state in the network.

9.4 Trigger Mechanisms

This section describes an approach for utilizing triggering events from any part of the system, which consists of both trigger collection and trigger processing (based on classification rules). The focus is on events that trigger mobility management actions (such as context-dependent, security-related upper layer requirements and other system-, application- or user-dependent events), but this approach could easily be extended to cover additional types of events or to help in triggering other actions (i.e. any ACS function could benefit from those collected and processed events).

9.4.1 New Concepts Introduction

In order to cope with the challenges discussed above, several new concepts have been introduced: trigger grouping, on-demand trigger filtering and classification, trigger correlation, mobility dimensions and filtering/classification based on criteria. For each of them, a short explanation is given within the following sections. Further details can be found in [191] Sections 4.3 and 6.1.

9.4.2 Trigger Grouping Concept

Triggering events originate from several different sources. Exploration of communication standards [107–133], research papers and past projects [134,135], mobility scenarios of AN project [190] and requested contributions from specialists in other AN working packages give over 200 different events that might be considered useful as mobility triggers. Certainly, even more such events exist and will be emerging in future. Mobility management mechanisms cannot efficiently handle such amount of triggering events. Therefore, grouping and classification of the events is desired. Independently of the communication technology used, it is possible to group similar or related triggers together. The identified trigger events have been grouped into 30 event groups in order to simplify the handling of the events. The number of groups is large enough to allow still guarantee sufficient diversity for event separation and classification. The event groups contain triggers related to changes in network topology, available accesses, radio link conditions, user actions and preferences, context information, operator policies, QoS parameters, composition and routing events, security issues, etc.

9.4.3 On-Demand Trigger Filtering and Classification Concept

The main idea is that any function (called a trigger consumer) can request a specific trigger classification and filtering service based on its own need. Thus, such a function may decide to receive a set of triggers and may also decide to stop receiving triggers or may decide to change the request on the fly.

9.4.4 Trigger Correlation Concept

It might be useful to consider causal relationships between some triggering events in order to avoid generating too many transient triggers based on the same physical event (e.g. a single link break). This concept is called the trigger correlation mechanism and it will help in reducing the number of events to be processed.

9.4.5 Mobility Dimensions Concept

The actual handover decision and the selection of appropriate mechanisms for HO execution can partially rely on a suitable set of triggers selected according the classification. From the triggers received, it is possible to derive the information on whether the mobile entity is moving with a single network, or crossing different technology boundaries, and whether the addressing scheme, trust domain and/or provider domain should be changed as well. Triggers may also provide a hint about whether or not the communication endpoint should be moved

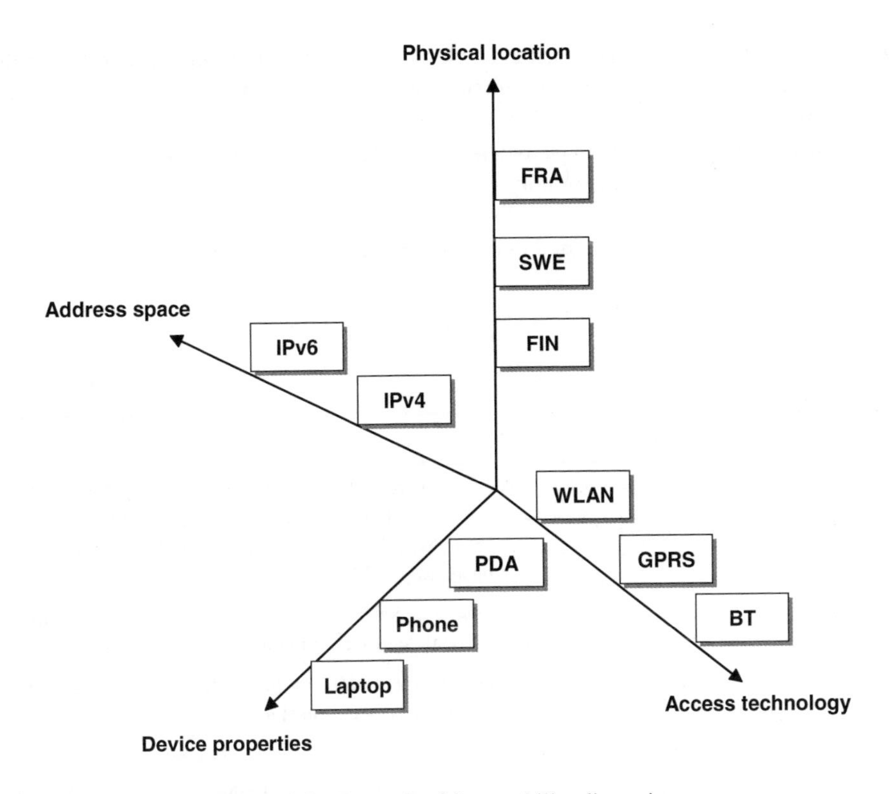

Figure 9.8 Example of four mobility dimensions

from a device to another. These different kinds of movements may occur independently or as related together. Hence, mobility events can be seen to take place in multiple orthogonal dimensions. They are referred to as mobility dimensions, seven of which has been identified: physical location, access technology, address space, security domain, provider domain, device properties and time. Figure 9.8 depicts four mobility dimensions. Triggering events may relate to one or more mobility dimensions (coordinates). Thus, mobility management mechanisms can be seen as updating these coordinates. Consequently, mobility dimensions give a broader view of traditional mobility.

9.4.6 Filtering and Classification Criteria Concept

In order to perform any filtering or classification, it is mandatory to agree on the criteria that will be used. Five criteria are proposed in this section.

1. *Type of the event*: The idea is to identify the triggering events according to their possible type (predicting, triggering or forcing).
2. *Source of the event*: The idea is to identify the triggering events according to their possible origin within an Ambient Network (within ACS, via ANI, via ASI, via ARI).
3. Frequency of the event: The idea is to identify the triggering events according to their type of occurrence (periodic or asynchronous).

4. *Persistence of the event*: The idea is to identify the triggering events according to their persistence (volatile or nonvolatile), answering to the question: Is this event transient or everlasting?
5. *Time constraint of the event*: The idea is to identify the triggering events according to the moment they occur (real time or non-real time).

Obviously, new criteria may be added in the future. Table 9.1 details the criteria classes.

Table 9.1 Trigger classification criteria

Criteria	Classes	Description
Event source	ACS	An event is received from the local Ambient Control Space, i.e. from any of the functional entities in the local ACS. Further subclassification could be done based on the FEs
	ANI	An event is generated in some other AN and is delivered to local ACS via Ambient Network Interface
	ASI	An event is generated on higher layers and received via Ambient Service Interface
	ARI	An event is received via Ambient Resource Interface
Event type	Predicting	An event that does not imply the need for a handover yet, but might be a hint of the need for handover in near future and consequently allow anticipation
	Triggering	An event that alone may lead to a handover. Other concurrent or earlier events may affect the handover decision and procedure
	Forcing	An event that mobility management has no way to optimize or negotiate. This forces a handover execution. Otherwise, communication will be interrupted
Event frequency	Periodic	An event occurs periodically with a constant time interval
	Asynchronous on demand	An event may occur at any time, but is generated on demand due to a request by the system
	Asynchronous self-generated	An event may occur at any time in a self-generated manner
Event persistence	Volatile	An event may occur and disappear in the sense of a transient event
	Nonvolatile	An event that must be raised within a certain time frame
Event time constraint	Real time	An event that must be raised within a certain time frame
	Non-real time	An event that can be raised at any time without time constraint

9.4.7 Requirements for TRG FE

For carrying out the specifications and the design of triggering FE, a list of requirements has been raised. Triggering FE should meet

- functional requirements (e.g. trigger collection, classification and distribution);
- performance requirements (e.g. scalability);
- security requirements (e.g. trigger integrity, prevention from a malicious ANTrigger consumer).

A detailed list is given in [D4.3 Annex] Section 4.4.

9.4.8 Functional Description

For Ambient Networks prospective network architecture, triggering framework has been designed for

- collecting triggering events from any sources (within a particular AN);
- classifying them according to dynamically definable classification rules;
- performing event correlation (may be an advanced mobility management function);
- providing classified ANTriggers to all registered ANTrigger consumers (e.g. handover, routing group, network management functions);
- permitting that each ANTrigger consumer receives only those ANTriggers it is interested in. The ANTrigger consumer should be able to define its preferred classification rules to be used by the TRG FE.

Please refer to [D4.3 Annex] Section 4.3.

Figure 9.9 shows triggering components (TEC, ANTCE and ANTR), internal/external interfaces (TCI, TDI, TCCI and TII), triggering events sources and ANTrigger consumers. Those will be defined and described in the following sections. Note that the same functional entity may act as a trigger source and an ANTrigger consumer at the same time.

9.4.9 Triggering Events Collection

Triggering events collection (TEC) is a function in TRG FE which receives triggering events from various sources in the network system via trigger collection interface (TCI). TRG FE may contain several TECs, which may be distributed and which may be responsible of collecting different types of triggering events. For example, each device driver could implement its own TEC, which would be capable of handling triggering events produced by the specific device only. The output from any TEC is an ANTrigger.

For scalability reason, this function can be aggregated or split into subfunctions (additional capabilities for different network access). It may be possible to have several instances of TEC within the same AN. Thus, a collaboration of those different TECs is needed in order to gather a larger amount of events. Figure 9.10 shows how different TECs within different ANs could be cascaded.

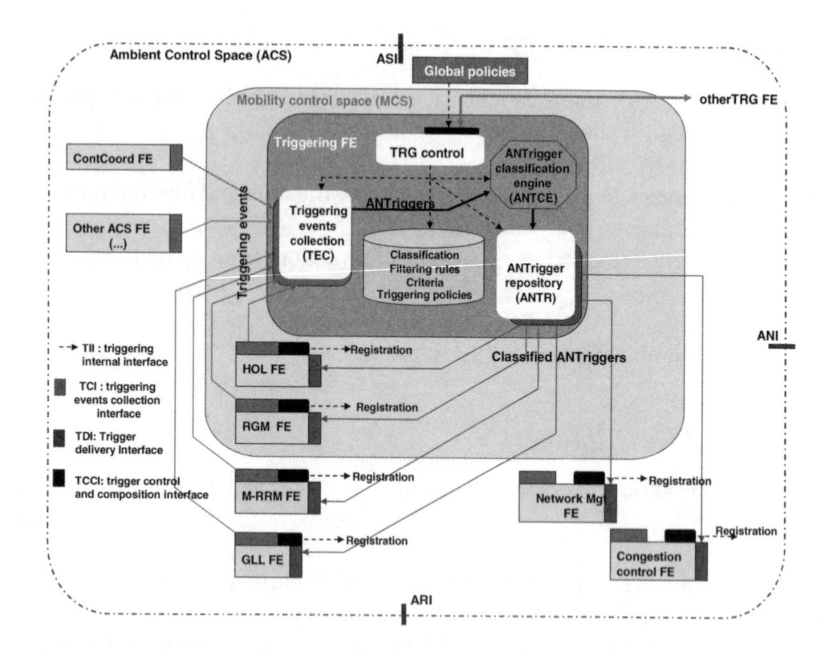

Figure 9.9 Triggering functional entity overview (TRG FE)

9.4.10 ANTrigger Classification Engine

The main idea of AN triggering framework is to handle trigger using a common format, i.e. reformat any legacy triggers into ANTriggers once they come in through TEC. This may require the extension of the trigger with new fields (criteria and mobility dimensions). Any

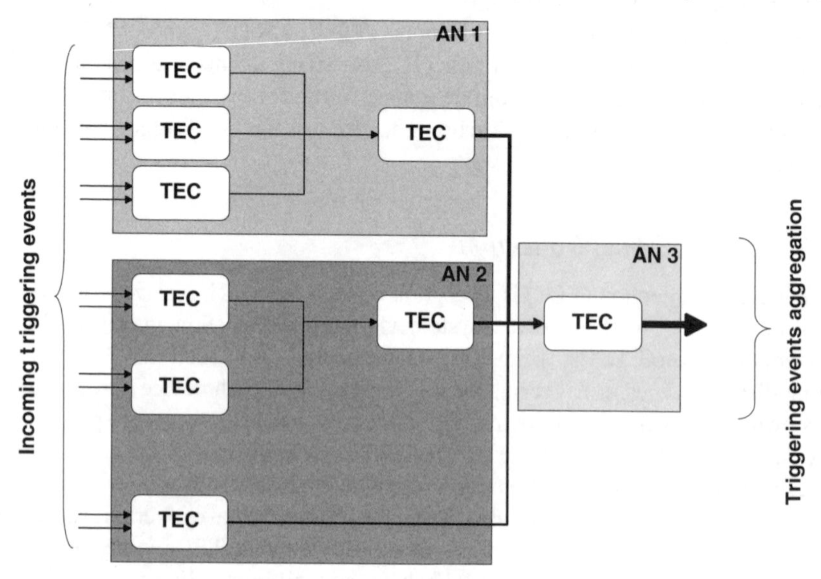

Figure 9.10 Cascade of TEC configuration

source of trigger should implement TEC and TCI in order to be able to make them available for other AN functions (trigger consumers).

Any ACS function can register (or de-register) to TRG FE to request a set (or subset) of ANTriggers. For instance, HOL FE might be interested in receiving link status and radio measurement report, RGM FE might be interested in RG advertisement and network supervision function may focus on link status. All those ANTrigger consumer functions can express their trigger requirements using filtering and classification rules. Some default rules are provided for consumers which do not need elaborated classification. Basic rules can also be used as basic bricks for building more sophisticated rules. The assumption is that consumer may provide redundant, wrongly formatted or conflicting rules. Therefore, the ANTrigger classification engine (ANTCE) should

- evaluate (on the fly) the request of each consumer and build an optimized filtering/classification algorithm (get rid of redundancy, conflicting rules, etc.);
- perform the actual filtering and classification for each individual consumer by evaluating the value of extension fields or by updating them when no value is set for such fields.

9.4.11 ANTrigger Repository

Triggering repository is used as a temporary storage for classified ANTriggers. Any other FE registered to triggering FE may use the classified triggers for further operations like the decision of the handover execution. ANTrigger repository is designed mainly for the use of mobility management triggers. But it might also be used for other FEs as well. One possible implementation for this temporary storage is the ACS common registry (e.g. CIB) that is originally designed to store the context information.

Requirements for the repository are

- It must have functionality to store and remove ANTriggers.
- Each trigger must be stored in a specified ANTrigger format.
- Each trigger that is stored in the repository has a lifetime and should be removed automatically when the lifetime expires.
- FEs that are using triggering FE (as a consumer) must be able to fetch the triggers from the repository.

10

Overlay Networks for Media Delivery

Acknowledgements

This chapter is based on the joint experiences and efforts of the researchers in the first phase of the AN project and particularly the following people listed as contributors and authors (i.e. in alphabetical order): Guido Gehlen (Chair of Communication Networks, Aachen University), Stephan Goebbels (Chair of Communication Networks, Aachen University), Uwe Horn (Ericsson EED), Thomas Petersen (TNO Telecom), Aude Pichelin (France Telecom R&D) and Aruna Seneviratne (National Information and Communications Technology Australia). This chapter has been edited by Frank Hartung, Jose Rey, Stefan Schmid and Thomas Petersen.

10.1 Introduction

In the dynamic and mobile environment targeted by Ambient Networks, there will be a broad heterogeneity of (radio) access technologies, terminals, signalling and transport protocols, applications and services, as we have seen in the previous chapters. In such diverse environments, it is highly desirable to enable differentiated handling of multimedia data streams – depending on the various end user's contexts and needs. As a consequence, it may be required to cache, transcode, split, synchronize or transform multimedia streams in some way or another before they can be delivered to the user or properly displayed by the user terminal.

With today's technology, this transformation of multimedia content is generally assumed to be provided by the end systems, either the user terminal or the media server. In both cases, this leads to either quite complex user terminals in terms of the required media processing and buffer capacity or redundant content transmissions from the server. Transformation of multimedia data and possibly signalling traffic may therefore be motivated by the service provider, the network provider or user preferences. In most cases, it would be unreasonable to place the burden of data transformation on the client device, as especially mobile devices are limited by performance constraints such as battery power, processing capability, memory

Ambient Networks: Co-operative Mobile Networking for the Wireless World Norbert Niebert (Ericsson GmbH), Andreas Schieder (Ericsson GmbH), Jens Zander and Robert Hancock

capacity, available media codecs and (radio) bandwidth. It is as well unrealistic to expect that service providers can or should be responsible for performing all the required transformation and adaptation operations as the variety of codecs and display formats increases together with the number of user terminals and content items on multimedia servers. As multimedia communications is entering the mass market rapidly, media processing and delivery might soon become a task for the infrastructure itself on top of packet forwarding. This translates into a need for network-side media processing capabilities and transformation services (which is also referred to as MediaPort or MP here) somewhere on the media path, between the sink (MediaClient or MC) and the source (MediaServer or MS). These MPs must be able to transform the multimedia data from the MS into a form that is acceptable for the MC. This transformation takes into account the available context information for the purpose of optimal service delivery, relying on an option to share the responsibility for the data transformation with the end users and the service providers.

The following sections outline the Smart Multimedia Routing and Transport (SMART) architecture as a part of Ambient Networks. Section 10.2 elaborates on the problem statement and puts the media delivery architecture into perspective within the Ambient Networks framework. An overview of current state of the art regarding media delivery techniques is also provided in that section. Section 10.3 describes in very detail the proposed media delivery architecture. Section 10.4 presents the validation efforts performed including some simulation results. Section 10.5 finally contains the conclusions and an outlook for further work.

Finally, we would like to thank here all people in the Work Package 5 of the Ambient Networks project that have made possible the development of the SMART architecture.

10.2 Why Media Delivery Support in the Network Infrastructure?

10.2.1 Alice on the Media in 2010 – A Scenario

Imagine the year 2010. Alice, while in the office, tunes into a multimedia conference using her laptop. The other participants in the multimedia conference are on the move and thus connect to the conference session via smart phones and/or PDAs. The conference session consists of an audio flow, a video flow as well as a shared application flow. All media flows destined to Alice are initially routed along the same path from the conference server to her laptop. The data is streamed to Alice across the wired company network and to the other users across mobile networks. When Alice leaves her desk, she wants to keep on participating in the ongoing conference session. She switches on her headset and manually triggers the delivery of the audio flow from the laptop to the headset. Later, Alice leaves the office to visit a customer and switches the conference session to her smart phone. After the conference session, on her way to the customer Alice listens to an audio clip containing important business details about the customer. The audio clip is directly streamed from her company network to her car network. Alice arrives well prepared at the customer location.

With today's technology, the realization of this example scenario would be difficult or more likely not feasible. In order to deliver the multimedia data in mobile environments, the media path between the end terminals changes constantly due to user mobility, i.e. the changing points of attachment of the mobile users. A consequence of the user mobility in heterogeneous environments is that bandwidth capacities might change drastically over a short time resulting in a need for media adaptation operations to be applied. As the heterogeneity of end devices and access technologies is expected to increase in the coming years, media adaptation functionality will also gain importance as a means to cope with the diversity of end systems and their capabilities (e.g. bit rates, media codecs and screen resolutions). Although media adaptation can often be performed by the end systems directly, the question is rather whether it is desirable to burden user terminals and also media servers with such functionality given that in many cases media adaptation is better performed in strategic locations inside the network.

When Alice leaves her desk, her laptop needs to split the media session and be able to re-route the audio flow to the headset using the laptop's wireless network (e.g. Bluetooth or WLAN). With today's technology, this might be a feasible task but it is questionable whether the session split can be performed seamlessly and in real time.

At the moment when Alice leaves her office and takes her car, the headset, which was attached to the laptop's wireless network, will lose its association with the laptop network and later even with the company network. Again, with today's technology the realization of the necessary handovers might prove to be a difficult task as handovers between heterogeneous technologies are often limited to networks of the same operator. Additionally, it is questionable whether the session handover can be performed seamlessly and in real time and whether this handover takes care of the different user context (e.g. terminal capabilities) and bandwidth constraints (e.g. WLAN or UMTS). In order to solve this problem, the smart phone could be first automatically added to the multimedia conference session through the wireless local access network that covers the whole company building. Later on, the smart phone could be automatically added to the car network with the consequence that the underlying network connections change again. In other words, Alice's personal area network would drop the laptop as its gateway and start to use the smart phone, respectively, the car network as its gateway. The headset then would compose with the wireless network provided by the smart phone, at which point the path of the audio flow would need to be adjusted and redirected to the headset's new network location. The video flow would then be transcoded en route and adapted to the lower resolution of the smart phone's display.

During Alice's car journey, the network connectivity has to suffer from significant coverage fluctuations. In the worst case with today's technology, this patchy coverage could lead to undesired and unpleasant service interruptions. By using several proactive media caches along the highway, the audio clip could be prefetched at locations with better network connectivity, so that Alice can experience a seamless media delivery service.

10.2.2 Overlay Networks to Support Media Delivery

Although we have sketched the technical barriers to implement the scenario given above, we should also notice that from the user's perspective this seamless and adaptive media scenario is very much straightforward and intuitive – its implementation should therefore be a natural task for future Ambient Networks. As a part of Ambient Network's Ambient Control Space (ACS), specific functionality for the distribution, adaptation and caching of multimedia content has been developed in order to make this or similar user experiences possible in the future. Media distribution, adaptation and caching have been very active areas of research in the last few years. However, most of this work has taken only a partial view of the overall problem of media routing and delivery, often limited to the application layer. Past research work was mainly dedicated to either caching architectures, media adaptation or multicast protocols separately. Furthermore, these solutions were usually optimized to solve only one specific problem such as a network congestion state, limitations of end devices or mobility. In contrast, the proposed solution for media routing and delivery of Ambient Networks takes a holistic view and supports media adaptation, distribution and caching altogether.

The solution is based on the concept of overlay networks in order to overcome the described shortcomings in an integrated way. This concept introduces a new abstraction layer in the network stack which enables flexible and application-aware addressing of all participating nodes for the purpose of media routing and delivery. This new abstraction layer is transparent to the underlying network (i.e. without the need to replace the existing infrastructure) as well as the end-user applications. One of the important advantages of the overlay concept is that it enables the in-parallel establishment of different types of overlay networks as needed. This allows, for example, tailoring the virtual addressing scheme and the overlay routing to best suit the requirements of a particular service and more advanced multimedia transport techniques that enable transparent integration of value-added media processing capabilities into the end-to-end media delivery path. To perform the routing and transport of multimedia content, the overlay nodes are configured with forwarding tables that reflect the next hop within the overlay level for each possible virtual destination address.

10.2.3 State of the Art on Media Delivery Overlays

Much research work has also been developed outside the AN project on the concept of 'overlay networks' coupled with multipath media routing or network-side media processing. X-Bone [117], and its enhancement GX-Bone, dynamically deploys and manages Internet overlays as a way to reduce configuration effort and increase network component sharing, but concentrates on routing aspects. Other research projects such as resilient overlay networks RON [110], QRON [112] and OverQoS [111] have introduced the concept of overlay networks aimed at improving the quality of service (QoS), mainly by routing around problem spots in the underlying networks, like congested network links or nodes. Together with the QoS provisioning, the inclusion of network-side media processing resources within the overlay network has been proposed by Nahrstedt *et al.* in different projects and concepts, specifically Spy-Net [113], QUEST [114] and Qualay [115]. However, such proposals concentrate on routing aspects, do not provide explicit support for mobility and assume fixed and known support nodes. Additionally, Spy-Net uses network flooding to maintain information about network topology and link characteristics, making the scalability of the system questionable.

3GPP has started a study item on network composition [3GPP SA1, TR 22.980], which may also look into media delivery related aspects as discussed here. This study however is still in an early stage and is mainly based on the results of the Ambient Networks project.

In the ETSI TISPAN architecture, a service-based policy decision function (SPDF) accepts service level policy requests from applications and translates them into IP QoS parameters. The access network is then asked if it can provide this QoS. What happens next will depend on the type of access network used. In TISPAN-based networks, the SPDF contacts the border gateway function (BGF) to enforce the policy. Policing can therefore occur on a per-media component basis. A similar, network-centric QoS architecture is currently developed by 3GPP for the evolved system architecture [SAE – System Architecture Evolution, TR 23.882]. These developments resemble the concept of the Ambient Service Interface (ASI) as described below; however, this concept also concentrates on QoS aspects and does not cover network-side media processing.

This chapter outlines a proposal that extends the concept of service overlays to allow the inclusion of supporting processing nodes, which we term MediaPorts, into the end-to-end transmission path. These 'service-specific overlay networks' are set up according to service requirements and include the needed MediaPorts for delivering media content adapted to the user context. As end users may have different contexts, streams for users will not always follow the same path. The routing of media data within an overlay network is based on a per-session or even per-flow [109] basis. Both the client components and to some extent service components of the overlay networks are assumed to be potentially mobile. As a result of this, media streams may originate, be routed through and terminate at different networked devices during the lifetime of a single multimedia delivery session.

10.3 Media Delivery Architecture

In the traditional end-to-end media delivery model, as illustrated in Figure 10.1, no interface exists through which service providers or end users can specify and request specific transport services to be provided by the underlying network – the network provides only a simple best-effort transport service.

This is fundamentally different from the end-to-end delivery model that is offered by Ambient Networks, as illustrated in Figure 10.2. The Ambient Service Interface allows user applications or services to negotiate and control Ambient Networks transport services. As such, it enables end users or service providers to take advantage of advanced media delivery or processing capabilities that may be present in the underlying networks to achieve a transport service more desirable than simply best-effort transport service.

The ASI allows applications to specify the service profile and service requirements to enable the Ambient Network to find the required processing capabilities for the transport of the service. The ASI also allows requesting information about advanced capabilities of the network in order to choose desirable resources for a transport bearer, and finally to transmit media across the bearer using these advanced functions.

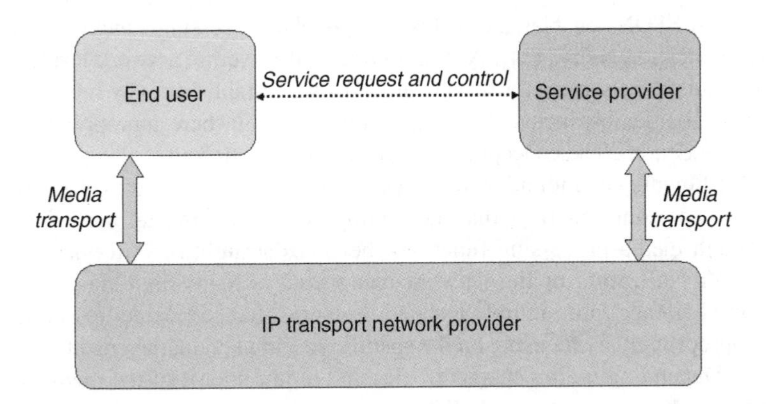

Figure 10.1 Traditional end-to-end media delivery model (prior to Ambient Networks)

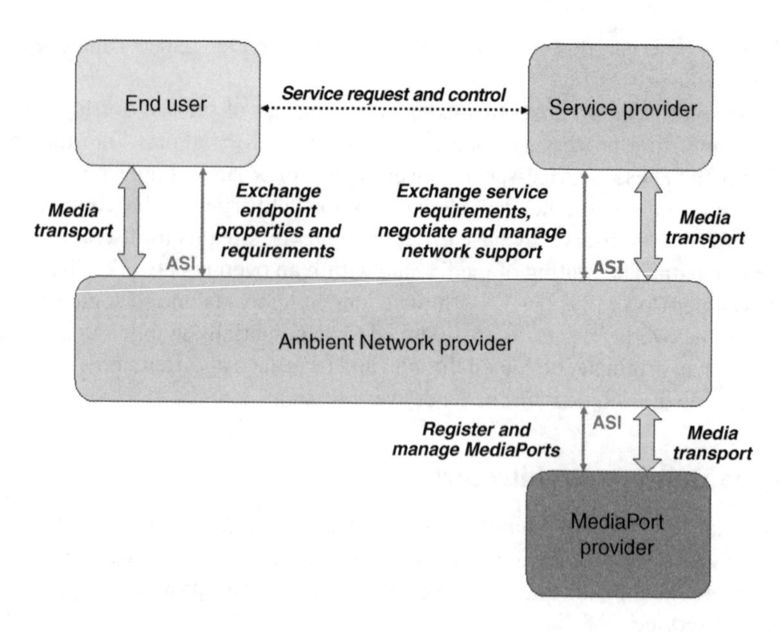

Figure 10.2 Ambient Networks media delivery with processing support in the network

Although the ASI exposes a rich service interface to the end-user applications and services that allows them to tailor the network resources and media delivery functions according to their specific needs, the complexity and actual mechanisms of achieving this 'network customization' remain invisible to them. To provide this customization functionality on a per-service basis, the control space configures separate virtual (or 'overlay') networks for each service, in the following referred to as Service-Specific Overlay Network (SSON).

10.3.1 The Concept of Service-Specific Overlay Networks (SSONs)

The use of such SSONs enables the inclusion of media processing functions into the end-to-end communication paths [104]. As the routing at the overlay network level can be fully controlled to satisfy the service for which the SSON was established, any support function in the network can be flexibly included into the end-to-end path where appropriate (independent of its physical location). The concept is illustrated in Figure 10.3.

The establishment of an individual overlay per service may seem excessive at first glance. However, an important benefit is that the routing of media flows for a service, including routing through media processing functions, becomes straightforward once the SSON is established. Virtualization of the network using SSONs helps dividing the routing and management challenge into multiple isolated problems that can be addressed more easily and more appropriately by focusing on the specific routing and management needs of a certain service. Depending on the characteristics and requirements of the media service, the SSON routes media streams through different media processing functions and/or network domains (e.g. WLAN or UMTS). Further, different components of a service (e.g. different media streams) can be routed over different paths, as proposed in [109]. The result is the

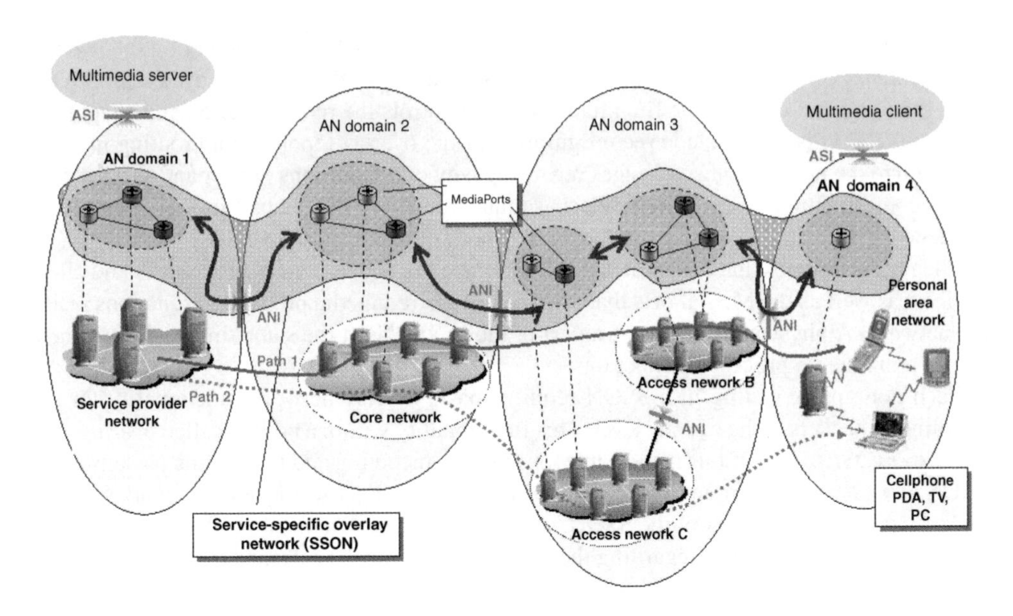

Figure 10.3 Service-specific overlay networks connect endpoints of the media delivery service and MediaPorts for network-based processing

seamless provision of value-added services inside the network beyond what is possible today.

A SSON is tailored to the specific service it has been established for and thus includes only those nodes that are required to provide the requested service. Among all MediaPorts in an Ambient Network, the most suitable ones are selected using appropriate metrics, which consider capabilities (e.g. transcoding between different media formats), costs of using the processing function, transport costs to route data to and from the MediaPort and the current processing load of the MediaPorts.

For this purpose, the Ambient Control Space encompasses a *MediaPort information* (MPI) function that enables efficient lookups of MediaPorts based on media processing capabilities, location, cost and load, among other characteristics. The MPI function maintains a directory of existing MediaPorts and has the capability to dynamically search for MediaPorts, for example, in a restricted neighbourhood near MediaPorts that are already involved in a given SSON [106,108].

SSONs are dynamic and can be reconfigured in response to changing network conditions (e.g. end-to-end delay or congestion), changes in network context (e.g. handovers) or changes to the user context (e.g. user profiles, device capabilities, media/content types). Adaptation of SSONs can happen on different time scales. 'Fast' adaptation is required to respond to critical changes that would otherwise disrupt a service, such as changes to the underlying network (e.g. if a link is congested or has failed). 'Slow' adaptation can optimize operation in response to noncritical changes, e.g. when MediaPorts join or leave an Ambient Network or when users enter or leave a SSON. For example, a SSON may adapt the virtual network topology after a technology or provider handoff of a mobile node in order to optimize the routing at the overlay level.

This functionality for the establishment and control of SSONs is provided by the *overlay management* (OM) functional entity of the ACS. It manages the entire lifecycle of a SSON, from the initial set-up until the final removal, and controls the reorganization and adaptation of the SSON. This may include reconfiguration of the overlay topology and routing in order to optimize the network when changes regarding network conditions or user and service context become apparent. It adaptively selects suitable routes for the media flows of a service by influencing the routing decisions in the overlay nodes that constitute a SSON.

The physical nodes that constitute a SSON include the end systems (server hosts and client devices) as well as the MediaPorts that provide the desired media processing functions inside the network. A physical node can be a part of many SSONs at the same time and implement one or more media processing functions.

Each node participating in a SSON requires basic overlay network functionality for the handling of packets at the overlay level. This functionality within a node is called *overlay support layer* (OSL). The OSL is responsible for sending, receiving and forwarding packets at the SSON level. It provides a common communication abstraction (overlay level network protocol and addressing) to all nodes of the SSON, so that they can communicate with each other independent of their differences regarding the underlying protocol stacks and technologies.

10.3.2 The SMART Architecture for Media Delivery

This section provides a high-level overview of the SMART architecture. 'SMART' in this context refers to the fact that the described multimedia routing and transport architecture enables the delivery of enhanced media services by taking advantage of network-side media processing capabilities that can be offered by Ambient Networks.

A central concept of the SMART architecture is the idea of SSONs, i.e. a different overlay network ('virtual network') is established for every media service (or group of services) that is tailored to the exact network needs of the service. This allows the configuration of appropriate high-level routing paths that meet the exact requirements (for example, QoS, media formats, responsiveness, cost, resilience or security) of the media service.

To illustrate how the SMART architecture relates to the overall Ambient Network architecture, Figure 10.4 pictures the Ambient Control Space and its control interfaces, namely the Ambient Service Interface, the Ambient Network Interface (ANI) and the Ambient Resource Interface (ARI).

Overlay nodes can take on the roles of MediaClients, MediaServers and MediaPorts – or any combination of these. These entities are controlled by the overlay management function to form a suitable network for a particular media service, namely a SSON. The MediaPorts, as special network-side functions, have the role of providing value-added media processing functionality, such as caching or media adaptation facilities, flow synchronization or special types of routing capabilities. The overlay management includes suitable MediaPorts inside each SSON, where most appropriate for the media processing service, and decides where different functions should be provided. In order to convey media flows to the end user in the best way and according to the end-user and service provider preferences, the infrastructural components of the SMART architecture are provided with context information from other functional entities of the Ambient Network. This includes, for example, information regarding underlying network conditions, mobility and user context.

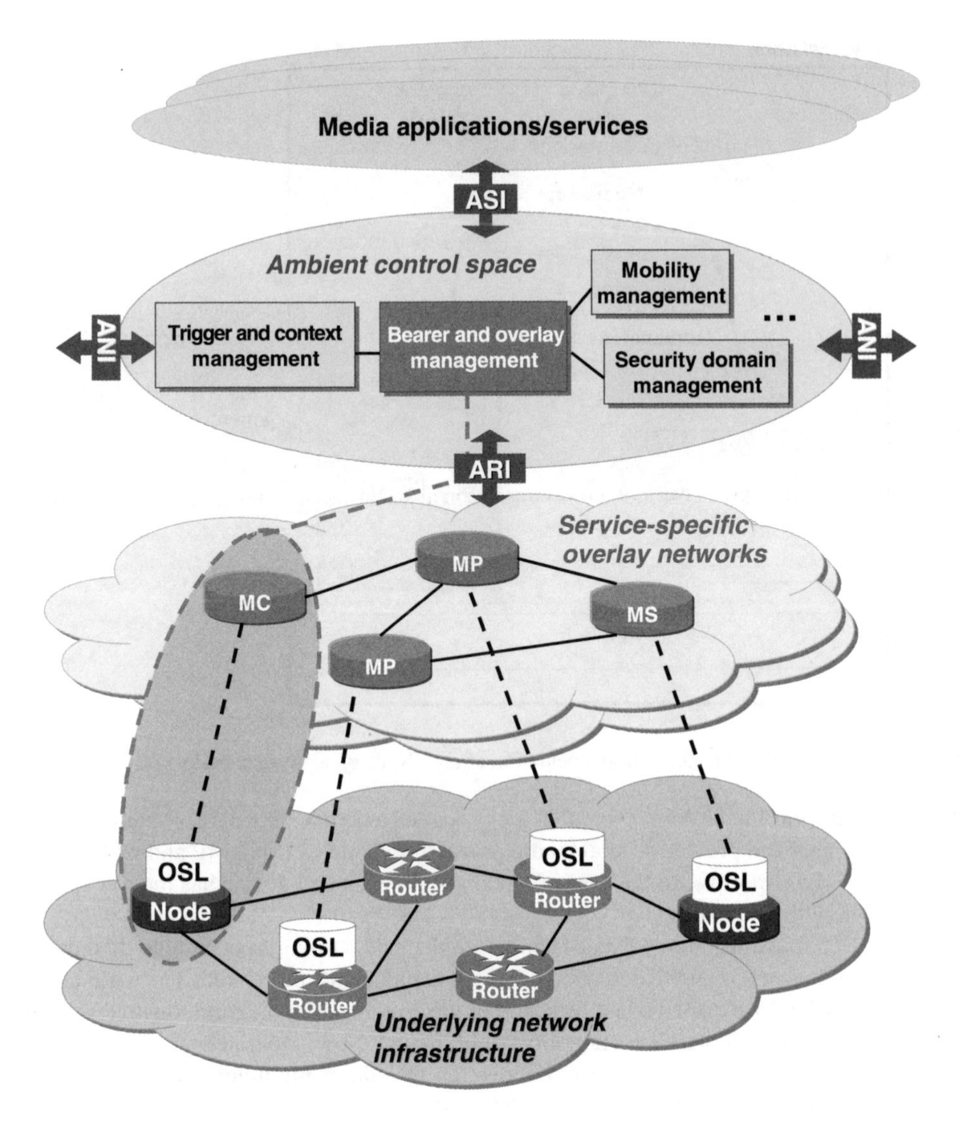

Figure 10.4 The SMART Multimedia Routing and Transport architecture

10.3.3 The Overlay Node (ONode) Architecture

An *overlay node* is a specialized Ambient Network node that implements the functionality required to join the SSONs by, for example, provisioning network-side media processing functionalities. ONodes can be described from the user plane perspective and the control plane perspective. For every SSON of which an ONode is a part of, separate MPs, MSs or MCs are instantiated. MPs, MSs or MCs are responsible for media routing in the control plane and, in the user plane, host the so-called application modules, each responsible for a particular network-side media processing functionality. Furthermore, and depending on the required media processing functionality, ONodes can take one or more of the roles of MediaClients,

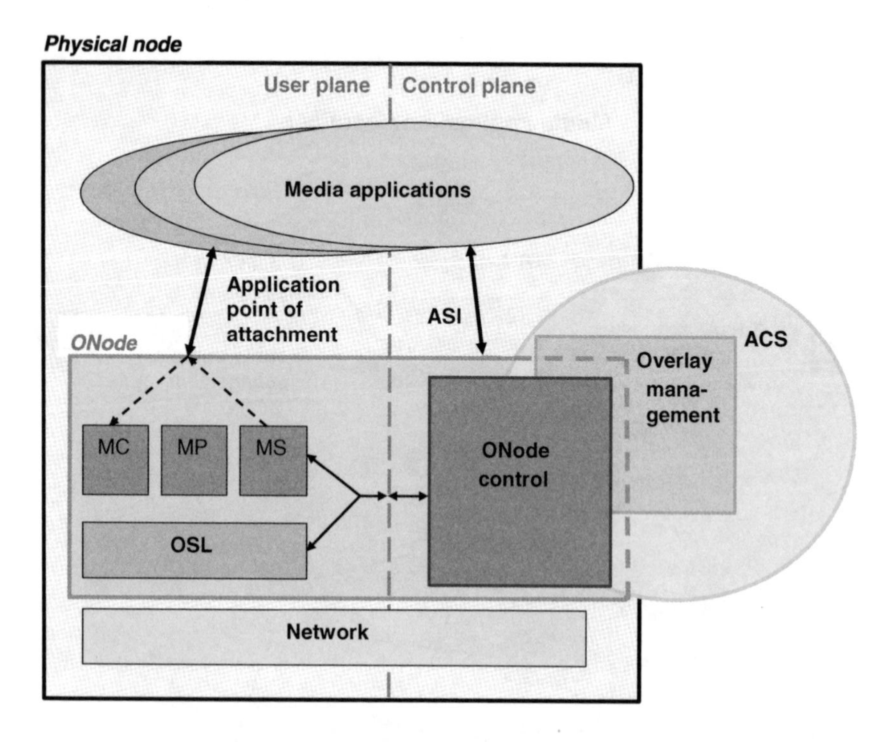

Figure 10.5 Implementation of an ONode on a physical node

MediaServers and MediaPorts. Note that a physical ONode can be a part of many SSONs at the same time or can provide several media processing modules (MC, MP and/or MS) in the same SSON. For example, in the case of a phone call, each end device has the role of a MC and of a MS at the same time.

Figure 10.5 illustrates how ONodes can be implemented on a physical node. It shows the internal components of an ONode and how these components are related. The main components of an ONode are displayed in the centre of the drawing. The diagram illustrates the user plane (left) and control plane functionality (right) of an ONode separately.

End-user media applications, located above the ONode, communicate with the Ambient Network and the SMART functions through the ASI in order to use and control the multimedia delivery capabilities provided by SMART.

The control plane of the ONode includes the ONode control entity, which is responsible for the general management of the ONode and the exchange of signalling messages (for example, it takes part in the routing of media flows in the SSONs). The ONode control consists of several components, which can be classified into those that deal only with the local control and management of the ONode and those that logically belong to the OM FE (Overlay Management Functional Entity), which is the functional entity residing in the ACS that controls the SSONs on Ambient Network wide level. The ASI is the control plane interface to the ONode through which endpoint media applications can communicate with the ONode control and thus the OM FE in order to request, configure and control the media delivery services offered by SMART through the SSON abstraction.

The user plane of the ONode encompasses the OSL and the application modules that take part in media processing actions. The OSL sits on top of the underlying network, which provides the functionality to include a node inside one or more SSONs and to forward and receive overlay packets. On top of the OSL, and using its services, there are application modules that implement the behaviour of a MC, MS or MP in regard to the data handling. MCs act as data sinks and send the multimedia data to the endpoint media applications through the application point of attachment, whereas MSs act as data sources and receive the multimedia data from the endpoint applications through the application point of attachment.

10.3.3.1 The Overlay Support Layer

The overlay support layer is the module within the SMART architecture that provides the functionality needed for the establishment of overlay links between the SMART nodes. Obviously, this functionality is mandatory for the introduction of any node into an overlay, and more specifically, into one or more SSONs. This makes the OSL a fundamental part of every ONode.

To provide 'overlay support', the OSL resides on top of the underlying network (e.g. IPv4 or IPv6 network) and makes use of it to exchange data packets with other OSL entities. The OSL also communicates and exchanges data with the node-local media processing modules or applications that are present in an ONode. These upper layer modules make use of the OSL to seamlessly communicate with other ONodes through the overlay links, regardless of the underlying network technology.

The OSL is responsible for the packet forwarding at the overlay level; however, it does not take any responsibility for the determination of the routing functionality. Thus, from an architectural point of view, it operates only within the user plane of the SMART scheme.

10.3.3.2 Addressing and Address Spaces

As a consequence of using overlay techniques, SMART introduces a virtual address space at the SSON level (i.e. at the OSL). Virtual addresses are needed to differentiate ONodes within a SSON. It is required that the virtual addresses of different ONodes are unique within a single SSON. As the SSONs constitute completely independent virtual networks, the format of the virtual addresses could be adapted to the specific addressing needs of the service for which the SSON was established. Given the growth of Internet applications, this could turn into a big advantage of the SMART concept which keeps a service-specific application realm embedded in a large universal Internet.

For the purpose of packet handling at the OSL level, some sort of a SSON ID is needed to differentiate packets belonging to different SSONs. The differentiation of SSONs allows that each service-specific overlay network can have its own, independent address space.

The use of globally unique identifiers as SSON IDs would avoid the need to account for SSON ID conflicts when two or more Ambient Networks compose, and thus simplify composition as well as routing through different Ambient Network domains. Considering that SSONs are a special type of bearer, it is foreseeable that the same identifier used to identify normal bearers, i.e. global bearer ID (see Section 4.4), could also be used to identify SSONs.

SSON IDs can be realized by means of an explicit identifier in the packets (e.g. by adding an extra SSON header). The disadvantage of this solution is that the SSON ID would have to be transmitted in every packet. It is therefore proposed to solve the problem by means of an implicit SSON ID. Here, the OSL associates a packet to a SSON based on the virtual link on which the packet was received. The particular set of headers used to classify packets might be extended to increase granularity. To allow the classification of incoming packets to SSONs, the OSL has to maintain the virtual link mappings for every SSON maintained by the ONode.

10.3.3.3 Forwarding Behaviour

The forwarding behaviour at the OSL is defined by the routing tables. These tables are dynamically defined by the overlay management FE (or more precisely the MRL; see the next section) and updated in the OSL when ONodes join or leave a SSON or when virtual links in the SSON change.

Upon receipt of a packet at the OSL, the forwarding process of the OSL looks up the next-hop ONode for a downstream packet (respectively the local processing module for an upstream packet) to which the packet needs to be forwarded based on the SSON to which the packet belongs and the virtual address of the destination ONode.

The routing table of a SSON must also reflect if packets need to be delivered to a local media processing module (e.g. MCs, MSs or MPs) in a certain ONode. In the case of MPs, this processing module is the appropriate caching, media adaptation or synchronization application. In the case of a MC or MS, the media processing module forwards traffic to the end-user client application or from the end-user server application respectively.

10.3.4 Overlay Management

The overlay management is the controlling entity in the ACS that provides a number of functionalities required for the management of SSONs. These functionalities include establishment of new SSONs, routing and addressing within the SSONs, alteration or adaptation of existing SSONs in order to account for changes in the network, changes in the service or the user context, and finally the termination of SSONs. In other words, the OM FE is responsible to handle lifecycle management of SSONs.

As illustrated in Figure 4.3, the OM is one of the control functions of the ACS. In order to communicate also with external control entities, the OM contributes to the external ACS interfaces, namely the Ambient Service Interface, the Ambient Network Interface and the Ambient Resource Interface. For example, the OM interacts with the user/application as well as the MediaPorts and with the service provider through the ASI interface to obtain information regarding the requested service profile, application requirements, user context, etc. (see Figure 10.1), or it communicates with other… OM FEs located in different Ambient Networks via the ANI (see Fig. 10.3). This enables SSONs to span several Ambient Networks and to be used and managed seamlessly.

The OM also needs to obtain the context information required for the establishment or adaptation of SSONs from the relevant control functions, namely the context management FE and the mobility management FE as well as any updates of this context information, which provides the basis for adaptation of the SSONs.

The OM FE is physically implemented in a distributed manner. It is composed of all the ONode control entities present in all the overlay nodes of the Ambient Network. It is important to note that the management and controlling decisions can be made by any combination of the ONode control entities, including a single node, a group of nodes or all of these ONode controls, i.e. the decision making can be implemented in a distributed, centralized or a semi-distributed way.

The following subsections describe in more detail the functionalities and responsibilities of the OM.

10.3.4.1 SSON Lifecycle

The set-up of a service-specific overlay network is triggered through the overlay management, upon determining the need to provide specific transport and/or media adaptation services to a communication session requested by a SSON user(s). In case a new SSON is not required, then the user device is just added to an existing one. The following subsections describe in more detail the SSON lifecycle.

10.3.4.2 Triggering SSON Set-Up

The minimum set of information required by the OM in order to judge whether a new SSON is required shall comprise the SSON service profile and the SSON user profile as follows:

- The service profile shall contain, at least,
  - A traffic characterization of the data to be transported by the SSON: data bit rate, allowable delay, etc.
  - Some type of service description, e.g. a description of the service in XML format including service name, a detailed description, including detailed session information like used transport protocols, whether the requested service is a conversational, a unicast or multicast streaming, whether it is interactive (e.g. web browsing) or it runs in the background (e.g. FTP downloads), a content distribution network or the computing resources of a grid network is requested. Important at this stage is also to know whether the users are mobile. It might also contain parameters (e.g. language codes, category codes, age rating, etc.).

This information is used to reserve network resources in the SSON. Many projects have dealt with the characterization of services and traffic, for example the IST Project Tequila [131].

- The user profile shall contain, at least, the user's identifier for authentication, authorization and accounting (AAA) purposes, the user's capabilities (hardware and software and even human capabilities) and the user's preferences for the given service, see [117,119]. Users' profiles may or may not be available at the SSON set-up stage, though

  - In subscription-based services, user profiles are made available to the service providers.
  - In a more dynamic scenario, users are added to a SSON dynamically or the SSON itself has been set up in an ad hoc manner. In this case, it is the users who can directly exchange their context end to end and resolve capability or preference mismatches (such as is done in the SIP Offer and Answer model [120]).

This information is used by the OM to select the appropriate overlay nodes to process and route the media.

A detailed specification of the primitives used for exchanging the aforementioned information is given in the SMART-related ASI specification (see Section 12 of [121]).

10.3.4.3 Selecting the ONodes for the Base SSON Topology

The service provided by the SSON must satisfy two different kinds of quality-of-service guarantees. The first guarantee refers to the 'network QoS'. The second is the so-called 'session QoS'.

The 'network QoS' relates to network attributes such as bandwidth, delay and jitter and all those metrics related to the packet data transport. In particular, network QoS refers to the traditional conception of QoS as, until now, the networks are unaware of the type of data being exchanged. On the contrary, the 'session QoS' refers to a more abstract kind of QoS: It relates to the required media adaptation procedures that may hamper two or more users from establishing a communication instance (a 'session') with each other. These adaptation measures include transcoding and caching of media data. The session QoS also encompasses the QoS as perceived by the user, i.e. although the transport path for the data fulfils the network QoS requirements, above the network level there may be other factors that reduce the QoS such as applications, the interaction between the different flows that form a session, etc.

Therefore, in order to select the MPs that shall perform the media adaptations, it is necessary to be able to describe the requested session QoS using some format. Typically, SSON clients include the requested session QoS in the session initiation exchange, e.g. in the payload using the Session Description Protocol [118]. Therefore, by comparing the requested session QoS to what the SSON is able to provide (the available session QoS) the required media adaptations are found, such as transcoding required streams to be synchronized, etc.

After this comparison process, we have a set of network QoS constraints and a set of session QoS constraints (or media adaptations) that shall be fulfilled. However, the problem of finding a set of network nodes that simultaneously fulfils a set of constraints might turn to be a difficult one. In particular, a multiconstraint problem where two or more constraints are additive[1] is known to be NP complete [126]. An example of this is the cost optimization delay optimization problem.

To circumvent this complexity and search time, the SMART architecture uses a sequential approach:

1. The OM first uses the INQA framework to find a base topology: a set of nodes that fulfils the requested network QoS for all the SSON users that shall use the SSON. We call this the base SSON topology and it represents a reduction of the problem space by prefiltering those overlay nodes that would anyway not fit the network QoS requirements. Once this topology is found, the next step is that the communication be possible among all users of the SSON, or just among a reduced set thereof, depending on the communication type (unicast, multicast or else as explained above). A communication matrix is thus required in which the potential service paths[2] are included. With this communication matrix in mind, it becomes apparent that in some cases the necessary media adaptations across the com-

[1] In general, if a network path has more than one 'additive' constraint (e.g. where the cost of the path equals the sum of the individual link costs), then the problem is not solvable without an extensive recursive search (i.e. NP problem).

munication paths[3] between two users are not supported by the chosen overlay nodes and hence gets clear which media adaptations are additionally needed.

2. The media routing logic (MRL) now assigns a 'weight' to each of the links connecting the overlay nodes in the base SSON topology. Using the 'weighted network graph' that results from this process, the MRL will now try to find the missing media adaptations by using two different strategies together with the information available from the MediaPort information (see [121]). The strategies depend on where the information on the MediaPorts is stored and what the routing capabilities of the overlay nodes are. The strategies may be
   - Running a routing algorithm upon the 'weighted network graph'. This is the traditional approach used in established routing algorithms such as link state routing or distance vector routing. After this step, all required media adaptations are found and the service paths are defined. This process is centralized, in the sense that it is the OM that runs this algorithm and not each overlay node. This is because the SSON itself does not yet exist; it is only an abstraction in the form of a weighted network graph.
   - The second strategy finds application in scenarios where the network knowledge is very limited, such as in ad hoc communication. In this case, we run a path-directed search for MediaPorts. Using the default routes[4] of the base SSON topology, the MRL starts searching in the 'proximity' of the overlay nodes on the path to the communication peer or peers using the path-directed pattern-based MediaPort information discovery service of the MPI (see Section 10.3.5). In this case, the search is distributed as each visited overlay node shall answer the query and forward it to those overlay nodes that it is aware of. Also, note that the amount of information that a given overlay node has about its 'neighbour' depends on many factors such as mobility or processing and storing capabilities. In the simplest modality, the 'proximity search' is performed by querying first those overlay nodes that are one hop away, then those two hops away and so on until the requested MediaPort is found or the hop limit has been reached. Of course, in this manner a suboptimal topology is found, but the complexity and search time is considerably reduced. In the case that the inclusion of the required media adaptations (in the form of MediaPorts hosted by other overlay nodes) worsens the network QoS considerably,[5] the MRL may opt for reducing the available network QoS and so they perform a new topology search. This process is iterative and network QoS may be downgraded as long as the SSON users accept it. Section 9.6 of [121] details a possibility to implement this strategy using NSIS building blocks. NSIS [127] stands for Next Steps in Signalling and it is a signalling framework under development in the Internet Engineering Task Force (IETF).

[2]The term service path is used in this context to denote an ordered sequence of processing or media adaptation steps required to enable the communication between two (or more) users.

[3]At this point, the communication type is very important; for example, source multicast users will never have to communicate with one another over the multicast network, so it is not necessary to cope with QoS mismatches. The same occurs with a content distribution network, in which only the QoS mismatches between the servers and each of the users shall be overcome. However, in a multiparty conference, the SSON shall provide for adaptation between all users.

[4]The 'default route' is understood as the one that optimizes the multiconstraint problem.

[5]The term 'considerably' depends on the weighting strategy used.

In both strategies as per above, a cost weighting function is required to perform the service path search and this function shall consider both the fulfilment of the network QoS and session QoS parameters. The weighting function used in this case shall take, at least, the following cost factors into account: the link cost, the MediaPort cost and the utility of the MediaPort processing, which depends mainly on its position on the service path. This cost weighting function is used to assign 'weights' to each of the links connecting the available overlay nodes in the base SSON topology. After this process, a weighted network map is available and traditional routing algorithms such as Short Path First (SPF) may be used. Thus, a generalized cost weighting function would be as follows:

$$\text{Cost(media path)} = \alpha \sum_{i=0,\dots,n} (\chi\, L_i) + \beta \sum_{j=0,\dots,m} (\delta \times \text{MP}_j) + \varepsilon \sum_{k=0,\dots,P} (\phi \times \text{Utility}_{\text{MP}_k}),$$

where

- The weighting factors (α, β, ...) are used for shaping the way each flow of the session makes use of certain MPs or links, e.g. depending on the user's service contract.
- The terms L_i and MP_j represent weight factors that consider the link's and MP's QoS parameters and costs. The term $\text{Utility}_{\text{MP}_k}$ represents a utility value of the MediaPort. A good example where utility functions are useful is multicast, where one MediaPort might be preferred to another depending on how many users benefit from that MediaPort's caching or routing. In this case, a simple utility function would be the ratio of served users over the total number of users. The higher this ratio, the better the MP for multicast.

This formula includes the three cost components that we have identified to be involved in the selection of the most appropriate:

- The link costs: in terms of QoS metrics like bandwidth, latency, jitter and packet loss.
- The MP costs, which include the costs in terms of usage cost (money), added latency, etc.
- The utility of the MP in the particular location. As explained more in detail below, a rate reducer is in general more useful close to the source due to the reduction of the link costs, but in the case of multicast a rate reducer is much more useful close to the set of users that have a reduced available bit rate, as the rest (a larger number of users) would benefit from the higher bandwidth.

Note that both strategies can be combined and the weighting factors might both be functions or constants.

Other solutions could have been used to address the problem of multiconstraint routing; an overview can be found in [128,129].

10.3.4.4 Using Utility Functions for Optimum Placement of MediaPorts

The question is now: How do we find out the optimal location of these additional overlay nodes? As we have seen in the weighting function used in the last section, the utility of the service provided by the MP is an important factor for choosing one or the other. This section explains the utility in more detail.

For example, in streaming, the need for a transcoder is determined by comparing the codec capabilities of the endpoints and the transcoder's position is irrelevant along the path. That is the utility does not change with the location. However, the need for caching along the end-to-end media path depends on the type of service because the added delay is tolerated only in browsing, streaming or CDN service classes. In those services, caches are placed close to the data sinks for reducing the latency of responses. On the contrary, the need for rate reducers is typically found by examining the QoS parameters of senders and receivers and the location depends on the number of users that are served by the high and low bit rate streams.

Thus, as we can see, the way media adaptation is best applied depends on the communication type, on the costs associated with the MP and on the media adaptation itself.

For illustrative purposes, we show how the utility of the MP, $\text{Utility}_{\text{MP}_k}$, could be used to estimate the optimum location for MediaPorts. For simplicity, we consider a reduced set of media adaptations and determine which location on the path is most appropriate for both point-to-point and point-to-multipoint media flow types.

Performing this exercise for each of the media adaptations possible would give us a method for obtaining a general formulation for the utility function. A more academical example of how to use utility functions can be found in [130].

Some remarks on Table 10.1

- Transcoding refers not only to the particular codec (MPEG, DivX, etc.), but also to media adaptation MPs that perform, for example, voice-to-text and text-to-voice transcoding, MPs that translate from one language into another or modify other codec parameters and preferences which do not modify the input rate, thus in general, any MP that modifies the codec or any of its parameters of the codec used.
- MPs that modify the bit rate such as packet droppers or transcoders that reduce the frame rate are grouped under the term 'rate reducers'. In the case of point-to-point streaming, they are best placed near the source to save bandwidth. In the point-to-multipoint case, the rate reducers are best placed where they can serve most users, hence in brackets.

Table 10.1 Service media adaptations for the streaming service

	Best locations		
Communication type	Close to sink(s)	Indifferent (i.e. where available)	Close to source(s)
Point-to-point streaming	Caching	Transcoding (no rate change)	Rate reducers
Point-to-multipoint streaming	Caching (rate reducers, if most users have low bw)	Transcoding	(Rate reducers, if most users have high bw)

- We have specified the caches near the sinks to be best as they have to cope with the losses on the wireless links and because the requests are quicker served if the content is available near the sinks.
- Although flow splitting and synchronization can also be described as media adaptations, these are not considered here for simplicity. In most but not all cases, they can be solved locally at the endpoints. For example, if one endpoint wants to have the audio using codec A and video using codec B and the other wants audio and video bundled using codec C, we could place a MP in the middle that splits the flow in audio and video and transcodes to the respective codecs, A and B. In the other direction, we would need a synchronizer that outputs codec C. However, it is much easier that we specify the session such that there is no need for splitting and/or synchronizing the flows in the network, e.g. the endpoints just agree on different codecs for audio and video on each side (in the worst case) and transcoders are introduced by the MRL, where needed. However, if the audio and video endpoints reside in different devices, as the example scenario with Alice depicts, some splitting and synchronization in the network is needed.

10.3.4.5 SSON-Level Routing

Once the set of overlay nodes hosting the required MediaPort functionalities are found, the next steps are

1. Set up the links to connect the selected overlay nodes. This implies making the necessary reservations of resources, both network and media adaptation resources.
2. Installing the routing and addressing instances at the routing MediaPorts inside the selected overlay nodes, bootstrapping the overlay nodes at overlay level (i.e. providing them with a SSON address, a virtual address, besides the physical address) and filling in the routing tables while observing the route ordering constraints that may apply to the distinct media adaptations. Note that virtual addresses are stored and handled by the overlay support layer.

For the first step, the OM resorts again to the help of the INQA and network composition, which are in charge of performing the necessary changes or additions to the composition agreements for setting up the links between overlay nodes. The set of selected overlay nodes to be connected is given by the abstract network graph obtained by the MRL, as explained in the above section.

For the second step, the OM has to initiate the addressing of the SSON entities in the ONodes and also evaluate which routing protocol is best, and then install (or just kick-start) the routing daemons in the MediaPorts performing the routing functionality. The selection of the routing protocol depends on many factors, the most significant being the number of nodes, the route convergence, the dynamicity of the route changes due to mobility or other factors, and finally the rate at which users join or leave the SSON. Traditionally, link state routing protocols (e.g. OSPF [122]) have proved to be more appropriate for big networks with relatively low dynamicity (in routes and join/leaves), whereas distance vector protocols (e.g. RIP [123]) are appropriate for small and dynamic environments such as moving networks. Hierarchical protocols (e.g. ATM's PNNI) are very valuable in the case of especially large networks.

On the contrary, if the route convergence and frequent join/leave is an issue, peer-to-peer DHT-based routing protocols such as those used in Chord [124] or Pastry [125] guarantee deterministic route convergence as an upper bound in the number of hops, independent of the number of ONodes and high resiliency of the routing information under frequent join/leave events. Finally, also hybrid approaches such as the hierarchical DHT routing used in Skype are possible.

A note on scalability: Although the SSON-level routing may need to maintain per-flow state, the Ambient Networks approach is scalable as this is not the case for all ONodes along the path. More details can be found in Section 7 of [122]. Specifically, only those ONodes that are implicated in the modification of the default overlay route have to maintain such per-flow state. These nodes are typically the edge nodes in the different QoS domains and are the same ones that negotiate the composition agreements.

10.3.4.6 Maintenance

As we have seen, the OM FE manages the creation of SSONs upon Ambient Network users request and controls the reorganization or adaptation of existing SSONs based on changes in the underlying Ambient Network (e.g. hardware failures), changes in the context of a user (e.g. mobility, link conditions, etc.) or changes in service policies (e.g. service downgrade, resource management).

In order to react to these changes in the MediaPorts, there shall be a SSON-wide monitoring and notification mechanism implemented into the distributed OM. At the end, a lot of information on the capabilities and status of the MediaPorts has to be compiled and made available to the rest. The MediaPort information provides this service as explained below.

10.3.4.7 Termination

Finally, after the SSON has provided its service, the overlay management shall ensure that resources are freed.

10.3.5 MediaPort Information Management

As mentioned before, for efficient lookup of media processing functions located in the network, the OM needs a MediaPort information function.

In this section, two approaches are described that can be used in conjunction with an ACS MediaPort information, namely a centralized MediaPort database (DB) and a distributed pattern-based MediaPort information discovery technique. The decision to work towards two MPI approaches was mainly driven by the fact that the Ambient Networks architecture is intended to be flexible enough to cover a wide range of possible networking scenarios. Thus, it cannot be expected that there is a ‘one size fits all’ solution for the addressed problem space. In this context, we concentrated on two solutions covering a wide spectrum of possible scenarios. In ad hoc network and opportunistic communication scenarios, the pattern-based MediaPort information discovery technique can be the only possibility to cope with the dynamic and ad hoc structure of the resulting networks. In addition to this, the returned sequence of MediaPorts can already be considered as an approximation of the optimal media delivery path. In more static scenarios, a MediaPort DB based on a DHT principle provides a platform with provable communication characteristics and success guarantees in case a requested media processing capability is available. By using the principle of range queries combined with a recursive search, a MediaPort DB can provide a value-added search function, which is supporting the media routing logic by reducing the complexity of media delivery path calculation.

This MP information is divided into two parts:

1. A MediaPort database (MPDB) is used to store information about available MPs in the administrative domain of an OM FE. The MPDB maintains information such as available processing modules, current load or usage costs per registered MP. This information can be accessed using the MediaPort information interface (MPII) by other ACS FEs, especially by the overlay management FE. Conceptually, a MPDB can be centralized (a directory service approach) or distributed (a peer-to-peer approach).
2. A pattern-based path-directed MediaPort information discovery service is used for the on-demand discovery of MPs 'close' to the media delivery path between the MediaServer and the MediaClient(s). This mechanism is necessarily distributed as it requires the processing and forwarding of queries by each visited overlay node.

In summary, either the MP database or the pattern-based MP discovery service of the MP information FE is used to perform the following tasks:

- Storing in objective terms the specific set of MP functionality required for the service delivery. The information is preferably stored using an extendable format, such as in SDPng [119].
- Retrieving from the MP information function the required operational information about potential overlay nodes (e.g. cost of using the MP) that is used to select the best nodes.

Although the pattern-based path-directed MP discovery has its main advantages in ad hoc or dynamic environments, the MPDB provides a scalable and guaranteed service that is well suited for managed networks. The MPDB is effectively a centralized or distributed database that lists the available MPs of a single or composed Ambient Network, as well as their characteristics (i.e. processing capabilities, transport address of the hosting ONode, location, connectivity, cost, etc.). The MPDB obtains this information through MP registration, i.e. ONodes register available MPs at the MPDB during start-up, using the ASI.

The role of the MP information FE in a scenario where transcoding is required is illustrated in Figure 10.6.

10.4 Concept Evaluation and Demonstration

As a starting-point of the development of the SMART architecture, seven new concepts had been defined. For the purpose of assessing the new concepts and technologies used for the smart delivery and transport of multimedia as presented above, the consideration of many different aspects is required. However, not all of these concepts and technologies can be assessed by one single evaluation approach or implementation. For aspects like system consistency

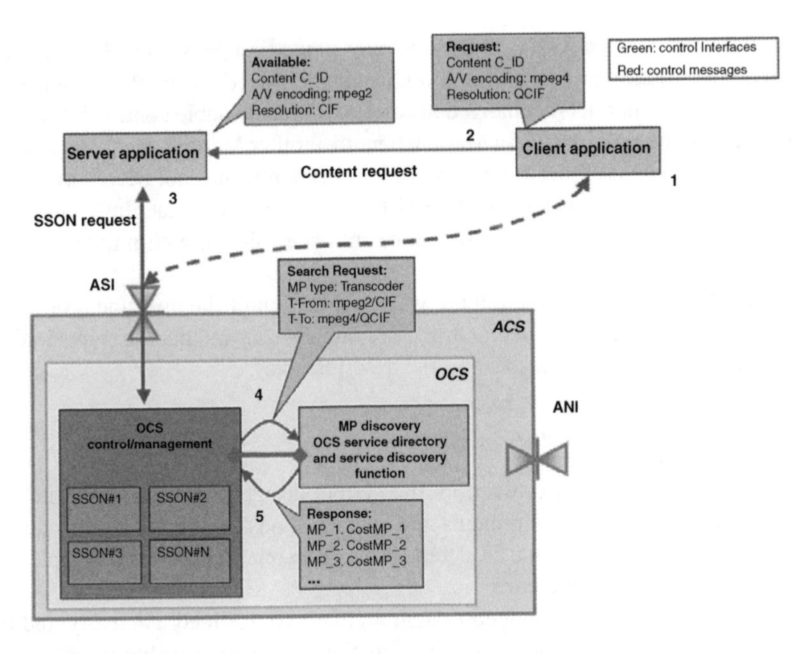

Figure 10.6 Role of the MPI in a transcoding scenario

and completeness, a conceptual analysis is the only suitable method of evaluation. For the verification of the scalability of the approach, it was decided to simulate selected scenarios including a huge number of network elements. And for other consistency checks and protocol verifications, some testbed implementation has been performed.

10.4.1 Concept Demonstration

The testbed implementation of SMART shows the new concepts for media delivery and transport working in a prototype environment. It proves the viability and usefulness of the developed media delivery architecture including its essential component, the media routing logic. It also gives an idea of the working principle of the overlay management FE when being a part of a more complex Ambient Control Space, even though only a few functionalities of the ACS were implemented.

The evaluation scenario of the testbed implementation involves three parties, namely a service provider that offers a streaming service to customers, a network operator that provides the Ambient Networks enabled transport platform for the service provider to deliver the service to its customers and the mobile end user consuming the service. In this sense, the benefit of the media delivery approach for these three parties could be proved.

In more detail, the demonstrator has shown to a somewhat limited extent the feasibility and usefulness of the first concept, the use of a service-specific overlay network for media delivery purposes. Network-side routing and transcoding MediaPorts specifically introduced for a streaming service can be included into the end-to-end communication path in a dynamical way. The second concept, adaptive overlay networks, has also been verified by the testbed implementation. The media overlay network can be dynamically adapted by the inclusion and release of transcoding MediaPorts and mobile clients as required for the seamless delivery of the streaming service. The third concept, multi-user services for heterogeneous environments, has not been verified with this demonstrator in the first phase yet. However, the fourth concept, dynamic routing at the overlay level, can be performed using the demonstrator. The media paths can be dynamically changed according to the available context information, e.g. information about the proximity of a MediaClient to another MediaClient. The fifth concept, media flows of the same session can be routed independently, has not been verified with this demonstrator yet. The same is true for the sixth concept, SMART caching. And last but not the least, the seventh concept, media adaptation, has been shown within the prototype using a transcoding MediaPort.

Overall, the testbed implementation allows a verification of the usefulness of the selected concepts as well as of the developed architecture and also shows that the provisioning of enhanced multimedia services is possible in the described way.

10.4.2 Conceptional Validation

For the purpose of conceptual evaluation, the system design was compared with and justified against specific system requirements and specifications. By verifying all requirements, interface objects and service flows in detail, it was possible to check the completeness and consistency of the target architecture.

An in-depth model of the developed system architecture for media delivery and transport is provided by performing an object-oriented analysis according to the object engineering process of the oose.de GmbH and the rational unified process. The analysis includes a recap of the requirements and features making sure that the required functionality is present in the model. One of the core functionalities of the media delivery framework, the session handling functionality, is presented as a standard sequence of actions. This sequence shows how a multimedia session is created and maintained within Ambient Networks making use of flows traversing a SSON. Also, through appropriate concatenation of the various offered functionalities of the system design, it was possible to cover all investigated essential functionalities that were required. That is why the developed architecture can be considered complete and why it can be claimed that the media delivery architecture offers a framework for the creation of a multitude of overlay networks.

However, some critical issues have been identified during the evaluation. Standard sequences of actions were presented that show how important the creation and update of sessions are. This issue has to be a part of future investigations especially when it comes to a higher frequency of context information updates or retrievals.

10.4.3 MPDB Simulation Results

As described above, the MediaPort information FE should provide information such as available processing modules, current load or usage costs per potential or registered MediaPort to the OM FE in case a SSON needs to be established or adapted. The more static part of the MPI FE, the MediaPort database, stores permanently information about MPs that are available in the administrative domain of an OM FE.

The simulation of such a MPDB reveals results that prove that such an approach can scale. For example, in the case when multiple MPs are required to realize a SSON, the search for these MPs can be issued in parallel and therefore the response time for a distributed MPDB (DMPDB) search query does not have a major impact on the time required to set up a SSON.

The query response time is used as one metric in the experiments with the result that the average response time for a DMPDB with a size of 570 ONodes maintaining information about up to approximately 8620 MPs is below 500 ms for the used test network. As a main result of the simulation, it can be stated that especially small-to-medium size content addressable network (CAN) based DMPDBs have very accurate query response times even in the case 'just' a three-dimensional CAN address space is used. Moreover, in the case of a DMPDB consisting out of 10 000 nodes storing information about 103 000 MPs, the expected average response time for a query can be below 400 ms in the case of a seven-dimensional DMPDB address space and an average IP layer one-way delay of 49 ms. In the case that information about the expected number of DMPDB nodes is available, the corresponding dimension of the DMPDB address space can be chosen with regard to the response time requirements and the measured average IP layer delay in advance.

This fact and the obtained simulation results are an indicator that a DMPDB approach based on Distributed Hash Tables (DHTs) can be considered as a realistic concept for the delivery of multimedia content over Ambient Networks.

10.4.4 Pattern-Based Discovery Simulation Results

The second part of the MPI FE, the pattern-based path-directed MPI discovery service, is used for the dynamic discovery of network-side media processing functions 'close' to the media delivery path between the MediaServer and the MediaClient(s). This function is used by the OM FE to identify the most suitable overlay nodes for a service-specific overlay network and to manage the configuration and set-up of this overlay network in case the MPDB is not able to fulfil the requests of the OM FE.

For the purpose of evaluating the detection of potential overlay nodes using the path-directed search pattern and the scalability of such an on-demand discovery of MPs, a set of simulations has been performed to observe the message complexity and the total number of nodes visited in a single-destination scenario. We measured the percentage of the number of nodes visited in the single-destination pattern as compared to a pattern, which makes a full search based on a straightforward flooding approach. The percentage of visited nodes is computed out of the total number of nodes. The simulation results show that the path-directed search pattern uses around 1–10 % of the number of messages and visits only about 1–10 % of the nodes. This evaluation gives an insight to its performance. By not visiting all the nodes, the discovery algorithm is quicker than a flooding approach, produces less overhead and the quality of the overlay network is enhanced as only MPs are selected that are close to the media path anyway. The simulations show that the percentages of the number of nodes are proportional to the inverse of the radius of the network (assuming the source is in the centre and the destination at the edge of the network). This result depicts one of the main advantages of the path-directed pattern. Because the search is made only along the end-to-end routing path, the number of visited nodes does not considerably increase with the size of the network.

The quality of the service-specific overlay network with respect to the stretch introduced by the overlay and the price incurred were also studied in a multiple-destination scenario. The result shows that even with densities of potential overlay nodes as small as 1 %, the quality of the network constructed using the pattern-based system can be more than 70 % (for sideway expansions above 1) of the quality achievable if the overlay was constructed based on full information about the network. The quality with larger densities of potential overlay nodes is comparable with the maximum possible quality. For instance, with 95 % confidence, the average quality achieved with a sideway expansion of two hops and a potential overlay node density of greater than 7 % is nearly 90 %. The average quality will be even greater than 95 % for this density with sideway expansion greater than two hops.

Concluding from these simulation results, the on-demand establishment of service-specific overlay networks with high quality is possible with only a small traffic/message overhead. The ability to construct high-quality overlay networks without prior knowledge of the existing physical network topology and the available network-side processing capabilities makes the pattern-based approach most valuable for ad hoc and dynamic environments. Being a scalable discovery mechanism of potential overlay nodes, it allows for periodic execution of the scheme as required by the concept of media delivery and transport within Ambient Networks.

10.4.5 Proactive Caching Simulation Results

Session continuation over heterogeneous user devices and access technologies should be possible with enhanced multimedia delivery and transport. To increase network performance (e.g. lower latency and/or load) and to deal with discontinuous coverage of wireless links, the network should be capable of proactive and context-aware data caching. The principle of proactive caching is based on ONodes which temporarily store multimedia content on its way to the end user. The benefit relies on the performance improvement of the wireless link, which usually represents the so-called last mile in the end-to-end connection. The higher variance in throughput and reachability of those wireless links requires fast reactions from the backbone network. Due to the high number of hops between the server and the client, a fast reaction towards link losses and session re-establishment is hard to achieve. Also, the maximum throughput of wireless links might never be reached by the overall end-to-end connection. Therefore, capacity might be wasted due to some bottleneck link on the backbone path for an ongoing session. By always providing enough streamable data close to the end user, it is possible to maximize the utilization rate of the wireless link which in turn improves the overall system performance.

This can be done by the caching of data close to the wireless hop at the edge of the core network to partly overcome coverage problems. The communication between a data source and the consumer is separated by a caching entity so that the first segment is exclusively confined to wired networks and therefore manageable by legacy technology. All issues regarding the radio data transfer are then restricted to a very limited region and are much easier to handle. By grouping access points (APs) and base stations into clusters, a user moving in such a clustered area would be continuously served by one caching MP. This implies that after switching over to another cell of the same cluster, the same caching MP can be employed and accessed.

Four different scenarios were compared by simulation. The maximum cumulative throughput, which could be reached if 100 % network coverage would be provided to the end device, was used as reference. A scenario with a theoretically reachable throughput for a coverage area of 50 % without proactive caching was compared to a scenario where the proactive caching enables much higher cumulative throughput. Due to the clustered area with proactive caching, the throughput almost reaches the level of the reference case despite of the area of only 50 % coverage. Packets, which cannot be transmitted if the terminal is between two coverage areas, are simply buffered and forwarded as soon as a WLAN connection is provided. Due to the larger bandwidth of a WLAN connection, it is possible to transmit all stored packets on top of the usual traffic while connected. However, a small gap remains between theoretical throughput and the final value of the simulation, which is caused by protocol overhead.

The simulation results illustrate the performance enhancement due to the caching of packets close to the wireless link and the division of the end-to-end connection. For non-real-time services, a throughput close to a fully broadband coverage can be reached in theory, although only 50 % of the area is actually illuminated by the various access points. Hence, the employment of proactive caching in Ambient Networks allows an improvement for multimedia delivery services.

10.5 Conclusion, Outlook and Further Work

This chapter introduces a concept for media delivery in Ambient Networks that moves away from a traditional end-to-end paradigm, in which the network supports the media delivery (only) by best-effort datagram transport, to an integrated concept for network-assisted media delivery in which the network provides QoS and media processing capabilities to the service to be transported and distributed.

This is facilitated by setting up service-specific overlay networks for each service. The main features of SSONs are as follows: First, a SSON is set up to optimize the routing transport between media servers and media clients for a particular service. Thus, it helps in providing and controlling the required QoS for the service, as it controls the routing topology (on the overlay level) that is used for the delivery of the services. Second, a SSON does include not only routing nodes, but also processing nodes, as required by the service. Thus, it allows media processing of any kind in the network that means between media server and media client. Examples for such processing include not only transcoding, but also caching, flow synchronization, spam filtering or virus filtering and many more.

The functionality for the establishment and control of SSONs is provided by the *overlay management* functional entity of the ACS. It manages the entire lifecycle of a SSON, from the initial set-up until the final removal. Also, it influences the routing decisions in the overlay nodes that constitute a SSON and reconfigures overlay topology and routing tables in order to optimize the network when changes occur. Each node participating in a SSON requires basic overlay network functionality for the handling of packets at the overlay level. This functionality within a node is called *overlay support layer*. The OSL handles packet forwarding on the overlay level and handles overlay addressing. In order to establish a SSON, all MediaPorts need to be known that may have to be included into the SSON. This is enabled by providing a MPI function that enables efficient lookup of MediaPorts based on media processing capabilities, location, cost, load and other characteristics. The MPI function maintains a directory of existing MediaPorts and has the capability to dynamically search for MediaPorts, for example, in a neighbourhood near MediaPorts that are already involved in a SSON.

SSONs are dynamically reconfigurable in response to changing network conditions (e.g. end-to-end delay or congestion), changes in network context (e.g. handovers) or changes to the user context (e.g. mobility, user profiles, device capabilities, media/content types). Adaptation of SSONs can happen on different time scales: fast adaptation to respond to critical and disruptive changes like link congestion or outage, and slow adaptation in response to noncritical changes, e.g. when MediaPorts join an Ambient Network or when users enter or leave a SSON.

The concept presented here allows for the first time the deployment of generic network-assisted service delivery, in which the network adds more value to the service than datagram transport. Specifically, it adds network-assisted media processing capabilities. This simplifies the deployment of scalable multi-user services in heterogeneous systems with diverse device capabilities and access network properties.

The concept has been proven by various simulations and has been partly implemented in demonstrators that were shown publicly. It will be further developed within the project and be introduced for standardization in suitable standardization bodies such as 3GPP and TISPAN.

11

ContextWare – Context Awareness in Ambient Networks

Acknowledgements

Editors of this chapter are Alex Galis (University College London), Raffaele Giaffreda (British Telecommunications Plc.) and Theo Kanter (Ericsson AB). This chapter is further based on the joint experiences and efforts of the researchers in the first phase of the Ambient Networks project and particularly the following people listed as contributors and authors (i.e. in alphabetical order): Dineshbalu Balakrishnan (University of Ottawa), May El Barachi (Concordia University), Fatna Belqasmi (Concordia University), John Dang (British Telecommunications Plc.), Roch Glitho (Concordia University), Hamid Harroud (University of Ottawa), Kerry Jean, Annika Jonsson (Ericsson AB), Anders Karlsson (TeliaSonera AB), Ahmed Karmouch (University of Ottawa), Mohammed Khedr (University of Ottawa), Heimo Laamanen (TeliaSonera AB), Mikko Laukkanen (TeliaSonera AB), John Mattam (Concordia University), Roel Ocampo (University College London), Christoph Reichert (Fraunhofer FOKUS) and Mikhail Smirnov (Fraunhofer FOKUS).

11.1 Introduction

This chapter describes the architecture, functional components and properties of ContextWare (CW). ContextWare is an information network that enables the automatic collection, dissemination and management of network context information within Ambient Networks.

The novel aspect introduced with this chapter is the exploitation of network context information aimed at the reduction of network complexity and the ability to dynamically and autonomically adjust the behaviour of Ambient Networks to be increasingly responsive to users needs. The goal of making Ambient Networks control functions context aware is therefore essential in both guaranteeing a degree of self-management and adaptation and

Ambient Networks: Co-operative Mobile Networking for the Wireless World Norbert Niebert (Ericsson GmbH), Andreas Schieder (Ericsson GmbH), Jens Zander and Robert Hancock

supporting context-sensitive communications that best exploit the network resources available without compromising unintended disclosure of private information from the various context information owners. The aim is not only to make the network itself context aware, but also to offer context information about the networks to high-level applications and services. This is a new and challenging area of research that has only just begun to be investigated by the research community. The context-aware Ambient Networks work therefore represents research that results in novel contributions to the study of network context awareness in general and to context-aware networks in particular.

We introduce the new term ContextWare to describe the system architecture for network context management. In essence, ContextWare is a mediator between context sources and sinks and its design resulted in a flexible architecture whose building blocks can be distributed to address scalability while breaking down complexity as the system usage and size grow. The scope of our approach to the deployment of ContextWare ranges from a single context lookup, through context associations (stable context exchange), to context dissemination strategies being defined dynamically, and finally to purposeful context processing, e.g. aggregation of network-level context information. It represents an approach to network context handling that combines already known advances in the area of context-awareness support in restricted end-user single-administrative domains with new techniques for extending the exploitation of context awareness to network control functions and across administrative domains.

11.1.1 Structure of the Chapter

The structure of this chapter is as follows: Section 11.2 introduces network context awareness, Section 11.3 elaborates context awareness in Ambient Networks and Section 11.4 introduces ContextWare, the main architecture and system design for management of network context in Ambient Networks. Section 11.5 describes the prototypes that have been developed to validate the ContextWare architecture and concepts. Section 11.6 provides work conclusions.

11.2 Network Context Awareness

11.2.1 Expanding the Notion of Context Awareness

Context awareness, as a process, system and concept, is based on a group of interrelated research domains: mobile, ubiquitous and grid computing and networking, service-aware networking, programmable networks and autonomic communications [138]. In each of these research areas, context has been used to enhance human–computer and computer–computer interactions, thereby helping achieve the vision of providing seamless computing and networking anywhere, anytime.

Traditionally though, context awareness has been studied as a required ingredient to enhance the user experience in pervasive computing environments [138]. In fact, many publications documented in the literature (see [132,144–147,150]) show how it is possible to support a user and adapt his applications to reflect his context.

Furthermore, we notice that most of the time, context information has been a synonym for user-related information (location, environmental conditions, mobility, situation, etc.) and was used to meet very specific application requirements. This means that from a networking perspective context awareness has been exploited only within islands of connectivity, in other words, well below its full potential.

The major deficiency in current systems we address can therefore be summarized as the lack of an infrastructure for the management and provision of different types of context information beyond user-related context, which makes it difficult to develop general solutions in the field of context-aware applications, resulting in proprietary mechanisms for context awareness developed from scratch most of the time.

The contribution documented in this chapter aims at developing an infrastructure for collecting managing and disseminating network context information across administrative boundaries and constitutes an important enabler for realizing more comprehensive context-sensitive communications.

The prospects deriving from enlarging the availability of context information beyond information about the user to include network-related information are not yet widely acknowledged. In fact until quite recently, data communication services have been available only via wired home or corporate networks or through wireless wide area coverage technologies such as 2.5G/3G. Most of the time, different types of service have been delivered on very distinct devices with a few opportunities for interworking between devices or for swapping services across devices.

In such environments, users currently have no choice but to use communication services within connectivity islands, unless they are particularly knowledgeable and confident with handling the latest technologies enabling interworking available in the market. Seamless use of services in such heterogeneous environments is achievable only in a very patchy way. Even then, the user needs to be proactive in gathering the right information in order to decide what service is delivered to which device through which network and to manually configure settings accordingly.

Recently though, the increasing availability of access technologies and related access networks, together with an increase of the number of devices each user has access to, is eventually leading us to work in pervasive networking environments which generate new opportunities and make much more desirable the availability of an enlarged context information base spanning across networks and devices.

In other words, it is becoming evident that a lack of an appropriate infrastructure for collection and dissemination of *network-related context information* is hindering further evolution or autonomy of communication services preventing exploitation of advantages that pervasive networking environments could bring for optimized service delivery.

Dey's review [143] provides the following key context definition: 'context is any information that can be used to characterize the situation of an entity. An entity is an object, place or person that is considered relevant to the interaction between a user and an application, including the user and applications themselves.' This definition does not cover all aspects of context as it presents only an external observer's viewpoint, revealed in 'characterizing the situation of an entity'. An internal observer's viewpoint would need to be added to this definition, identifying the structure of context, its domain, range, qualities, functionality and control.

Human user context characteristics include information representing the user's surroundings (user's location, available devices, networks, etc.) as well as his/her physical being (e.g. identity, presence, preferences, history, etc.).

Network context characteristics include network description (e.g. network identity, location, access types, coverage, IP address per machine, IP masks per subnetwork or address per domain), network resources in general (e.g. bandwidth, supported services, available media ports for media conversion, available quality of service (QoS), security levels provisioned) and flow context characteristics. Flows are a possible embodiment of the interaction between the user and networks. Context information that characterizes these flows may be used to optimize or enhance this interaction including the state of the links and nodes that transported the flow, such as congestion level, latency/jitter/loss/error rate, media characteristics, reliability, security; the capabilities of the end devices; the activities, intentions, preferences or identities of the users; or the nature and state of the end applications that produce or consume the flow. Because of the ephemeral nature of flows, flow context has to be handled differently than user or network context.

A full analysis of the context definitions and the state of the art in context awareness can be found in [132,151,172].

The work documented in this chapter illustrates ContextWare, a novel system architecture for network context management that bridges the gap illustrated above and whose features allow to reduce human intervention and provide Ambient Networks with the ability to self-adapt and reallocate resources based on network context changes.

11.3 Context Awareness in Ambient Networks

The challenge ahead is to make networks context aware, where software and hardware sensors are used to capture context information, in order to make increasingly heterogeneous infrastructure appear homogeneous to applications and users by enabling dynamic arrangement of connectivity and service delivery paths, minimizing the need for user intervention as well as gaining increased utilization of available network resources.

11.3.1 ContextWare in the Ambient Networks Architecture

This section positions ContextWare functionality within a wider background by illustrating its interactions with other entities in the overall AN architecture.

Looking at the various solutions created for user-related context-aware computing, it is clear that the objective of adding context awareness to any application is to reduce human intervention by making it adapt to the changing circumstances while it runs. The objective of adding context awareness is therefore to add automatic adaptability possibly without worsening the core functionality carried out by the application. This is clearly also the case when control functions in the ACS are made context aware, and in order not to affect their core functionality, there must be straightforward mechanisms for interaction with the ContextWare architecture. Also, different functions will have different requirements that in turn determine how they retrieve and use context information. For example, for multi-radio access decisions any context lookup needs to be performed very quickly, as opposed to the situation in which a context lookup is only needed to decide whether a better overlay network can be created for the delivery of a particular service.

These types of consideration have been taken into account when designing how other functions interact with ContextWare. In fact, a number of different protocols will be implemented allowing context information to be pushed, pulled in a generic way or in a specialized way (i.e. following a particular dissemination strategy and achieving an agreed quality of context (QoC) tailored to the requirements of the function acting as a context client).

11.3.2 The Goal of Ambient Network ContextWare

The goal of the ContextWare architecture is to provide a common framework for context awareness across all functions in the Ambient Control Space with the resulting advantages already illustrated in the introductory section. In this architecture (Figure 11.1), ContextWare

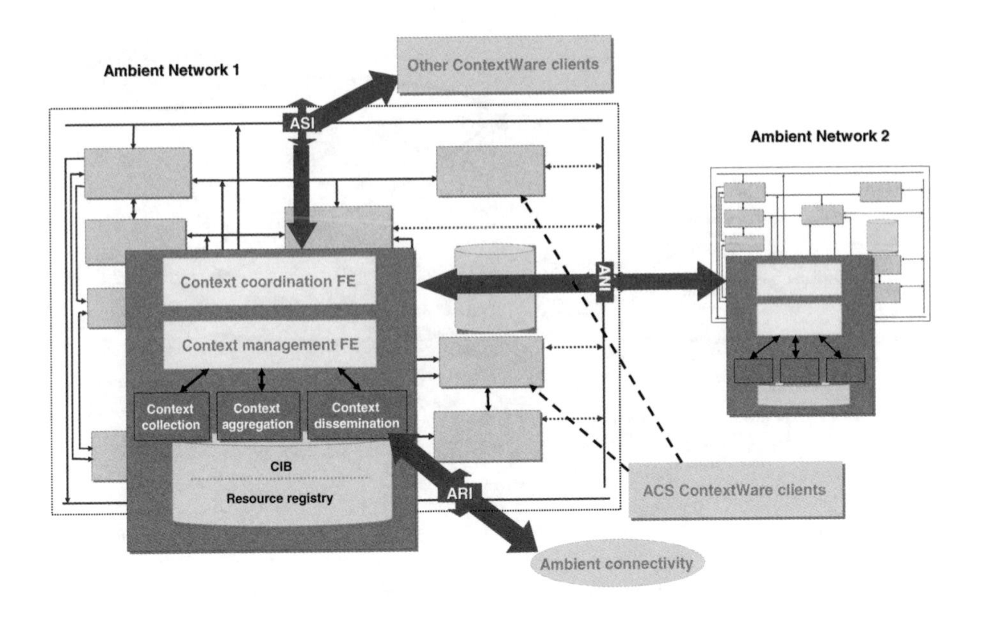

Figure 11.1 Ambient Networks ContextWare architecture

is realized by two context-specific functional entities (FEs), one interfacing other FEs of Ambient Networks (and other context clients) and the other implementing the core internal operations required in the context provisioning system. These two functional entities are the context coordination FE (ConCoord FE) and the context management FE (CM FE) and are further elaborated in the following subsections. The ConCoord FE was introduced to meet the requirements on creating a gateway into the ContextWare architecture and deals with indexing, registering, authorizing and resolving context names into location addresses. The CM FE was created to meet the requirements on creating appropriate context associations between clients and sources, managing the contents of the context information base and providing support for access control.

11.3.3 How to Interact With ContextWare

A graphical representation of the ContextWare interaction with other FEs inside and outside the ACS can be found in Figure 11.2. As the picture shows, the 'context provisioning' box synonym for the ContextWare architecture has links inside and outside the ACS to represent the fact that it not only can be used to make other ACS functions context aware but also is a key in achieving context-sensitive communication in the upper layers (context-aware services).

As shown in Figure 11.2, these interactions are achieved through appropriate signalling, which will be described later in this chapter and which will provide the ContextWare contribution to the definition of the three main AN interfaces, the Ambient Service Interface (ASI), the Ambient Networks Interface (ANI) and the Ambient Resource Interface (ARI).

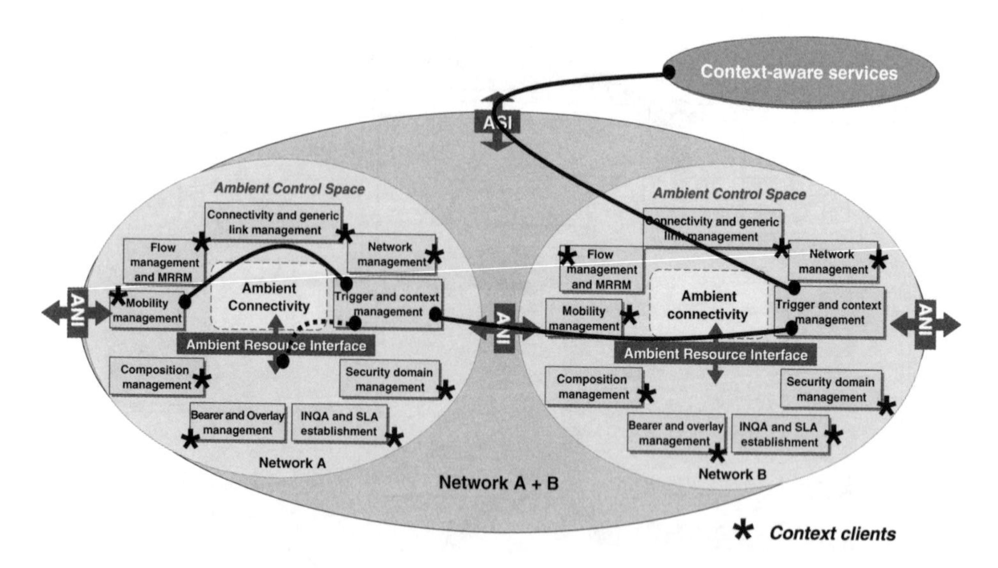

Figure 11.2 Interactions with other systems within and outside the ACS

- *ContextWare view point of ASI*: The ASI interfaces towards service infrastructures and allows applications and services to issue requests to the ACS concerning the establishment, maintenance and termination of end-to-end connection between functional instances connecting to the ASI. Contextualization of user facing services and applications (i.e. making them aware of network context) is performed through ASI.
- *ContextWare view point of ARI*: The ARI is an interface located inside an Ambient Network between the ACS and the Ambient Connectivity. It offers control mechanisms that the ACS can use to manage the network resources residing in the 'Ambient Connectivity'. Contextualization of network resource facing services is performed through ARI. In addition, the CIB functions of storage, updates, aggregation and dissemination are using ARI.
- *ContextWare view point of ANI*: The ANI interconnects different Ambient Control Spaces and facilitates the network composition process [139–142] as well as the exchange of control messages to agree on a shared control space. The interaction between ContextWare functions resulting from this role is performed through ANI. In addition, the multidomain management of context for composition/decomposition is performed through ANI.

11.3.4 ContextWare Interworking with Other Functional Entities

ContextWare plays a crucial information coordination role with respect to making other FEs of the Ambient Control Space context aware by collecting and distributing information needed for the autonomic operation and decision making of other FEs. The overall Ambient Network Control Space and its interactions are presented in Chapter 4. This section focuses on this interworking between ContextWare and the other FEs. Figure 11.3 gives some examples of how various FEs and other context clients in general can interact with ContextWare.

The design of the ContextWare architecture described in the previous section was mostly performed in a top-down fashion, working from scenarios, concepts and requirements. As a

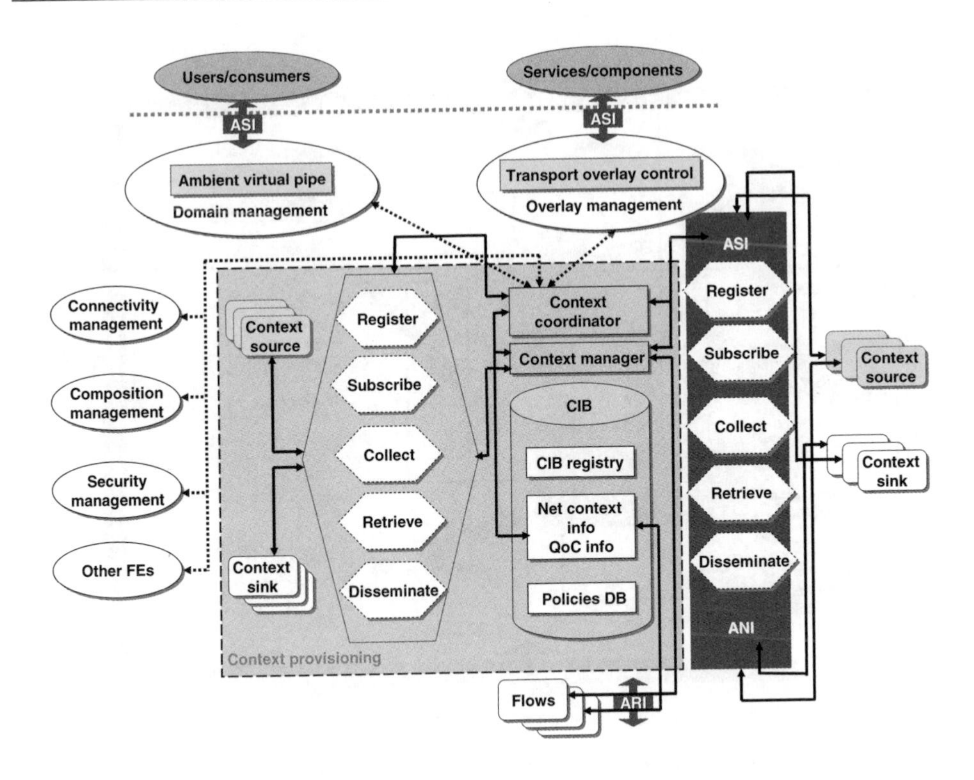

Figure 11.3 ContextWare interactions with other FEs

complement to that work, a bottom-up approach has also been implemented, where the starting point was to try to list context information that would be relevant to the different AN FEs. This work has resulted in a number of scenarios related to the other work packages in the project as well as their interaction with ContextWare. In this section, we briefly discuss in summary some of those interactions between ContextWare and the other FEs.

Figure 11.2 illustrates amongst other things, how for example the transport overlay control function, a part of the overlay management FE, can be made context aware. This allows the creation of appropriate overlays that automatically adapt to the change of network context (e.g. a particular network gets overloaded). The picture also illustrates the ambient virtual pipe (AVP) function that, contributing its core capability in the domain management FE (see Chapter 12), can be notified of context changes that influence its operations.

Both these functions contact the ConCoord FE to request the availability of particular context information, and if necessary establish well-defined associations with elements in the CIB, which is managed by the CM FE. This request for context information can include QoC demands and the level of service that the CIB should provide to this client. To summarize, any context-aware networking function must implement context client functionality, supporting the interfaces to ConCoord FE and the CM FE, and the associations to the CIB. The context client can be more or less advanced, depending on whether support for QoC, service levels, negotiations [149] is needed.

ContextWare has the role of managing the context information in the Ambient Control Space, including distribution to context clients. Context clients are context-aware services, either user facing applications or network services, which adapt themselves based on context

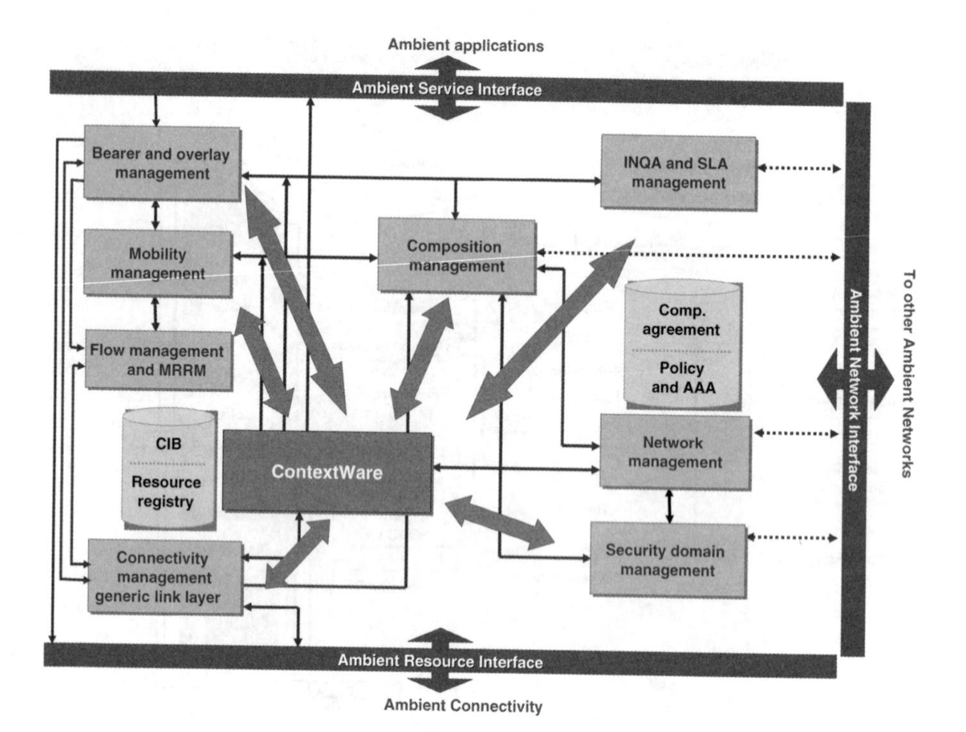

Figure 11.4 ContextWare as mediator of context information exchange

information. Network services in the Ambient Networks project are described as a number of functional entities, which act as context clients.

ContextWare enables a simpler and more efficient interaction between the different context sources and context clients. It acts as a mediating unit and reduces the number of interactions and the overhead control traffic. This is illustrated in relation to the AN FEs and other context sources outside ACS, in Figures 11.3 and 11.4. It is important to understand that the ContextWare FE plays a central role in the ACS as far as context information dissemination and collection is concerned. We also expect that most context sources and context clients belong to the other FEs of the ACS (and often even to other Ambient Networks). However, aggregation [170], correlation [169], indexing and other processing functions are typically located within the ContextWare architecture.

11.3.5 Context Identifiers

The atomic element concept of the architecture is the context association. A context association is a unidirectional relation from a context source to a context sink, i.e. the direction is that of the context information flow. The context source is the component providing/producing the context information, and the context sink is the component using/consuming it.

A context association has certain attributes, among which are context level agreements (CLAs) [171] and QoC specifications. The context association protocol is a client–server protocol, where the context source acts as the server and the context sink acts as the client. Modes of retrieval (server push versus client pull) are also attributes of a context association.

Context sink is any entity embedding a context client (making the entity context aware). Otherwise, we do not make any assumptions about context sinks, like when or why they request, how they are affected by or what decisions are made based on context information. A context source provides context information in the form of data objects. The content and structure of a context object are determined by its type, which may be, for instance, an SNMP MIB [163], an XML document type definition (DTD), a file format, etc. A particular context object of a given type is identified by its universal context identifier (UCI). The information represented by a context object may be dynamic and change over time, but only the context source can update its context objects. The basic operations to retrieve context information are that a context client fetches the content of a context object by means of a client–server protocol or subscribes for event notifications delivered when the context object changes its state.

UCIs are a new type of uniform resource identifiers (URIs) [160] and uniquely identify a given context object, but not its location within the network. A UCI is the conceptual rendezvous point between client and sources, i.e. whenever a client wants to get specific context information, it has to know the UCI of object representing this information. Similarly, a context source is assumed to know the UCIs of the context objects it wants to publish.

A fully qualified UCI looks basically as follows:
ctx://domain.org/path?options, where

- 'ctx' is the new URI scheme;
- 'domain.org' is the DNS domain name within which the context object exists;
- 'path' is a sequence of words separated by slashes ('/');
- 'options' specifies further modifiers like the data encoding format on the wire.

Most often, clients will query for local context information within their domain. In such cases, the domain name can be omitted: ctx:/path?options.

Fully qualified UCIs contain a double slash after the colon, whereas local UCIs contain only one slash. The main advantage of local UCIs is that context clients can retrieve their desired context information always by the same name, even when the network change aliases are allowed, so for example the following two UCIs, ctx:/path_one?options and ctx:/path_two?options, are equivalent if and only if they refer to the same context object. A source is therefore allowed to register several UCIs for a single object.

The structure of the path component is currently not yet specified. The difficulty here is that a given name hierarchy is basically as appropriate as another, as long as each object gets a unique name, and other requirements suggesting a proper choice could not be found. We therefore propose as the first path component for context objects specified within the Ambient Networks project the name 'ambient' and to further divide this subnamespace according to the functional entities or work packages. According to this convention, for instance, the meta-object of the ConCoord (see the next subsection) has the following local UCI: ctx:/ambient/wp6/concord.

11.4 Ambient Networks ContextWare: Architecture and System Design

11.4.1 Introduction

This section describes the AN ContextWare architecture, its internal structure, the primitives for communicating with it and a possible authorization and privacy framework. The design of the architecture is based on the concepts and requirements described in [134]. As described above, context information is modelled as data objects identified by UCIs. Given the nature of context information, its sources and its potential clients, it is envisaged that creating a system that scales well is of primary importance.

To fulfil these basic requirements, we based the design of ContextWare on the existence of two building blocks separating the tasks of registering sources and name resolving context information for clients (ConCoord FE) from the task of managing the actual context information in a distributed way (CM FE). Context information base (CIB) building block provides flexible context storage capabilities and it supports the operations of these two functional entities and in particular the uniform context dissemination from the vast diversity of network context information sources.

These three components, together with a proposal for an authentication and privacy framework, are illustrated hereafter.

11.4.2 ConCoord: The Context Coordination Functional Entity

The main purpose of the ConCoord FE is to coordinate information exchanged between different FEs in ANs. The ConCoord FE is the first point of contact for any context client.

All context sources register their context objects at a conceptually centralized entity, the context coordinator (ConCoord). The ConCoord is the first point of contact for a context client: Clients query the ConCoord in order to get the locations of context objects, and the ConCoord responds with the address of the source maintaining these objects.

In other words, the ConCoord does not store the context information itself, but pointers to it. This mechanism is quite similar to dynamic DNS [161]. The interface functions of the ConCoord are summarized as

- A registry where a context source registers the UCIs of its object with its contact information. Context sources are authenticated.
- A function to authenticate and authorize source registrations and client access to context objects.

The registry of the ConCoord is itself a context object, the meta-context object of all other context objects. The context source for this object is the ConCoord itself, and the meta-object is essentially the set of registered UCIs.

This meta-object should also be accessible like any other context object by the protocol primitives described below, as this enables context clients to subscribe to events like ‘notify me whenever a new object of this type registers’. This is important information for context clients to detect new sources of context information or to learn that currently used context sources are no longer available.

11.4.2.1 Locating and Distributing ConCord Functionality – A Realization with P2P Technology

The registry of the ConCoord has to map context UCIs to contact information of the objects identified by these UCIs. The registry can be realized by Distributed Hash Tables (DHTs) using multiple schemes from the area of structured peer-to-peer (P2P) overlay networks like Chord [164], content addressable networks (CANs) [165] or Tapestry [166]. In order to simplify the explanation, we assume that every context source, sink and manager join the P2P overlay, which makes up the distributed ConCoord.

After joining the P2P network, context sources register the UCIs of their objects in the form of pairs (UCI, contact). The context source first applies a common, uniform hash function to the UCI. The resulting hash key identifies the node on which the entry is to be stored. Then, the pair (UCI, contact) is sent through the overlay to this node and stored there.

When any other overlay node issues a RESOLVE request for a given UCI, the hash function is applied on the UCI to identify the node where the entry is stored, and the entry is then retrieved via the overlay.

Methods to self-organize the nodes of a P2P network into a resilient overlay topology are well known and guarantee upper bounds for the scope of a search (e.g. $O(\log N)$ in the case of Chord [164], where N is the number of nodes in the ConCoord), so that a distributed ConCoord is quite feasible.

Within an Ambient Network, both sources and clients have to locate the local ConCoord in order to register their context objects or to resolve UCIs, respectively. Additionally, any ConCoord must be able to locate any other ConCoord for interdomain resolution. A single mechanism enabling both intradomain and interdomain ConCoord locations is therefore desirable. We propose two mechanisms for this:

- *Locating ConCoords via DNS*: DNS can be extended by SRV resource records [161] or entirely new resource records [162]. As soon as a source, client or ConCoord knows the name of the target domain, it can resolve the domain name to that domain's ConCoord location.

The problem with this approach is that the namespace for Ambient Networks and domain names might be different.

- The other mechanism is via the Generic Ambient Network Signalling (GANS) protocol. GANS provides a mechanism to route control messages with identifiers like 'FE@domain.com' to the functional entity 'FE' in domain 'domain.com'. As ConCoords represent the context FE within an Ambient Network, this mechanism can be directly exploited for ConCoord location.

11.4.2.2 Context Protocol Primitives

The basic message sequence chart for ConCoord messages is given in Figure 11.5.

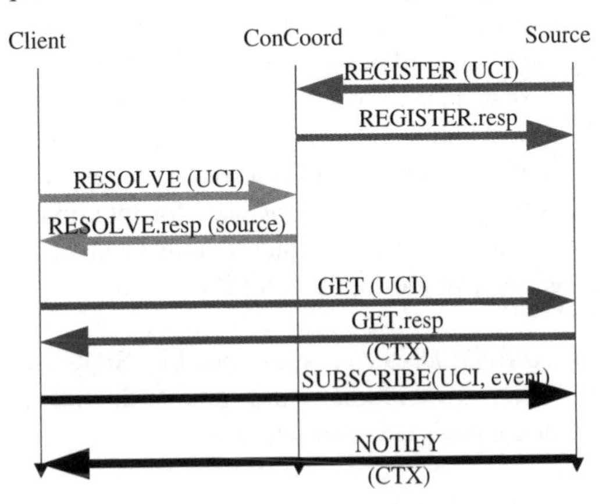

Figure 11.5 Basic message sequence chart of context primitives

A context protocol provides basically the following five primitives, depicted in

REGISTER:

- A context source registers the UCIs of its context objects and its contact information at the ConCoord.

RESOLVE:

- A context client requests context information in the form of UCIs from the ConCoord, which responds with the contact information of the corresponding context objects.

GET:

- A context client fetches the content of a context object from a context source.

SUBSCRIBE/NOTIFY:

- A context client subscribes to a context source object with an event specification. Thereafter, the client receives notifications whenever the object changes in the specified way.

Note that there is no primitive to update an object, as this is done exclusively by the context source, which 'owns' its objects. The motivation for the RESOLVE primitive is that a client can locate its desired objects once and issue GET requests or change subscriptions as often as desired without generating unnecessary activity in the ConCoord. The alternative would be to send all GET or SUBSCRIBE requests via the ConCoord, which then locates the objects and forwards the requests to them; this solution however has the potential of creating a bottleneck in the ConCoord especially for a growing number of queries. Further information about these primitives can be found in [138].

11.4.2.3 Interdomain UCI Resolution

A context client in one domain should be able to ask for context information about another domain, which requires querying the ConCoord of the other domain. Mechanisms to locate the remote ConCoord (or a representative node in the corresponding DHT) are described as follows.

Within an Ambient Network, both sources and clients have to locate the local ConCoord in order to register their context objects or to resolve UCIs, respectively. Additionally, any ConCoord must be able to locate any other ConCoord for interdomain resolution. A single mechanism enabling both intradomain and interdomain ConCoord locations is therefore desirable. We are investigating two mechanisms for this.

- *Locating ConCoords via DNS*: DNS can be extended by SRV resource records [161]. As soon as a source, client or ConCoord knows the name of the target domain, it can resolve the domain name to that domain's ConCoord location.

  The problem with this approach is that the namespace for Ambient Networks and domain names might be different.

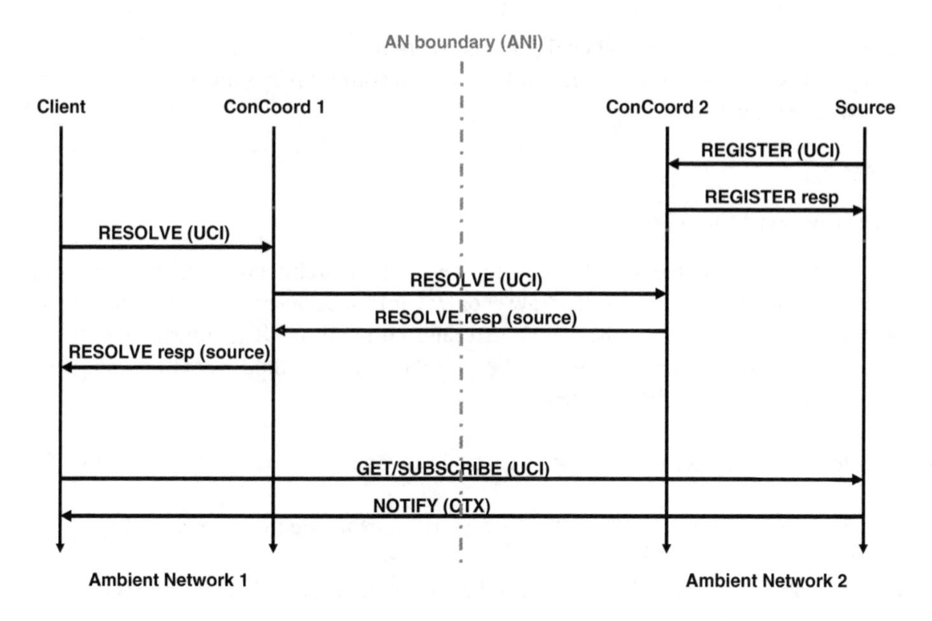

Figure 11.6 Interdomain UCI resolution

- The other mechanism is via the GANS protocol. GANS provides a mechanism to route control messages with identifiers like 'FE@domain.com' to the functional entity 'FE' in domain 'domain.com'. As ConCoords represent the context FE within an Ambient Network, this mechanism can directly exploited for ConCoord location.

A client always asks its local ConCoord even for remote context UCIs. It is then the task of the local ConCoord to locate and contact the remote ConCoord and to relay the RESOLVE request to it, maybe after a context level agreement has been established between both ConCoords (see Fig. 11.6). This design decision is motivated by the goal to keep context clients as simple as possible, and also to enable caching of both remote ConCoord locations and CLAs by the local ConCoord.

11.4.3 Context Management Functional Entity

The CM FE manages context within and across domains. This would involve operations such as collection, modelling and dissemination of context information to the interested entities, as well as managing the sharing of context information among different domains, i.e. cross-domain management. The CM FE is also responsible for scheduling interactions between context sources and context clients, monitoring these interactions, reallocating channels of interaction (in the case of context changes), and finally aggregating and composing context according to clients' requirements. The context management FE therefore represents the service provided by a number of distributed processes that can

1. be dynamically created based on context client requirements;
2. provide aggregation, translation, inference capabilities;

3. cache context information on behalf of context sources;
4. cache context information at different locations to address performance optimization and minimize retrieval time from clients.

11.4.3.1 Context Managers

The above-mentioned context protocol primitives and the architectural elements introduced so far allow clients to request only 'raw' context information provided by objects. These do not yet provide a way to filter, aggregate and correlate context information. This can be accomplished in a number of ways. We present here a method based on using context managers, which consist of three parts:

- A context client to get context information from one or more input sources of type T_1, T_2, ..., T_n.
- A processing function f: $T_1\ T_2 \ldots T_n\ T$, which transforms the input context of types T_i into output information of type T.
- An output object of type T that represents the processed information and makes it accessible via the context protocol.

A context manager is therefore a context client 'back to back' to a context source with a processing function in between. Context processing is then performed in a data-driven manner following the pipes-and-filters pattern by associating the output of a manager with the input of another.

Figure 11.7 depicts a directed acyclic graph (DAG) whose nodes are context sources (SRC), context managers (MGR) and context clients (CLT) and whose edges are context associations. Nodes with zero in-degree are initial sources and the original source of 'raw' context information. Nodes with zero out-degree are final context clients and the ultimate sink of context information. Nodes with nonzero in-degree and nonzero out-degree are context managers. The subgraph for a client, i.e. the subgraph made up of all nodes and edges having a directed path to the client, is called the multi-pipe for that client.

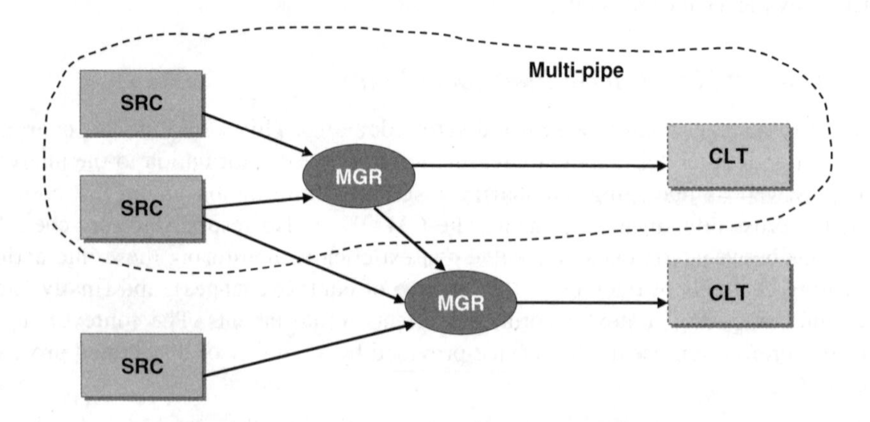

Figure 11.7 Directed acyclic graph of context associations

A multi-pipe must be type consistent in the sense that a context processors output type must match the input type of the next one. Polymorphic processor functions might also be advantageous in some cases. In essence, a multi-pipe with multiple context processors is the functional composition of its constituent processing functions. As an example, a linear pipe of managers with functions f, g and h provides the nested processing function $h(g(f(x)))$.

Context managers with a single input are either filters, extracting specific information from their input, converters, transforming the input into another format, or loggers, recording the history of context information. Context managers with multiple inputs act as aggregators or correlators on their inputs.

Context processing can basically be performed in two modes. When a client issues a GET to get context information from a directly associated context manager, the GET request is propagated back towards the initial context sources. Alternatively, a context manager might be able to serve the request from a cache. Recursive fetching and processing of context information is the client-driven mode (pull). In contrast, recursive subscription and notification is the source-driven mode (push).

Mixtures of both modes are also possible. For instance, a caching manager with a single input might exist solely for the purpose of transferring load from another context source that is experiencing a high rate of GET requests. The caching manager subscribes to the object and responds to GET requests on behalf of the initial context source.

Note that only the type, but not the UCI, of the output object of a context manager is determined. The actual content of the output object and its UCI depend on the object instances feeding the manager.

11.4.3.2 Creating Context Managers

Context managers are dynamically created based on the requirements of context clients. Once created, they register their output type and capabilities with the ConCoord. The ConCoord's registry therefore maintains initial context sources and context managers.

This enables recursive multi-pipe establishment of in a distributed way. The ConCoord locates the final context manager, which locates the managers for its input, which in turn locate the managers for their input, and so on, until the inputs are all initial objects. More precisely, the following steps occur:

1. A client sends a RESOLVE request with a UCI to the ConCoord.
2. The ConCoord checks whether the UCI identifies an initial source. If so, it returns the contact information of that source. Otherwise, the ConCoord infers the type of the required object, locates a manager with this type as its output type and returns a RESOLVE response with the manager's location.
3. The client sends a GET or SUBSCRIBE request to the manager.
4. The manager infers from the UCI in the GET or SUBSCRIBE request the UCIs of its input sources. For each of these UCIs, it sends a RESOLVE REQUEST to the ConCoord.
5. Steps 2–4 are repeated until all UCIs are resolved during step 2.
6. Figure 11.8 depicts the message sequence chart for the recursive establishment of a linear pipe between a client, a manager and a source.

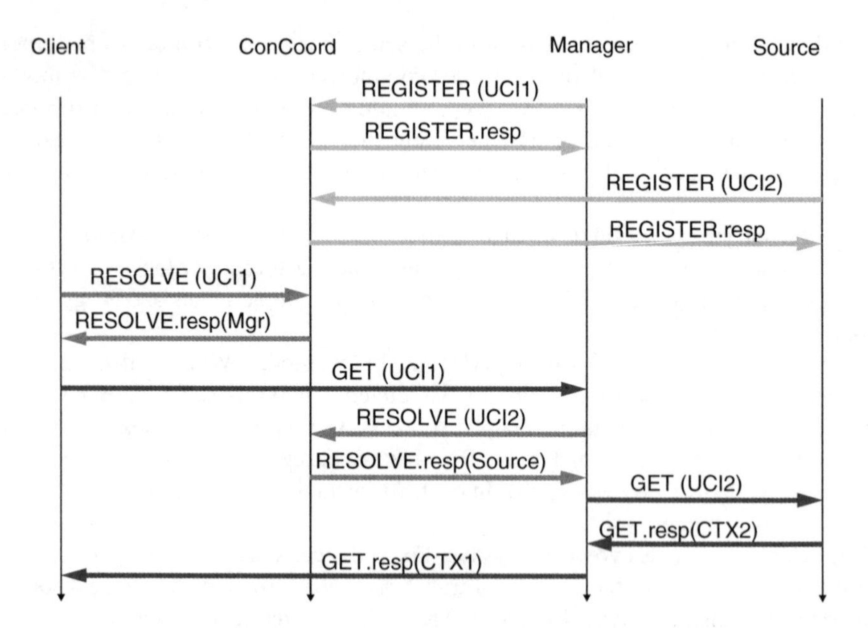

Figure 11.8 Recursive set-up of a linear multi-pipe with one manager

The critical requirement for the recursive approach is that a manager, knowing already its input types, is also able to infer which object instances of that type, i.e. their UCIs, are required.

There is another way for a manager to find its input sources. This alternative scheme is applicable if the manager inputs are not to be derived from a requested UCI and the manager knows in advance which input sources it needs. In this case, the manager subscribes to the ConCoords meta-object, specifying which type of source it is interested in. Whenever a new source of that type registers, the manager is notified and can SUBSCRIBE to it.

11.4.4 The Context Information Base

Alongside with the two functional entities described earlier sits the CIB, an architectural building block providing flexible storage capabilities, mainly in support of the context management FE. The CIB is meant to address the following points:

- Context information may need to be distributed for performance reasons (to allow both efficient retrieval from context clients and fast update from context sources).
- Not all the context information will be retrieved directly at its source, for various reasons (the owner of context information may want to publish it but not deal with all the requests from context clients, the owner of context information may not have the information available in the formats that the clients understand).
- The diversity of context sources makes it unlikely that information can be retrieved directly from the sources according to a common ontology. The CIB therefore gives the flexibility of storing context information according to a common ontology, which simplifies the implementation of context clients (by making it more generic rather than tailored to read from a particular sensor).

- Context managers described earlier need to have access to storage facility to make the results of their processing readily available to clients. Note that this is expanding the previous translation example, given it can be applied to context aggregation, interpretation, etc.

The context information base therefore adds on top of the ConCoord one more level of flexibility, which is twofold. On the one hand, it allows sources of context information to delegate publishing the information they own to the context management FE. On the other hand, it allows us to keep in the design the flexibility of moving around/pre-processing context information to accommodate better performance for both client requests and sources updates.

11.4.5 ContextWare Authentication and Privacy

11.4.5.1 Motivation

The main concern that use of context awareness usually generates is surrounding privacy considerations. The word 'privacy' in the context of this section is synonym for achieving control over who has access to context information. ContextWare is no exception and same concerns must be addressed given it can store information considered private by end users, but most importantly it can also grant access to network context information needed to improve the operation of an Ambient Network's ACS. This category of context is normally 'owned' by the network provider, which may or may not want to share it according to well-defined policies.

In the light of these considerations, a valid framework for protecting the information accessible through ContextWare architecture must be in place.

11.4.5.2 Security Framework Requirements

To achieve proper privacy protection and authorization for access to context information, due to the distributed nature of the context information base, the following goals must be met by the framework:

1. *It must work on a distributed architecture.*

When context information is forwarded from one node to another within the network, where authorization decisions might take place at multiple locations and with authorization policies being distributed at different places it is necessary to ensure that (a) policies travel with the context information either via reference or value and (b) the authorization framework is able to deal with failure cases and dynamically changing policies.

2. *It needs to have mechanisms that restrict distribution.*

Authorization policies are a convenient way to express the context provider's intent regarding the distribution of context information to other entities. The policies need to be extensible and rich enough to allow detailed definition of the conditions that govern the distribution of context information, the actions that are taken if conditions are met and the transformations (operated by the context manager) that need to be applied prior to returning information.

3. *By default no context information is distributed.*

This mechanism ensures that no information is provided to other entities without explicit indication (e.g. if the context provider did not attach any policy). It also ensures that disabled authorization policies do not harm the context provider's privacy. This situation can occur when policies are stored centrally and need to be retrieved in order to compute the authorization decision. Evaluating additional policies can only lead to more information being disclosed rather than less. As a consequence, exceptions cannot be expressed conveniently. We do not see this unnecessary restrictive as ephemeral identifiers are heavily used in Ambient Networks and can therefore be exploited to bypass blacklists by generating a new identifier. The unsuitability of the blacklist approach has been recognized with the current email system where new email addresses can be generated easily and they can also be spoofed.

4. *Simple access policy design to prevent unintentional information disclosure.*

Configuring policies is an error prone task. In order to avoid misconfiguration and a false trust in the security of the system, the policies need to be designed to be as simple to express as possible. This aspect is, to some extent, related to the user interface design but indirectly to the functionality provided by the underlying authorization policy language.

Given these high-level goals, a strong relationship with the work in the IETF GEOPRIV working group [167] is apparent. We argue for the application of the GEOPRIV privacy framework to the ContextWare architecture in order to control access to the information stored in or referenced by the context information base.

The GEOPRIV framework (see Fig. 11.9) aims at protecting location information against unauthorized access to preserve the privacy of the owner of that information. Although GEOPRIV is mainly focused on authorization of location information, the framework was extended to provide authorization functionality also for presence information as well. With RFC 3693 [168], a protocol-independent model for access to geographical location information was defined and is illustrated in. The model includes a location generator (LG) that produces location information (LI) and makes it available to the location server (LS). A Rulemaker (RM) uploads authorization policy rules to the location server. An authorization policy is a set

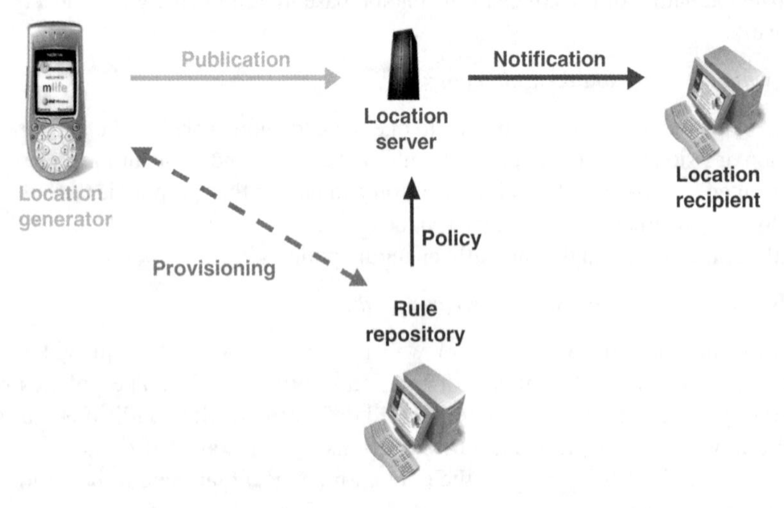

Figure 11.9 GEOPRIV architecture

of rules that regulate an entity's activities with respect to privacy-sensitive information such as location information. Each rule consists of a condition, an action and a transformation part. The LS stores LI and authorization policies and acts as an entity that authorizes access to a particular target. The request sent by a location recipient (LR) to access location information of a particular target receives either an error message (in the case of an authorization failure) or the location object that consists of LI together with authorization policies.

11.4.5.3 Mapping GEOPRIV to ContextWare

We illustrate in this paragraph the blocks in the ContextWare architecture, which are of particular relevance to the authentication and privacy framework. As illustrated earlier, ContextWare architecture has two main functional entities, the context coordinator and the context manager, that interface context requests with context providers.

Figure 11.10 shows the map of the existing GEOPRIV framework onto the ContextWare architecture created for storing context information in Ambient Networks.

Compared to GEOPRIV where the location server always acts as a policy enforcement point, we have identified a similar role in the context manager, which is also responsible for setting up multi-pipes [136] for context aggregation. We are also adding the flexibility of granting the source of the context information itself to act as an enforcement point.

Also it is beneficial to consider having two levels of authorization, one for coarse access control managed by the ConCoord and based on basic policies of the domain it represents (common to all context information) and the other for refined access and based on finer grain policies implemented by the various sources. Note that either of these may be void if there are no concerns about sharing context information widely.

As authorization policies are attached to the retrieved context information, further distribution can be allowed, restricted or denied. Some degree of trust has to be placed in the entity that is granted access to context information in order not to violate the provided authorization policies.

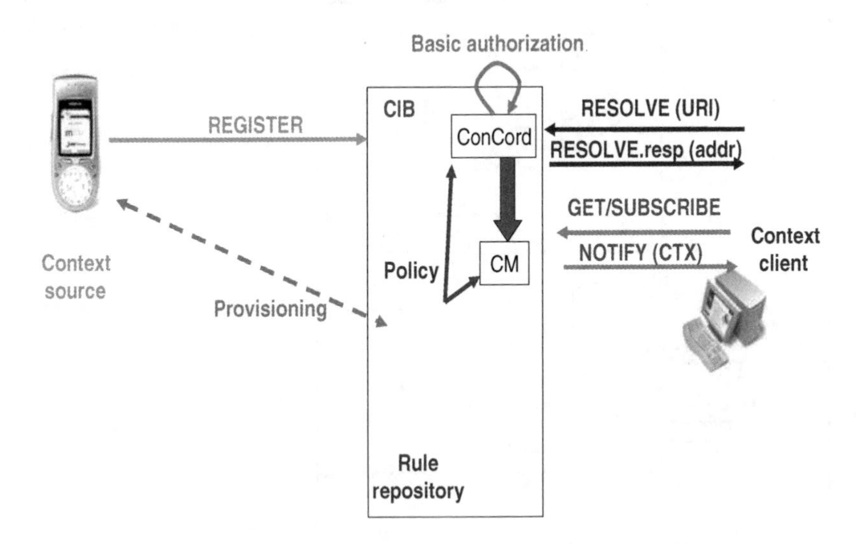

Figure 11.10 Mapping GEOPRIV onto ContextWare architecture

A client can request context information from both within and outside its own domain, though authorization policies may be in place to restrict its onward distribution to other domains, based on the structure of the identifier or based on various traits and roles.

Analogously, a context source might register with a ConCoord of another domain; in this case, the context source is external to the ConCoord's domain.

11.4.5.4 An Example of a Secured Context Request

Based on the above architecture and discussion, Figure 11.11 illustrates the sequence of events that characterizes a secured context request. This scenario also covers the case whereby context associations are established between a context source and a client when multiple queries from the client are envisaged (such as desired in a subscription-based environment).

First, the context source registers the URI for its context information with the ConCoord; it also notifies whether or not it wants to delegate the context manager as a proxy to deal with client requests (CM = YES in the case considered). Next, a context client asks (Figure 11.11) the ConCoord where the context information identified by the RESOLVE URI is located. The ConCoord performs (2a in Figure 11.11) the basic policy check. This basic authorization (based on a common policy for all the context requests) enables the ConCoord to decide whom to provide this 'context lookup' service to. For example, a provider may want to deny access to context information for all foreign clients. The query is then resolved (2b in Figure 11.11), and the address of the requested context (CM_addr in this case) is sent back to the client.

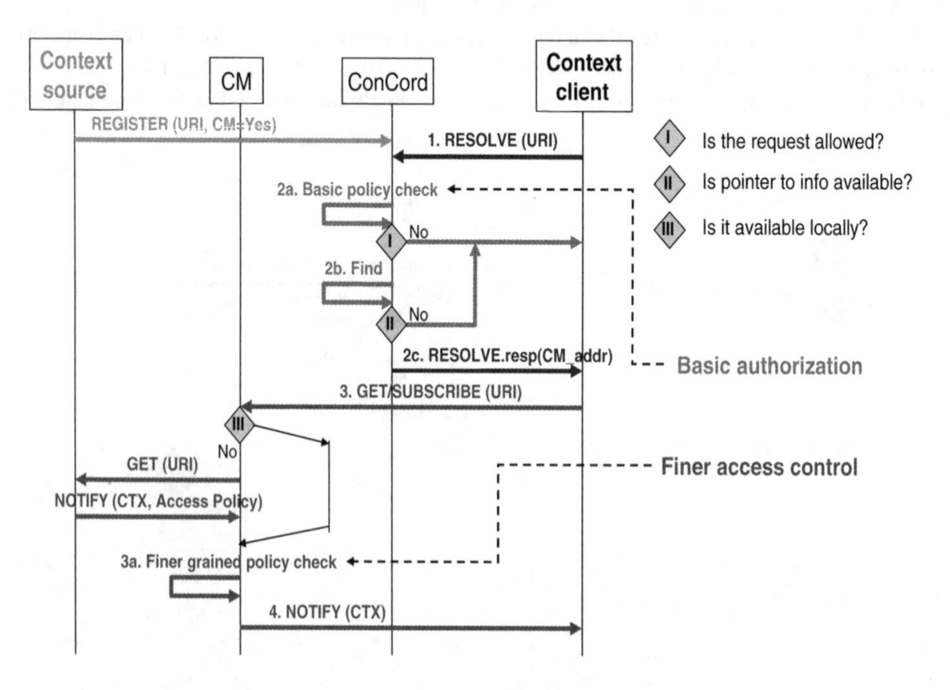

Figure 11.11 Context request example message flow

The client now has the address of the CM. It issues a GET request to pull the information or alternatively it can establish a context association issuing a SUBSCRIBE request. The context manager, acting as a proxy in this example, checks whether the requested information is locally available. If not, it recursively fetches the information from the original source, authenticating itself as a context manager. The source responds with a NOTIFY message containing the context information and its access policies. It then checks the finer grained policy for the requested information on behalf of the source and returns (NOTIFY) the context information to the client.

Note that further GET or SUBSCRIBE requests from this client will involve neither the ConCoord nor the context source, but only the context client and the CM.

As it can be seen, the behaviour following a context request is similar to what happens in the GEOPRIV architecture with the context manager playing the role of the location server (checking finer grained access control policies that might be set by the context source itself).

11.5 ContextWare Prototypes

Three ContextWare experimental systems were designed as proof of concepts prototypes [172] and two of them are described below. The first prototype illustrate the use of ontologies for structuring data retrieved via the context information base, whereas the second prototype focuses on the self-contextualization of an adaptive management service overlay in Ambient Networks.

11.5.1 An Context Ontology Demonstration System

Thinking about the definition of context quoted earlier in this chapter, it is easy to acknowledge that any type of information that can be used to identify the situation of an entity is in its nature very diverse as well as originated from the most disparate sources. As a consequence of that any system or framework dealing with context information (network context is no exception) needs to have a commonly agreed (between clients and sources) approach to modelling context information.

The first demonstration scenario shows how structuring context information according to well-defined ontologies provides a very simple context aggregation functionality which then facilitates a meaningful comparison of data from different networks for decision-making purposes. For this demonstrator network, throughput, bandwidth and cost were used to dynamically select, amongst a number of available Ambient Networks, the one for which the result of a composition process would best suit the requirements of the services to be provided. Based on the above selection of the context information, three ontologies were developed:

- an Ambient Network root ontology;
- a QoS ontology;
- a cost ontology.

As we anticipated, through the use of these different ontologies the demonstrator provides network context information which is then appropriately compared for automatic decision-making purposes. In this section, due to space constraint, we illustrate demonstrator results related to the use of the QoS ontology only (interested reader can find more detailed and complete information in [172]).

The AN ContextWare QoS ontology is made available in the context information base by the context manager and is illustrated in Figure 11.12.

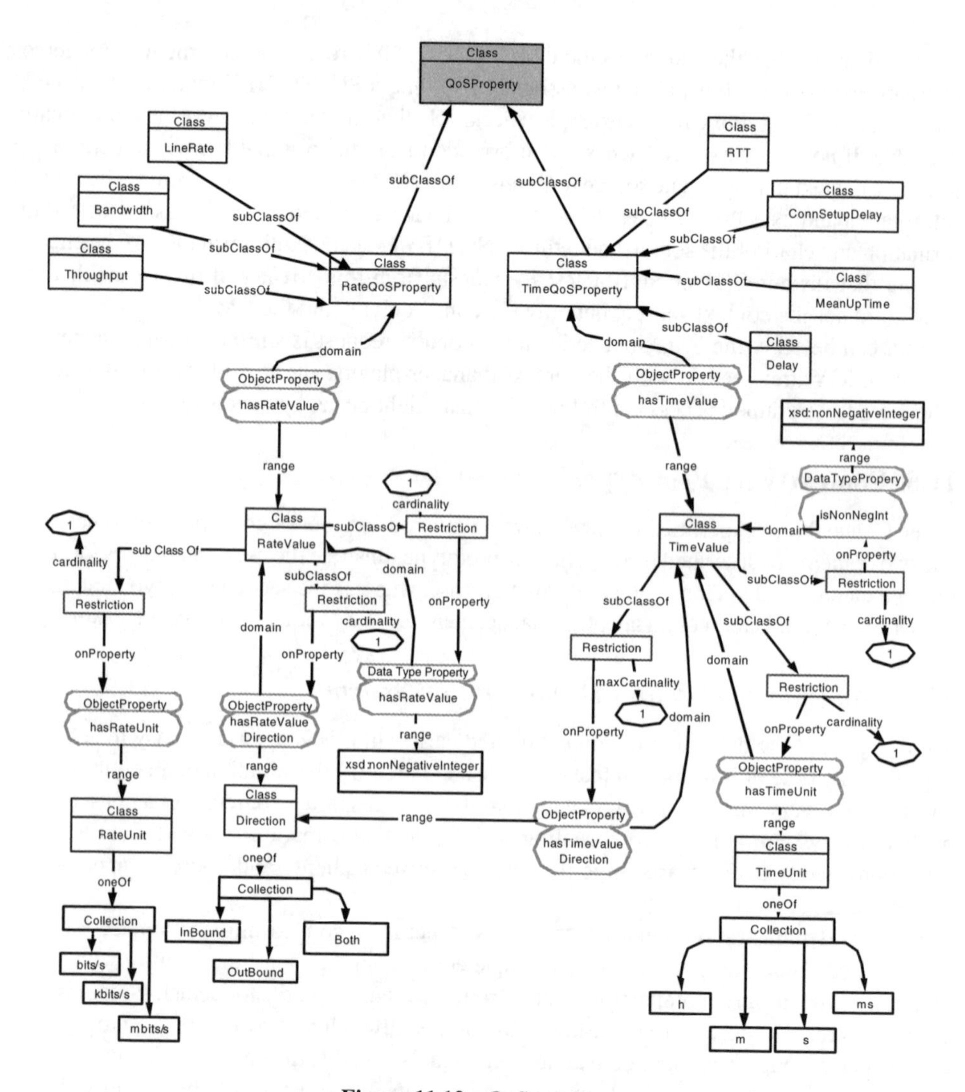

Figure 11.12 QoS ontology

Focusing on the request for QoS information, a generic query sent using an agreed UCI for network QoS is sent from to the ConCoord FE which returns the address of the context management FE capable of dealing with such requests for composite context information.

In order to reply to such a query, the context management FE parses the OWL query and mapping it onto the QoS ontology it initiates simpler queries to sources of context information that contribute then to the aggregation of 'network QoS' information. In the demonstrator, throughput and bandwidth (i.e. subclasses of RateQoSProperty, only a subset of the overall QoS ontology) are retrieved from the appropriate sources and then aggregated and formatted in an OWL reply which is then sent back to the client for comparison with replies from other networks and automatic decision-making purposes.

Previous work on suitable data modelling [137,138] on context modelling and dissemination highlighted the ontology-based modelling technique as the most promising approach to use in a network context management framework. Our demonstrator has shown its ability to address some of the challenges in retrieving, aggregating and comparing context information. While addressing most of the requirements of a context modelling framework, ontology-based modelling suffers of one drawback which was also highlighted in the demonstrator and relates to the high verbosity which would limit its applicability when, for example, composite queries are sent over links with limited resources.

11.5.2 ContextWare Demonstration System

The main aim of the second prototype is to validate ContextWare concepts and technologies in the context of the train scenario. It highlights

- service-level composition, i.e. network service composition with QoS contracts;
- dynamic reconfiguration of services triggered by changes in network context.

The ContextWare demonstration system is a scenario-based demonstration of the ContextWare concepts. It implements a network service called the ambient virtual pipe providing secure, QoS-assured, context-aware and adaptive management service overlays autonomically built on top of the underlying composed peer-to-peer AN management hierarchy [152]. AVP managers subscribe to ContextWare components and use context information for dynamic provisioning and self-adaptation [168] of the management overlay. Ambient Networks nodes are developed as programmable DINA nodes [159].

11.5.2.1 ContextWare System Components

Two general classes of ContextWare components were implemented, namely context sources (CSs), which are basically software-based context sensors and aggregators, and the context coordinator. Interfaces are provided for CCs, which may be applications or FEs within the ACS that need context information. Figure 11.13 shows the software components of the CW prototype. The modules currently implemented in the demo map are a straightforward way to the design indications found in Figure 11.13. The ConCoord software modules in the demo handle the registration of CS objects (context sensors) and supports queries from context clients, providing basic functionality described in the architectural specification. The various sensors implemented and described in the previous sections, on the contrary, implement context monitoring functions by interfacing with the actual physical context sources and disseminate context information through Notify primitives. The AVP manager on the contrary provides an example of a context client within the ACS.

Figure 11.13 illustrates the mapping of the prototype to various components in the ContextWare architecture. The mapping is shown by enclosing the implemented software components in the corresponding architectural block, for example, the ConCoord software component from the demo is enclosed in a block labelled 'context coordination FE', which is its corresponding architectural component. These software components, namely the AVP manager, ConCoord, the various context sensors and the PAP composition management software, may in turn be mapped back to their corresponding equivalent blocks in the software architecture diagram shown in Figure 11.11.

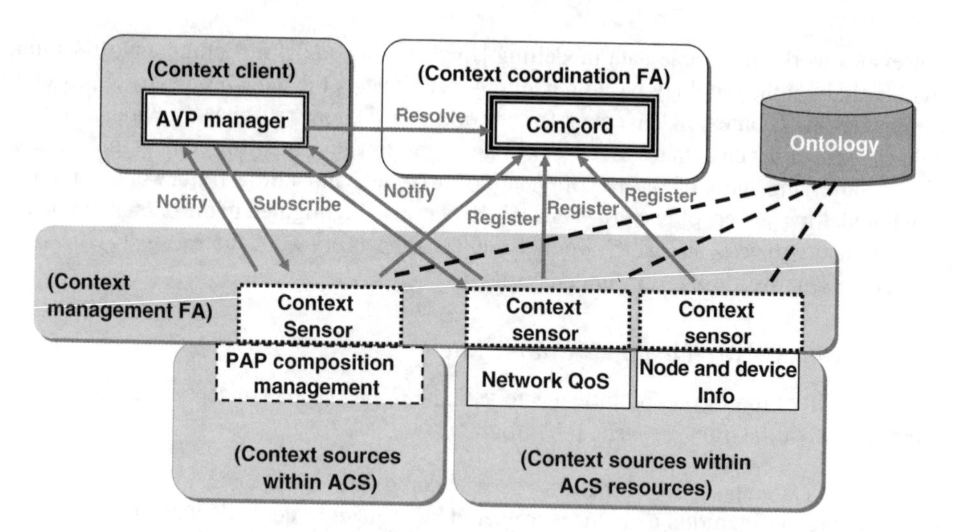

Figure 11.13 Mapping the prototype to the ContextWare architecture

The ContextWare components implement primitives described in [135]. When CS objects are deployed and instantiated, they register with the ConCoord using the Register primitive, supplying a context object's name and a requested time to live (TTL). In the current version of the demo, the IP address and UDP port number serve as a reference to the CS object and are supplied to the ConCoord during registration. The TTL is used to set a timer for registration entries within the ConCoord; entries may be renewed through a Refresh message prior to time-out, otherwise they would have to be re-registered. The ConCoord replies with a RegisterAck message confirming the registration, supplying the remaining TTL value, which may be different from the requested TTL and a unique identifier within the ConCoord for the CS object. Secure messaging between ContextWare components is provided by the AVP service.

ContextWare components in the demo prototype were dynamically deployed and executed within the Ambient Network prototype using an active and programmable networking [156–159] approach. Components such as the node and device context sensor were typically implemented on the DINA active networking platform [159] and dynamically deployed within the network as DINA active packets.

Self-contextualization is the process by which an entity whether a service, system or overlay autonomically becomes context aware. The prototype is self-contextualized and makes use of context information in management and in the provision of services to its customers. The key to self-contextualization of the AVP is the ContextWare infrastructure and the programmable network. This context exchange allows optimization of the functionalities of the overlays themselves, by taking into account the aspect of dynamism as required by mobility and changing networking conditions and context. It also allows the dynamic creation of overlays, composition and decomposition as a result of context changes.

11.5.2.2 Context Sensors

Context Sensors software sensors obtain context information by interfacing with the composition monitoring functions [152] within the ACS and context events related to management-level composition and changes in the peer-to-peer management topology. They also obtain network-related context from other sources, including the SNMP [163], management information base (MIB) or other low-level sources of network information. In our implementation, we initially limited ourselves to three main classes of CS: one that receives context event notifications from the composition management application, another that monitors network QoS and a third that obtains node- and device-related context information. (We henceforth use the term 'sensor' to refer to CS objects.) The details of these sensor classes are as follows:

- *P2P context event sensor*: The P2P context sensor listens for events from a peer-to-peer overlay management platform (details of which may be found in [152]) that manages network composition. Our current implementation of this class uses SNMP traps defined through a MIB object to support event notifications to context clients.
- *QoS context sensor*: This class of sensor provides information on the QoS state of an Ambient Network. Our prototype uses a subclass that dynamically monitors bandwidth utilization on an interface, or more specifically, the traffic traversing a particular Linux IP chain [153]. Unlike the other two sensor classes whose implementations were written fully in Java, the bulk of the code for this sensor was written in C, for two major reasons. The first reason was that sensors of this class required low-level access to node services and resources not normally accessible via Java, such as high-resolution timers and per-chain packet and byte counters. The second reason was that performance and code execution latency were seen as crucial factors in generating real-time QoS measurements.
- *Node and device context sensor*: This class of sensor provides information on node-related context. The sensor can also provide information on the geographic location of the host, although the data currently supplied by the sensor is only a simulated location.

Figure 11.14 shows a part of a reply to a context query sent to a node sensor; the information shown in the diagram is displayed through the CIB browser, a small application we have developed for debugging and diagnostic purposes.

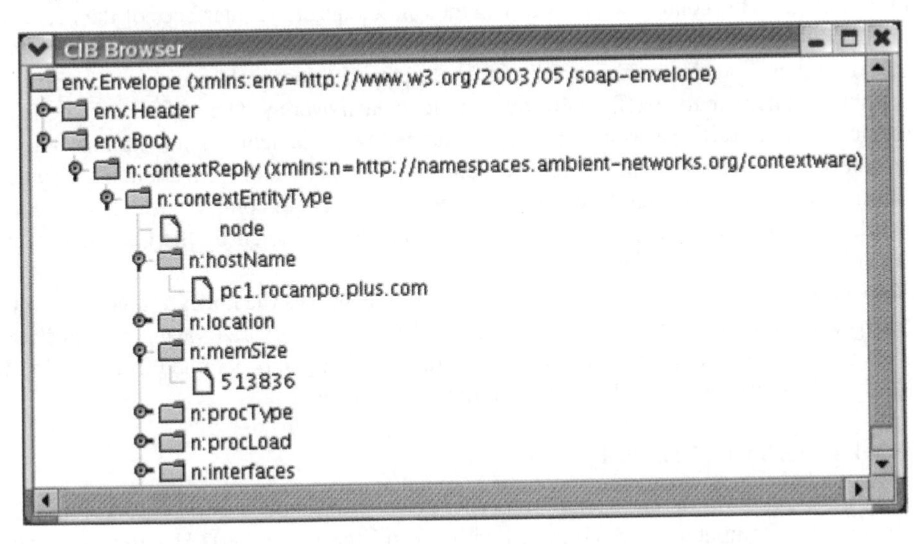

Figure 11.14 Fragment of a context reply from a sensor object, as viewed by the CIB browser

Typically, a context client such as an application or a functional entity within the ACS would query the ConCoord using a Resolve message for CS objects that can provide a certain type of context information. The ConCoord searches its internal registry and replies through a ResolveAck message with a reference (IP address and UDP port number, in the current version) to the appropriate context sensor.

Context replies from node sensors are typically formatted in Extensible Markup Language (XML) [174], although other proprietary formats are also supported. It should be noted that although the sensor reply shown in Figure 11.14 implies the use of Simple Object Access Protocol (SOAP) [175] message formatting, the protocol itself is not supported in our current implementation. In general, all of these sensors support event notifications to context clients; however, due to the nature of the context information they provide, node sensors are typically used in Get-Notify mode. On the contrary, due to the highly dynamic nature of network QoS, clients of QoS sensors either subscribe to continuous information streams (with a specified time interval between Notify updates) or request a Notify based on specified context events such as congestion or when traffic levels exceed a certain threshold.

11.5.2.3 Context Information Used

- *Peer-to-peer network context*: This context information concerns the activity involving members of the peer-to-peer network. It details when members join and leave an overlay network, the identity of the super peers and normal peers in each overlay. This context information includes the peer IDs of each peer along with the corresponding overlay ID and IP address. The P2P context information is stored in an SNMP MIB. SNMP traps are used to send context event notifications that arise when peer-to-peer network context changes. These context events are captured by the peer-to-peer sensors. Examples of these context events are newOverlayMemberJoined, overlayMemberLost, becomeSuperPeer, newMember and superPeerLost.
- *QoS context*: This context involves traffic measurements at the interfaces of the networks nodes. Filters are defined using iptables to do traffic measurements across an interface based on these iptables entries. Each overly is mapped to an iptables entry and hence we are able to calculate the traffic patterns specific to each overlay. The same can be done for services. These traffic measurements are combined with the actual bandwidth capacity of an interface to create computed context detailing the bandwidth utilization. This context information is used to determine congestion enabling peers to compose with higher bandwidth networks when congestion arises. This context is handled by the QoS sensor and delivered to the context clients.
- *Node and device context*: This context information concerns the nodes and devices in the network such as host name, processor type and load, memory size, available interfaces, display capabilities and others. In most cases, this information is obtained from the SNMP MIB by the node sensors and delivered to the context clients.

11.5.2.4 Experimental Testbed

As a means of partially validating our prototype, we set up an experimental testbed consisting of six notebook computers, each equipped with wired Ethernet and 802.11 wireless network interfaces. All ran the integrated prototype on the Linux operating system. Using the train scenario, we mapped the devices in our testbed to some of the scenario actors: two of the computers represented PAN Ambient Networks of Alice and another person, one computer represented a railway staff's AN, another two represented the Ambient Networks of two train passenger cars, and finally one computer represented the rail company's hotspot at the railway station.

We then walked through selected parts of the scenario using our testbed. The ANs of Alice and another passenger first composed with each other based on their AN identifiers and their respective composition policies. We then simulated their joining the train's Ambient Network by switching the two passengers' 802.11 ESSIDs to that of the train's 802.11 ESSID; as soon as they detected each other, the composed AN of the two passengers composed with the train's AN. The interconnection of the passenger cars and the train station network were simulated in sequence using Ethernet (wired) connections. Finally, we simulated the railway staff boarding and disembarking from the train using the same ESSID switching technique.

Figure 11.15(a) shows the state of the testbed from the point of view of the peer-to-peer management application, at that point in the scenario where the two passengers and the train staff are on the train while the train is still in the rail station. The application shows each Ambient Network as a node in the graph and graphically illustrates the topology, hierarchical structure and composition state of the AN management overlay.

During each composition, the election of a management super-peer [152], the identities of the composing ANs and overlay join/leave events were sensed as context, triggering the deployment or self-adaptation of AVPs interconnecting the management peers. When the railway staff boards the train and composes with the train AN, the associated context events trigger policies on a super-peer that determine the amount of bandwidth to be allocated to the AVP. As the railway staff member disembarks from the train and composes directly with the railway company's AN, this change in context, i.e. movement from an AN owned by a third party operator to his home network, triggers an adaptive reallocation (increase) in the AVP's bandwidth. The graph in Figure 11.15(b) shows the comparative bandwidths allocated to the AVP connected to the railway staff member's AN as he moves from one AN domain to another. The graph is provided in real time by a management application called the AVP bandwidth monitor and the data represents the amount of actual video traffic transmitted through the AVP. The data is provided to this management application by QoS sensors we discussed in previous sections; this is another concrete example of an application acting as a client of ContextWare.

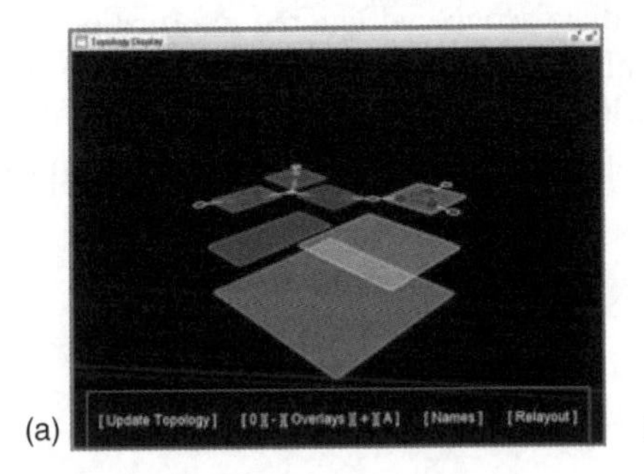

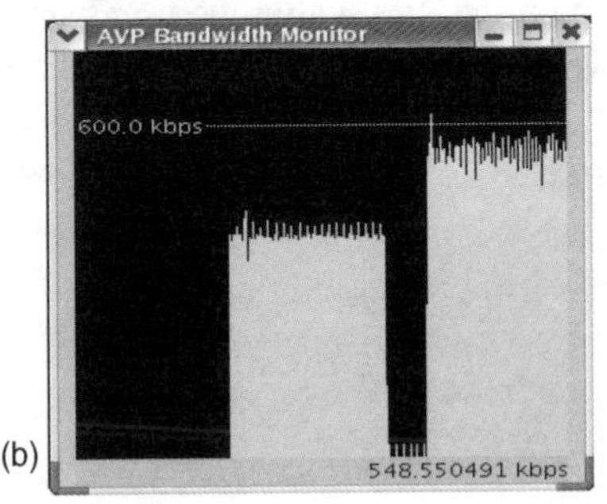

Figure 11.15 (a) Topological view of the peer-to-peer management overlay. (b) AVP bandwidth adaptation in response to context change

As the prototype in its current form uses the composition process primarily as a source of contextual events, our future work will focus on the complementary scenario where the composition process itself is context driven, thus completing the other aspects of the train scenario. For instance, the data collected from the individual node-based QoS sensors may be aggregated to provide AN-level QoS context descriptions that may aid the selection of a suitable uplink as described in the scenario. Node and device context sensors may be used to provide information on the capabilities of Alice's PAN so that the media stream may be dynamically adapted based on device capabilities and the available QoS.

11.6 Conclusions

Ambient Networks is new concept that generates new key requirements, which must be implemented in order to achieve the overall objectives of the project. The work undertaken by the context-aware networks work package has defined a number of context-based Ambient Network scenarios and requirements that clearly demonstrate the motivation, the value and the pivotal role of the context awareness in Ambient Networks. Each of the requirements defined in the Ambient Networks project paves the way for novel and distinctive research ideas in context-aware networks and some of those were further developed during the second phase of the project.

The goal of making Ambient Networks control functions context aware is seen as essential in both guaranteeing a degree of self-management and adaptation and supporting context-sensitive communications that best exploit the network resources available. The work detailed in this chapter has been undertaken to achieve this goal through the creation of ContextWare, an infrastructure for network context information collection and management within an Ambient Network.

The aim was not only to make the network functions context aware but also to make network context information available to high-level applications and services. This is a new and challenging area of research that has not yet been investigated by the research community.

Early research work on context awareness mainly focused on the use of location information to allow applications to adapt to the user's environment. Later on, researchers shifted their focus towards developing context-aware applications that use both user and environment physical contextual information to provide specific services to users as they move (e.g. guided tours and adaptive maps). However, these approaches were restricted to the use of context information that lies within the user's immediate scope and ignored information related to the underlying network. Now research recognizes that network-centric contextual knowledge can play a major role in providing the user with guaranteed services that can best utilize the underlying network. This chapter provided a novel contribution towards achieving this goal.

12

Towards Ambient Networks Management

Acknowledgements

Editors of this chapter are Alex Galis (University College London), Róbert Szabó (Budapest University of Technology and Economics) and Marcus Brunner (NEC Europe Ltd). This chapter is further based on the joint experiences and efforts of the researchers in the first phase of the AN project and particularly the following people listed as contributors and authors (i.e. in alphabetical order): Henrik Abrahamsson (Swedish Institute of Computer Science AB), Jorge Andres (Telefonica Investigación y Desarrollo SA Unipersonal), Eskindir Ayallew Asmare (NEC Europe Ltd), María Ángeles Callejo (Telefonica Investigación y Desarrollo SA Unipersonal), Lawrence Cheng (University College London), Márk Erdei (Budapest University of Technology and Economics), Alberto Gonzalez (Kungliga Tekniska Hogskolan), Anders Gunnar (Swedish Institute of Computer Science AB), Péter Kersch (Budapest University of Technology and Economics), Zoltán Lajos Kis (Budapest University of Technology and Economics), Balázs Kovács (Budapest University of Technology and Economics), Gergely Molnár (Ericsson Hungary), Johan Nielsen (Ericsson AB), Giorgio Nunzi (NEC Europe Ltd), Roel Ocampo (University College London), Csaba Simon (Budapest University of Technology and Economics), Simon Schuetz (NEC Europe Ltd), Rolf Stadler (Kungliga Tekniska Hogskolan), Ambrus Wágner (Budapest University of Technology and Economics) and Kai Zimmermann (NEC Europe Ltd).

12.1 Introduction

This chapter describes the management technologies needed to realize the vision of Ambient Networks. It focuses on the different management approaches that enable efficient management of Ambient Networks.

Network management systems of Ambient Networks must work in an environment where heterogeneous networks cooperate and compose, on demand and transparently, without the

Ambient Networks: Co-operative Mobile Networking for the Wireless World Norbert Niebert (Ericsson GmbH), Andreas Schieder (Ericsson GmbH), Jens Zander and Robert Hancock

need for manual (pre or re) configuration or offline negotiations between network operators. To achieve these goals, Ambient Network management systems must become scalable, self-managing and responsive to the network and its environment. This requires decentralized management architecture and algorithms.

From a management perspective, Ambient Networks differ from traditional networks in having dynamic configurations and topologies. ANs dynamically compose and gain connectivity through rapid establishment of internetwork agreements, bringing new challenges to all network management tasks, such as interaction among administrative domains, monitoring, end-to-end control, etc. Traditional network management tasks, such as configuration and establishing connectivity to other domains, usually involve human intervention. Although this is a valid approach for small-scale and static networks, it is infeasible for ANs due to their dynamicity. A human manager cannot react to changes produced by dynamic composition in a timely manner. Therefore, ANs require new management approaches that permit them to be autonomously managed. Management systems for ANs must dynamically adapt to changing network conditions. Moreover, such an adaptation must be robust (to cope with topology changes), efficient (in terms of management overhead) and scalable (to cope with large networks).

12.1.1 Ambient Networks Management Approaches

The results of our investigation show that the specific challenges of AN management can be met through the realization of four complementary approaches, namely peer-to-peer and pattern-based management, plug-and-play and closed-loop traffic control. Integration of the four management approaches into the overall AN architecture is also discussed and analysed.

Peer-to-peer (P2P) systems are characterized by the interaction among equal partners, called peers. This makes them to be particularly applicable in highly dynamic and heterogeneous scenarios, like ANs. In such environments, cooperation and internetworking can be dynamically established using P2P negotiations. Specifically, the characteristics of P2P approaches make them a good candidate to manage network composition as discussed in this chapter. This chapter presents an integrated solution composed of two P2P-based interworking proposals for the management of network-level composition. The first proposal is a framework for the organization of nodes and ANs into a dynamic hierarchical structure. This hierarchical structure is created and maintained by means of P2P negotiation processes among nodes and ANs based on physical network topology, policy and context information. The second proposal is the so-called ambient virtual pipe. It is responsible for the dynamic establishment of an autonomic, secure and self-managed service management overlay network to support (new) AN management services or external applications/services. The functionality of both approaches is demonstrated through a common demonstration.

Pattern-based management is a distributed approach to provide scalability and robustness to management tasks. It is based on the use of graph algorithms to control and coordinate the processing and aggregation of management information inside the network. Pattern-based applications map network-wide operations into local operations that will be performed by the managed nodes. Fundamentally, the graph algorithms are clearly separated from the local operations performed on the nodes, and both can in many cases be exchanged without changing the other one. This chapter presents three results in this area. First, a pattern for the set-up and maintenance of service overlay networks. Its evaluation proves that it is efficient and scalable in terms of management overhead. Second, a pattern enabling real-time continuous monitoring with accuracy objectives in an AN environment. Our evaluation shows the feasibility of controlling the fundamental trade-off between accuracy and overhead in a large-scale dynamic network. Third, this chapter also discusses concepts to extend patterns designed for fixed networks to work in wireless and mobile environments.

AN plug-and-play (PnP) aims to achieve full autoconfiguration of AN elements, which is mandatory in such AN dynamic scenarios. The PnP approach enables new AN elements to become integral parts of AN domains. It also periodically reconfigures the network elements to reach optimality in the face of changing network conditions. The approach followed is that each AN node is responsible for configuring IP connectivity on its own and to integrate into existing ANs. This means that management tasks are performed locally on each element, in contrast with traditional centralized management systems. This chapter discusses two aspects of PnP management. First, the autoconfiguration of base stations in wireless networks. For this purpose, a distributed flat approach is proposed and evaluated. A key aspect of the proposal is that stations autoconfigure based on state information exchanged only with immediate neighbours. A prototype of the base station configuration has been developed. Second, the provisioning of IP addresses to network nodes and routers. This mechanism assigns unique IP addresses to composing network elements. A hierarchical approach is proposed.

The *closed-loop traffic control* approach explores traffic engineering mechanisms as a way to optimize the network resource usage in AN environments. The objective of traffic engineering is to avoid congestion in the network and to maximize use of available network resources by adapting the routing to the network state (traffic profile and topology). This is a key task so that service provides can meet the subscribed SLA with their customers. This chapter presents and evaluates different approaches to handle dynamic traffic loads, a key characteristic of ANs scenarios. First, traffic engineering algorithms are evaluated for AN scenarios. Results show that the deployment of such algorithms outperforms current practice. This work has also identified the need for new mechanisms in order to cope with the appearance of disruptive traffic patterns. Second, a novel distributed approach to traffic engineering is proposed and evaluated. Its key characteristic is that it requires only local information at the routers instead of network-wide traffic statistic. This increases its scalability and robustness. Our experiments show that the proposal permits more traffic to be carried than current practice and also reduces packet loss. Finally, traffic engineering aspects are discussed from an interdomain point of view. The focus is on connectivity provision.

12.1.2 Structure of the Chapter

The structure of this chapter is as follows: Section 12.2 presents the requirements and challenges of Ambient Networks management, Section 12.3 presents the four management approaches to Ambient Networks and Section 12.4 discusses the integration of our proposal into Ambient Networks architecture and presents an outlook on further work.

12.2 Ambient Networks Management Challenges

The Ambient Management System (AMS) contrasts with traditional management systems mainly in the area of dynamic heterogeneous management layer composability and self-manageability.

Generally, traditional management systems are homogeneous and have a clear separation between the roles of management systems and network elements. SNMP or CMIP are generally used, but they do not scale well in large networks, in particular dynamic (de)composing networks. AN management approaches must be robust, scalable and adaptable and require less management traffic, less processing load on the management station, shorter execution times, etc. Automatic configuration management is also not supported in traditional approaches. Traditionally, manual configuration has been deployed due to the small scale of network involved. In AN management, however, manual configuration is impractical. Policy-based management was proposed to ease configuration, but the deployment of policy-based management is inhibited by the difficulties in understanding of the concept of policies by practitioners and the difficulties in standardizing configuration management.

Handling the composition of heterogeneous ANs is the major challenge of AN management. Prior to composition, each (heterogeneous) AN has its own Ambient Control Space (ACS) that is controlled by its own AMS. The latter is governed by a set of predefined policies that are known to the AN. AN AMSs are heterogeneous. Different ACSs controlled by AMSs may (de)compose, resulting in a common ACS AMS to manage heterogeneous network elements (that were previously managed by their own management system). Network elements may also manage themselves, i.e. self-management. Figure 12.1 shows the contrasts.

During AN network composition, a set of policies control the composition process, and a set of common policies for the new combined ACS must be agreed on. Once the networks are composed, the individual ACSs are also composed, resulting in a common ACS. From a management point of view, the common ACS is controlled by a common AMS that is governed by

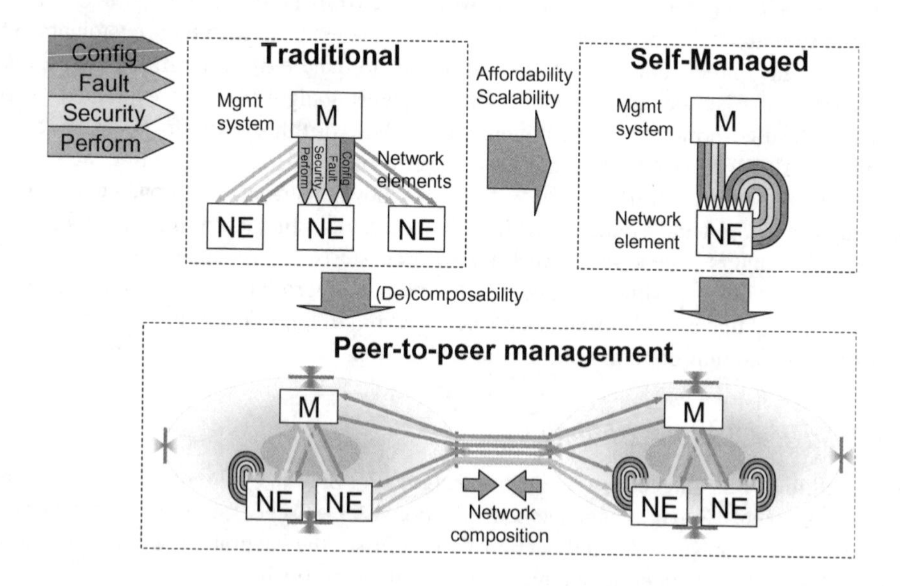

Figure 12.1 Traditional management systems versus Ambient Management System

common policies. The term 'composition', therefore, does refer not just to network and control layer composition, but also to *management layer composition*.

Management system composition, through interdomain interfaces, has already been demonstrated in the TMN model. However, management layer composition in AN is much more dynamic and requires a simple, efficient and scalable mechanism for pooling and sharing management information within and across heterogeneous composed networks. The composition mechanism itself should be autonomous and self-managed. This is needed because large- or small-scale ANs may leave or join on the fly, i.e. arbitrarily compose and decompose. The common AMS should be service oriented, that is, capable of supporting new services on demand. The common AMS should consist of distributed AMSs that are coordinated through Ambient Network Interface (ANI). The AMSs should be capable of accessing context policies within the AN(s), so that the common AMS may adapt itself to its context based on network conditions or administrative policies. The AMSs should coordinate policy negotiation between ANs during composition processes. This also implies that the AMS should resolve conflicts in ANs based on the composition agreement.

From a business perspective, network composition is regarded as a temporary agreement between independent networks (operators) to achieve a common business goal. Thus, the AMS should support automated creation and administration of network composition. Also, services that were available only in individual AN may now be provided to end users anywhere throughout the composed AN, so increasing business opportunities. The challenge of providing services across continuously composing and decomposing ANs is that autonomic reconfiguration of services is needed. Our results suggest that a highly flexible information infrastructure is needed to support automated AN management. This infrastructure should provide on-demand quality-guaranteed connectivity services between AMSs, as well as support for automated management. Although today's virtual private networks (VPNs) provide some QoS guarantees, they suffer from a low flexibility to adapt to the rapid changing, i.e. composing and decomposing ANs. Thus, an infrastructure that is secure, dynamic, QoS guaranteed and programmable [188] for rapid service deployment is required to support the common AMS.

This chapter provides some insight that the dynamic management layer composition challenge and the self-management challenge of AN may be dealt with four complementary approaches presented in the next section that develop and adapt different technologies for the purpose of dealing with these challenges.

The four complementary approaches are P2P, pattern-based, PnP and traffic engineering (TE) (also known as closed-loop-based approach). The P2P and pattern-based approaches target the management layer composition challenges, whereas the PnP and TE approaches target the self-management challenges. In the next sections, our approaches towards AN management challenges are presented.

12.3 Ambient Networks Management Approaches

Ambient Networks are under development and they are based on novel networking concepts and systems that will enable a wide range of user and business communication scenarios beyond today's fixed third generation mobile and IP standards. Central to this project is the concept of Ambient Control Space and the domain manager control function, which manages the underlying data transfer capabilities and presents a set of interfaces towards the supported services and applications. This section describes the different management research challenges and four complementary solution approaches (i.e. pattern-based management, peer-to-peer management, (un)PnP management and traffic engineering management application approaches) that enable efficient management of Ambient Networks, and the relationships between them, and presents the main results achieved so far.

12.3.1 Peer-to-Peer Management for Dynamic Network Composition

Peer-to-peer-based management is aiming at offering a scalable and flexible answer to the management challenges posed by the worldwide dynamic and heterogeneous Ambient Networks. The main feature of peer-to-peer systems is the immediate interaction among equal partners that are called 'peers'. This section focuses on providing dynamically established cooperation and interworking using peer-to-peer negotiations and still benefiting of the peer-to-peer systems' self-organization characteristics.

From a network management point of view, the most important potential application of P2P concepts is configuration management, more specifically, the provision of distributed network self-organization mechanisms based on peer-to-peer negotiations. The peer-to-peer self-organization framework presented in the next section organizes nodes and Ambient Networks into a hierarchical structure described by an overlay network model. The proposed network self-organization model addresses network composition, one of the most important features of Ambient Networks.

This section, however, presents an integrated solution composed of two interworking approaches for management layer network and service composition that were developed in the AN project. The *peer-to-peer Ambient Management Overlay (p-AMO)* function structures AN network resources and their topology according to dynamic composition rules and policies (i.e. topological resource composition). The *ambient virtual pipe (AVP)* is responsible for the establishment of an autonomic, secure and self-managed service management overlay network to support (new) AN management services or external applications/services. The AVP is a network service facing the resources (i.e. service resource composition).

The p-AMO is responsible for the topological management layer network composition. An AN is formed by one or several AN nodes (which are known as peers). This system organizes different peers into the AN at different hierarchical levels. It also provides a basis for every higher-level management task through the provisioning of messaging and peer discovering functionalities. The AVP is created dynamically between AN management entities that are organized in a P2P hierarchy by the p-AMO function. The p-AMO function organizes the underlying peers, and the AVP provides a secure and QoS-assured overlay network between management entities in/across the composed ANs. This overlay network is a flexible, self-adaptive and context-aware management overlay network in/across ANs for service (de)composition management.

12.3.1.1 A Framework for P2P Network Self-Organization in Ambient Networks

The peer-to-peer network self-organization framework organizes nodes and Ambient Networks into a dynamic hierarchical structure [184]. This hierarchical structure is created and maintained by means of peer-to-peer negotiation processes among nodes and Ambient Networks based on physical network topology, policy and context information. The hierarchical structure itself is specified by a hierarchical overlay network model, whereas changes and evolution of this structure are given by a network composition model. The basic components

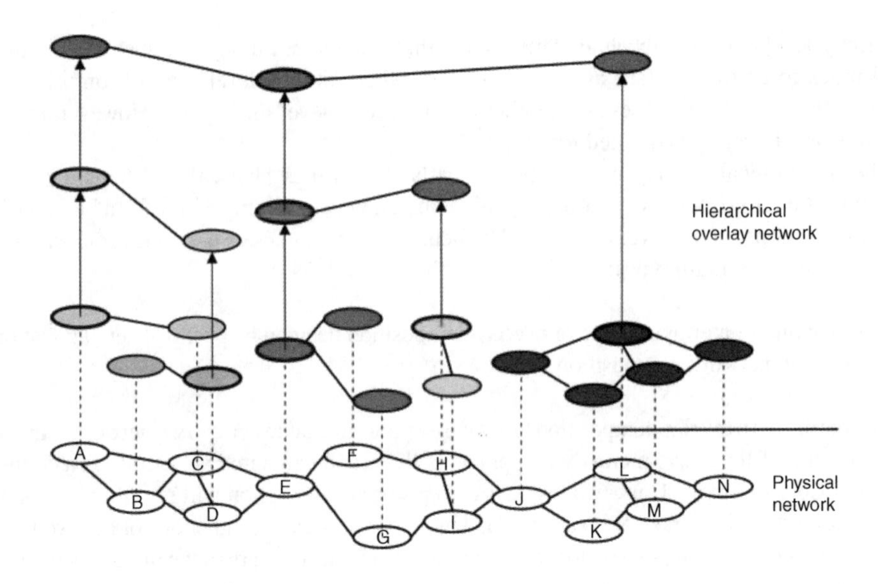

Figure 12.2 Hierarchical overlay network model

of the hierarchical overlay model are overlay nodes called peers, super-peers and overlays. On the one hand, an overlay is a set of peers belonging to one Ambient Network. On the other hand, overlays extend the AN with virtual connections – network – between its constituent peers. The AN itself is defined by the common Ambient Control Space shared by its peers.

Each overlay elects a super-peer to represent the overlay towards the outside world. It is important to note that this super-peer is solely responsible for negotiations with other overlays and has no other special privileges within its overlay. Super-peers may also form overlays at higher hierarchy levels, thereby creating a hierarchical overlay network structure. Figure 12.2 shows an example network topology and its associated hierarchical overlay structure atop this physical network. Peers are marked by coloured ellipses and super-peers are drawn by thicker lines. All peers belonging to the same overlay are filled with the colour of their super-peers.

As shown in Figure 12.2, the hierarchical overlay structure can be described by a complex graph model. Vertices of this graph are overlay nodes and neighbouring overlay nodes of the same overlay are connected by a nondirected edge if there is a neighbourhood relationship between them. Besides these neighbourhood links, there are also directed edges pointing from super-peer nodes to the parent overlay node at the next upper level.

Neighbourhood relationship between overlay nodes is determined by hierarchical overlay structure and physical connectivity. Two nodes of the same overlay are neighbours if it is possible to select at least two leaf nodes from the overlay subtree of the two nodes so that they are physical neighbours. In the example network in Figure 12.2, overlay nodes A and E are neighbours at the topmost level, because nodes C and D from the subtree of A (A, B, C, D) are physical neighbours with node E from the subtree of E (E, F, G, H, I). In contrast, nodes A and K at the topmost level are not neighbours because there is not physical connectivity between nodes of subtree of A and K.

Note that neighbourhood relationship is defined only between nodes of the same overlay at the same hierarchy level. Another characteristic of the presented hierarchical overlays is that hierarchy levels are not absolute. This means that one cannot assign an absolute hierarchy level index to an overlay (see again Figure 12.2, where the top-level overlay comprises two peers as third-level super-peer and another one as second-level super-peer). However, the bottommost level overlay is defined for all peers.

The hierarchical overlay graph unequivocally determines physical and logical network structure; therefore, network self-organization and all network compositions can be described as manipulations on this overlay graph. The behaviour and logics of network composition are defined using two main types:

1. Absorption or gatewaying type of overlay composition decided by peer-to-peer negotiations.
2. Bottom-up network composition.

Type 1 defines two overlay composition types: absorption and gatewaying. Two networks compose by absorption if they have mutually acceptable policies and can agree on setting up a common Ambient Control Space. Thus, two overlays composing by absorption will result into one single overlay represented by one single super-peer. This super-peer can be either one of the two former super-peers, but it is also possible to elect a new super-peer from peers of the unified network.

If the two networks cannot create a common Ambient Control Space – due to address conflicts, other control space problems or incompatible policy sets, for example – they will compose by gatewaying. During gatewaying type of composition, the two overlays will keep their own separate ACS, but an upper level overlay will be created whose members will be the two super-peers. The ACS associated with this upper level overlay is responsible for providing and regulating interworking between the two overlays.

The number of hierarchy levels in the overlay structure may increase as a result of gatewaying type of overlay compositions.

Composition between networks having multiple overlay levels is the second type of composition. When two previously separate networks get in contact, bottommost level overlays will detect each other by neighbour discovery procedures. After the recognition that they belong to different top-level overlays, the bottommost level overlays will try to compose. If they can agree on absorption type composition, they complete the absorption procedure. Otherwise, the two super-peers forward composition to the next upper overlay until either the two parties can agree on absorption type composition or the top-level overlay is reached. In the former case, the top-level overlay of one network will merge into an overlay of the other network at some level. In the latter case, the two networks will compose by gatewaying and the two top-level super-peers will create an additional overlay level.

Overlay type is decided as a result of peer-to-peer negotiations between the super-peers of the two composing overlays. It is important to note that direct network-level communication may not be possible between the two super-peers during composition; thus, if they are not direct physical neighbours, all composition messages will be relayed by overlay boundary nodes of bottommost neighbouring overlays.

12.3.1.2 Ambient Virtual Pipes

One of the AN management issues addressed is the definition and setting up of a self-adaptive and context-aware service management overlay network, i.e. the AVP [186]. As shown in Figure 12.3, the AVP forms a management overlay network within/across (de)composed

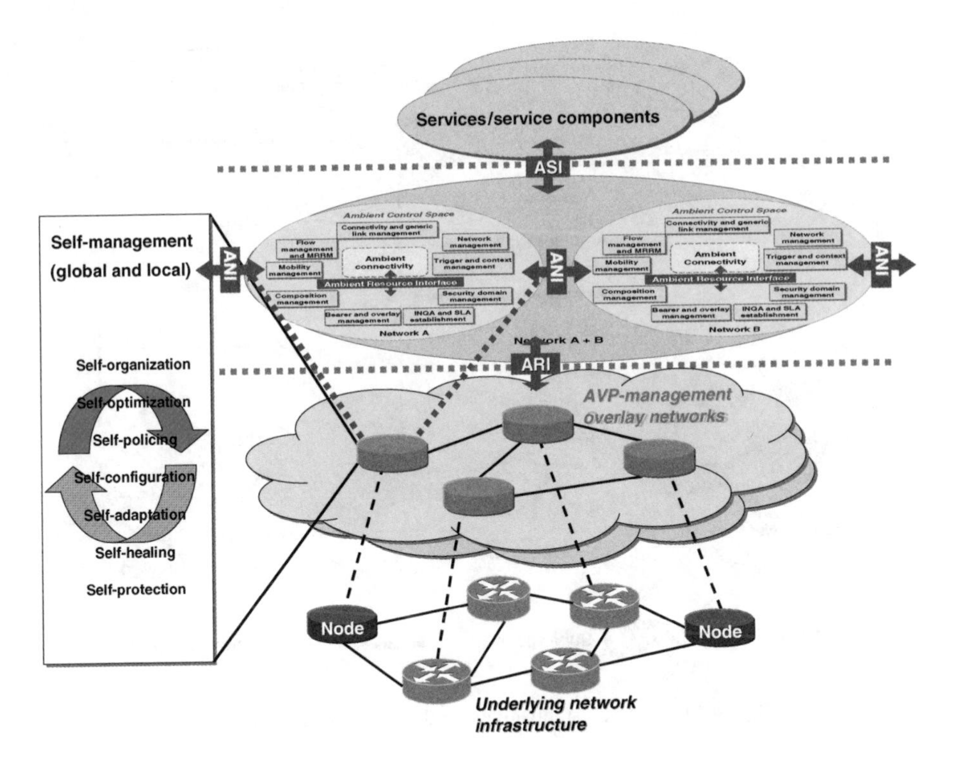

Figure 12.3 AVP as the management overlay in AN

ANs for supporting service (de)composition. In order to enhance scalability and efficiency, the AVP supports self-management, which is achieved through a programmable platform [188].

It is envisaged that a topological composition (conducted by the p-AMO function) would trigger a service (de)composition. Service (de)composition would in some cases trigger topological reconstructions/reconfiguration of network resources. The AVP is autonomically established and self-adapted such that new management services or external services/applications may be executed on demand in (de)composed ANs and is capable of dynamically reconfiguring underlying network elements on the fly. The AVP is context aware that the autonomic establishment and self-adaptation of AVP is triggered by the changes of the underlying network context. The AVP makes use of network context information retrieved from distributed AN context monitoring systems (CMSs) and context information bases (CIBs) for network information sharing within/across heterogeneous ANs.

A highly dynamic and flexible information infrastructure is needed in order to support (new) AN management services or external services/applications deployment and self-management over a composed AN management domain. This information infrastructure must be capable of providing secure and reliable connectivity across heterogeneous networks with guaranteed quality on demand. It may be arguable that conventional solutions like the currently available VPNs may be used to provide QoS guarantees to networks. However, the major drawback of the today's VPNs is their low flexibility to quickly adapt to the changing requirements. The AVP provides a programmable [179,188], flexible and network

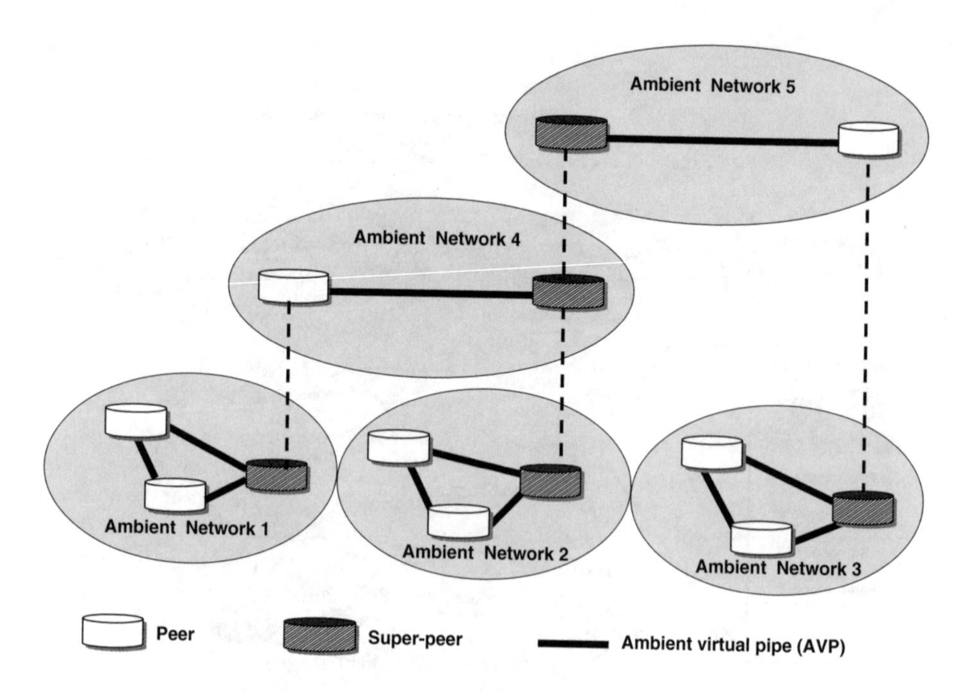

Figure 12.4 AVP establishment and AN network composition

context-aware information infrastructure with guaranteed QoS (for prioritizing management traffic) in the (de)composed AN management domain.

The establishment of AVPs is shown in Figure 12.4. It should be noted that the AVP resides on the management plane (rather than the network plane) for service (de)composition. AVP is formed among the AN management entities, i.e. between the peers and the super-peers, and it provides a secure and QoS-assured channel for P2P management information exchange and an environment in which (new) AN management services or external services/applications may be launched and executed. The secure and QoS-assured environment provided by AVP for service (de)composition is required to assure that the service (de)composition process is not disturbed. More specifically, distributed management information must be shared securely and in a QoS-assured fashion to enable service (de)composition to take place. Note that AVPs are not restricted to provide a secure and QoS-assured management service overlay network for (new) AN management services. The capabilities of AVP make it potentially ideal for the dynamic deployment of external (user-specific) services across heterogeneous ANs.

The dynamic creation of AVP is achieved by dynamically injecting active code to desired peers in order to instantiate AVPs. The active code carries executable programs, which result in security association establishment between peers. QoS in AVP is assured through the injection of active code to dynamically prioritizing AVP traffic flow.

12.3.1.3 Evaluation of Peer-to-Peer Management

During the research and design of the P2P management for ANs, the group recognized the possible validity of a comprehensive AN-wide P2P solution to the Ambient Control Space.

Therefore, this group developed a generic peer-to-peer-based Ambient Control Space Prototype (PAP) in which AN management functions were augmented. The p-AMO function structured the PAP into the previously defined hierarchical overlay structure for scalable and affordable autonomic management, whereas the AVP provided a secure, context-aware and self-adapted environment for protecting management traffic within ANs.

The experiences with the PAP showed that a scalable and robust overlay hierarchy of the ACS is feasible to manage and control the ANs. In order to manage the overlay structure in an efficient way, we made a decision on controlling the succession of the composition events. The main idea was the introduction of the so-called 'bottom-up principle' to execute the composition process as close to the physical network as possible. Further, for the sake of simplicity, the tree structure of the overlay hierarchy was strictly enforced by not allowing concurrent composition process, a limitation for simple implementation only.

It is widely accepted that a hierarchical architecture boosts the scalability of a system. The feasibility of using a hierarchical solution at all has been proved by the demonstrations with the PAP. In the case of hierarchies, the trade-off is between the gain obtained from the abstraction introduced by a hierarchical level and the cost of control traffic required to maintain the hierarchical structure. Detailed analysis of whether there exists any upper bound on the number of overlay levels above which the performance of the architecture decays is still required.

Scalability-wise, we must admit that one of the main reasons of introducing the hierarchical overlays was to confer scalability to the PAP. It is widely accepted that a hierarchical architecture boosts the scalability of a system. The feasibility of using a hierarchical solution at all has been proved by the demonstrations with the PAP. In the case of hierarchies, the trade-off is between the gain obtained from the abstraction introduced by a hierarchical level and the cost of control traffic required to maintain the hierarchical structure. Detailed analysis of whether there exists any upper bound on the number of overlay levels above which the performance of the architecture decays is still ongoing work.

Stability-wise, the hierarchical overlay network structure is maintained through the peer–super-peer relations. The function of super-peers is limited to inter-AN negotiations as representatives of their AN, and they are not dedicated nodes for intra-AN management of their Ambient Network. Using AN terms, it means that although functional entities of an Ambient Network might be distributed, the ANI of the AN is logically bound to the super-peer. Nevertheless, super-peers play central roles to maintain the hierarchical overlay structure; hence, their failure, which could be considered common, for example, in ad hoc environment, must be fast recovered. In order to keep the super-peer up to date and to re-elect a super-peer on failure, we can utilize the achievement of the pattern-based management (see Section 12.3.4) by using robust and scalable patterns for network queries or super-peer election.

Management information between distributed management entities in composing/composed ANs must be protected and QoS ensured to ease management layer composition in ANs. The AVP is a secure, context-aware and self-adapted management service overlay network, which provides a secure and QoS-assured environment for management information distribution and management service deployment and dynamic execution of (new) management services/external services/applications within/across heterogeneous ANs. Dynamic deployment and execution of new services is supported through a flexible and programmable platform.

The AVP consists of several components, each with a different task. The context manager (CM)/context monitoring system (CMS) is responsible for context source and client registration and management of network context pointers. The AVP manager retrieves network context information within the AN using network context pointers provided by the CMS and information from the PAP platform and build AVPs across ANs. The AVP managers are distributed and host key management tools for establishing and maintaining security associations between peers. AVPs are self-adapted, in that the current implementation the AVP bandwidth manager is implemented to provide real-time network traffic monitoring for adjusting QoS on specific paths. The AVP bandwidth manager consists of the QoS context sensor that provides information on the QoS state of ANs. The resource mini-broker provides access to low-level node resources.

Throughout the demonstrations we have evaluated the functionalities of the AVP. The AVP is capable of self-adapting in response to changes in network context. The design primitives of the AVP are justified through the implementation of the AVP functional entities. For instance, through connection with the ContextWare and the PAP platform, the AVP is context aware. Through the AVP's programmable platform, the AVP is capable of providing support for the deployment and execution of (new) management services/external services/ applications. AVP also provides provisioning to support secure distribution of management information as well as the support for QoS provisioning in order to avoid management traffic being disturbed.

12.3.2 Patterns for Dynamic Network Management

The pattern-based management paradigm is a distributed management method based on the use of graph traversal algorithms to control and coordinate the processing and aggregation of management information inside the network. From the perspective of a network manager, the algorithm provides the means to 'diffuse' or spread the computational process over a large set of nodes.

The paradigm achieves this through the development of two important concepts: the navigation pattern and the aggregator. The former represents the generic graph traversal algorithms that implement distributed control, whereas the latter implements the computations required to realize the task. A navigation pattern controls the flow of execution of a (distributed) management operation. It is described by an asynchronous network algorithm, which can be analysed for its complexity and scalability properties.

Pattern-based management aims at overcoming the limitations of centralized management: its poor scalability regarding (a) management traffic, (b) processing load on the management station and (c) execution times. Its goal is to build scalable, robust and adaptable management systems.

The main benefits of pattern-based management are that (i) it separates the semantics of the task from its flow control, (ii) it enables building scalable management systems, (iii) it facilitates management in dynamic environments and (iv) it does not require *a priori* knowledge of the network topology, in contrast to centralized approaches.

Figure 12.5 presents the simplest examples of navigation patterns.

Pattern-based applications map network-wide operations (such as identifying the top flows regarding traffic) into local operations that will be performed by the managed nodes. Then, they are distributed using navigation patterns, creating an execution graph. For monitoring tasks, local operations include the collection of statistics and the incremental aggregation of the collected data. This aggregation is done in parallel across the network asynchronously: all the nodes contribute in the calculation.

Operation	Typical application	Navigation pattern
Type 1: node-to-node	One node control/monitor (get/set of variables)	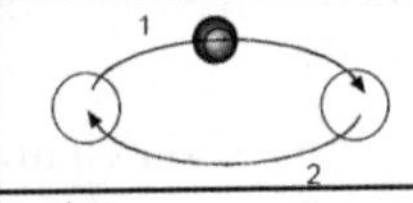
Type 2: visit all nodes along a path/flow	One flow/path control, e.g. traceroute, bottleneck detection, signalling, VPN operation	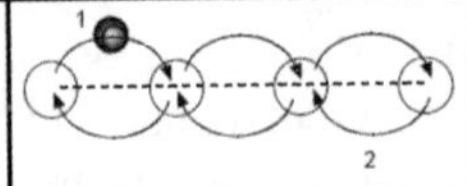
Type 3: distribute agent to all nodes in subnet (parallel control)	Subnet control, message broadcast, e.g. congestion location detection	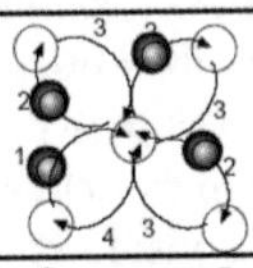
Type 4: visit all nodes in subnet (sequential control)	Subnet control, e.g. topology detection	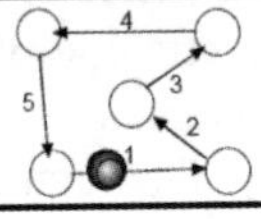

Figure 12.5 Examples of simple navigation

12.3.2.1 A-GAP: Distributed Real-Time Monitoring with Accuracy Objectives

The ability to provide continuous estimates of management variables is vital for management tasks, such as network supervision, quality assurance and proactive fault management. Generally, management variables, which are monitored in these tasks, are aggregates (obtained with using functions such as SUM, AVERAGE, MIN, MAX, etc.), which are computed from device variables across the network. Examples include the total number of VoIP flows in a network domain and the maximum link utilization.

For many of these tasks, such as quality assurance through monitoring SLAs, knowing how accurate such estimates are is crucial. However, network management solutions deployed today usually provide qualitative control of the accuracy, but do not support the setting of an accuracy objective.

Engineering continuous monitoring solutions for network management involves addressing the fundamental trade-off between accurate estimation of a management variable and the management overhead in terms of management traffic and processing load. Obviously, high accuracy comes at the cost of a high overhead, and similarly, low-accuracy estimations can be achieved with a low overhead.

In this work, we address the problem of continuous monitoring with accuracy objectives in large-scale network environments. Specifically, we want to achieve an efficient solution that allows us to control the accuracy of the estimation.

We present A-GAP, a generic aggregation protocol with controllable accuracy. A-GAP is based on GAP (Generic Aggregation Protocol), which allows for continuously computing aggregates of local variables using a self-stabilizing spanning tree and incremental aggregation [187]. A-GAP is push based, which means that changes in monitored variables are sent towards the management station. It controls the management overhead by configuring filters in the management nodes of the aggregation tree. The filters periodically adapt to the dynamics of the monitored variables and the accuracy objective, which can be set by the management station. All operations in A-GAP, including computing the aggregation function and filter widths, are executed in a decentralized and asynchronous fashion to increase robustness and achieve scalability.

12.3.2.2 Set-Up and Maintenance of Overlays for Multimedia Services

Creation and maintenance of overlay networks is relatively straightforward in fixed networks. However, in mobile and wireless networks, topologies and network characteristics dynamically change, causing maintenance to be much more complex. Additionally, the concept of overlay networks is a means of implementing service-specific routing, caching and adaptation functionality. So, the whole overlay becomes service specific and the topology of the overlay network depends on the service. Here, the usage of patterns for the specific network must be considered. Patterns are a very general communication method and can be applied to the creation of different types of overlays, and therefore allow for flexible creation and maintenance in different environments and with different requirements for a particular overlay. The requirement for the overlay mainly comes from the type of service installed.

In the following, we assume a specific type of overlay for the transport of multimedia data [180]. Three important factors when transporting multimedia in wireless and mobile environments are (a) heterogeneity, (b) network changes due to radio strength variations and (c) topological changes resulting from node movement. In such an environment, multimedia processing techniques in the network such as adaptation, caching and transcoding are used to provide the best possible service to a customer, taking into account the device capability as well as the network capabilities. As we assume that multimedia processing engines are relatively specialized network elements (potentially running on dedicated hardware), the multimedia flows must be forced to pass through those nodes. The nodes with such capabilities must potentially be visited in a certain sequence in order to offer a certain end-to-end service. So, only a relatively small set of nodes in the network is capable of running these expensive multimedia processing functions. We call these the potential overlay nodes.

The detection of potential overlay nodes is of primary focus here. Suitable nodes are selected by an optimization algorithm that aims to satisfy specific requirements such as using the cheapest overlay network or the overlay with the least number of hops, etc.

Concerning the detection of potential overlay nodes, we assume that each node stores the required parameters in a standard format. For instance, a node stores the set of functions it can perform, for how many overlay networks this node can perform each function and the cost of using that node for each function (these are the first assumed parameters of importance, others might be of relevance as well depending on the service and future capabilities). Detecting potential overlay nodes involves probing each node for the required resource and functionality for the multimedia service. This means only nodes capable of hosting a virtual node with a certain function are found in that process. Various ways of performing this function exist.

We show that the pattern-based paradigm is a nice tool for discovering network-side functions/resources. A pattern will determine which nodes should be probed. The pattern will also initiate the probing process as well as the gathering of information. Different patterns can be used for this purpose. However, the most suitable pattern for this application is the path-directed search pattern. The basic idea of this pattern is to limit the scope of the search to a configurable area along the end-to-end path between the communicating peers. The search pattern uses a parameter that defines the 'distance' (e.g. in number of hops, delay or any other measure, etc.) from the routing path that should be searched. This distance is also referred to as 'sideway expansion'. Depending on the type of resource or function that is searched, this parameter can be changed.

Our pattern assumes to know the source of a multimedia service and a number of destinations or regions where one or more receivers of the multimedia services are located. The path-directed search pattern starts from the source node and expands along the end-to-end routing path towards the destination nodes with a sideway expansion of a given distance (e.g. based on the number of hops, delay, etc.). After visiting the nodes defined by the pattern scope, the pattern contracts towards the source node gathering the requested information (depending on the resources/service we are looking for). The sideway expansion parameter of the pattern controls the scope of the search and thus limits the number of nodes probed during the detection. Above all, it allows the discovery of network-side resources along a close approximation of the routing path [183].

12.3.2.3 Evaluation and Robustness of Patterns

As has been shown previously, pattern-based network management introduces a promising concept for decentralized network management by separating the semantics of network management functions from the distribution and aggregation of this information. However, almost all work on patterns so far has assumed fixed, wired networks, with no or low failure rates of links and nodes, and where a node or link failure can be detected on the physical layer.

Robust patterns are required for the dynamic changing environment as envisioned in Ambient Networks; therefore, some extensions for dealing with the dynamics for above-mentioned patters have been developed. Initial simulation results verify that the proposed enhancements to the protocol make it suitable for a wireless environment.

In wireless environment, pattern message being sent out over a wireless network must be acknowledged by those receiving nodes that will reply with an answer to the request. This is because in a wireless environment the sender does not necessarily know who hears the message and will return an answer. On the contrary, a child node that already has received a copy of the pattern message from another parent does not have to reply to this copy of the pattern. Additionally, a child node will retain a copy of all potential parents it receives a pattern message from for a predefined time interval in order for the node to send the pattern reply message over an alternative link if the primary link no longer is valid, as described in the robustness section. Finally, a 'keep-alive' message sent and acknowledged at regular intervals between two nodes should be used to allow the nodes to know whether the link is still valid. If the link fails while a pattern is being executed, both nodes will recognize this event and deal with the situation, for example as the robustness area proposes.

In mobile environment, if a node or group of nodes move while a pattern is being executed, a node will drop its connection to another node and set up a link to another, new node; these patterns will be able to deal with this by adding extra functionality to the patterns, allowing them to return information multiple times to their parents in combination with return information being returned to other parents than those sending out the expansion message.

Concerning robustness, if a child node loses its connection to its pattern while a pattern is being executed, the parent must know of this in order to be able to report the information upstream that the parent receives from its other children instead of going into a deadlock waiting indefinitely for an answer from all its children. This is achieved by using the keep-alive messages described above. Additionally, a child will try to contact its alternative parents from the list of parents it has kept in order to forward the pattern reply message. If this fails, the node will ask its children if they have an alternative valid connection to another parent that could return the information. If a child has a valid alternative return path, the pattern reply message is forwarded in this way.

These ideas are not dependent on a specific pattern. Instead, these ideas are used by any pattern being developed for use within a wireless mobile environment.

12.3.3 (UN)Plug-and-Play for Ambient Network Components

Traditional management systems allow end hosts to configure themselves, e.g. through protocols like DCHP(v6) or IPv6 autoconfiguration. Still, the respective server parts have to be configured manually by an administrator, e.g. the DHCP server. Additionally, routers still have to be configured manually to integrate into a network domain.

AN PnP management is responsible for automatic and full configuration of AN elements. It enables new AN elements to become an integral part of AN domains. This includes not only the automatic configuration of end host interfaces, but also complete configurations of router elements when they are attached to AN domains and initialization/maintenance of a hierarchical peer structure for management purposes.

AN elements may arbitrarily attach to or detach from AN domains, thus incurring dynamics in network topology. Therefore, PnP management does not only have to configure newly attaching AN elements. In addition, constant or periodic verification and optimization of current configurations has to be performed.

PnP management allows automatic configuration of AN elements whenever they enter or leave one or multiple networks and thus addresses the challenge of dynamic self-configuring networks. Each AN element is responsible to configure IP connectivity on its own and to integrate into existing ANs. Consequently, management tasks are performed locally on each element instead of totally relying on central management stations like in traditional management systems. This reduces communication overhead and also reduces the risk of overloads or single point of failures in a central device.

We have identified basic set of components to configure AN elements in a PnP manner (compare Figure 12.6) that are presented in the following subsections.

12.3.3.1 Base Station Configuration

The base station configuration component provides a mechanism that allows base stations in wireless networks to configure themselves in an automated manner such that no or least administrative work has to be performed manually. This PnP principle includes not only the bootstrap process of the base station, but also its capability to react on system changes and events.

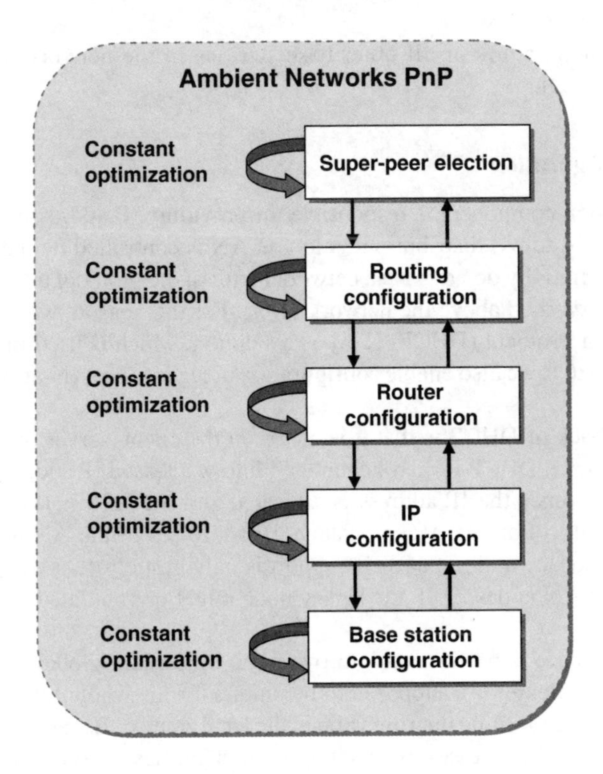

Figure 12.6 Overview of AN PnP

A base station is assumed to have at least two network interfaces. The first interface is a wireless one to provide service to client nodes, whereas the second one can be either wired or wireless to provide uplink connectivity.

The base station configuration uses a fully distributed approach. Therefore, the management capabilities have to be installed on all base stations within the network. Each base station retrieves configuration information from all its neighbouring stations. Based on collected information, a base station is then able to select an appropriate configuration. Notice that base stations are only provided with information, but the configuration is done completely autonomously, i.e. no base station can force another base station to use or not to use a specific configuration. Information exchange between base stations is done via the uplink interface, as other interfaces are usually configured to provide the base stations service.

The information that is maintained by each base station falls into three categories: private, local and global information. The system uses different information dissemination techniques for each kind of information. Based on these three types of information as well as locally monitored information, each base station constructs a local view of the overall network state and derives a consistent configuration from it. When the configuration changes – either because of a locally monitored change in the environment or because of reception of new information from a peer base station – a base station adapts its view of the network and may modify its configuration accordingly. In such an event, it may notify its neighbours or all other base stations in the network, depending on the specific changes [182].

12.3.3.2 IP Configuration

The IP configuration component is responsible for providing IP addresses to a node's active network interfaces. Address distribution within an AN is controlled by the AN's current super-peer. As nodes initially do not have a network address, the protocol used cannot work over the ACS (which is defined above the network layer). For this reason, we chose the Dynamic Host Configuration Protocol (DHCP) [178] as a solution, which is used in most today's networks. By using DHCP, we also enable configuration of legacy (not AN-aware) nodes through the ACS.

The only drawback of DHCP is that it is based on the client–server paradigm and uses a leasing mechanism, i.e. DHCP for IPv4 cannot withdraw a leased IP address of a node before it expires. However, once the IP address is allocated and the ACS is running, it is possible to use a special protocol on the ACS to change IP addresses of the AN nodes. This feature cannot be used by legacy nodes, but their support is only best effort as the primary aim is to seamlessly support AN nodes. Still, the legacy node might use outdated IP addresses, which need to be taken care of.

The IP address space is partitioned into two parts by default: global addresses and local addresses. Global addresses are supposed to be unique for the whole Internet, whereas local addresses have local scope only; they are valid in the local context. Because of the unlikelihood of having unique global addresses available in such quantities as to distribute them to each AN node, in the Ambient Networks local IP addresses are used, and the AN nodes capable of acquiring global addresses use NAT to provide external access. Thus, from this point the AN address space means local IP addresses, unless otherwise stated.

We have to differentiate three use cases for AN addresses:

- AN-wide permanent unique address for nodes to communicate with other AN nodes.
- Temporary local AN unique address for bootstrapping nodes in order to compose with an AN.
- Local AN unique address for supporting legacy (not AN-capable) nodes.

We have to differentiate between these address spaces because they have different use cases: an AN node with a permanent address can be requested via the ACS to change its IP address at any time during its action. In contrast, legacy nodes have to be leased addresses with legacy client–server-based techniques; thus, distributed addresses cannot be reclaimed from them. For the second and third use cases, we can use the same address space, so from here on the two address spaces will be referred to as permanent and temporary.

In the current Ambient Networks approach, composition takes place by protocols running above the IP layer. Thus, a network node has to possess an IP address in order to communicate its composition with other AN nodes.

The first process for a bootstrapping node is therefore acquiring an IP address that can be used for the time of the composition process. Because at this point of the life of a bootstrapping node the ACS might not be fully functional, a legacy protocol, DHCP, has been selected for the configuration of this IP address.

12.3.3.3 Router Configuration

The router configuration component is responsible for automatically managing address spaces of router elements. Each router may be assigned a certain address space for its network interfaces that it can then provide to its clients. Specifically, base stations typically give out IP addresses to their wireless clients (this holds at least for wireless LAN access points of a certain type of base station).

Router configuration uses a management hierarchy, where higher-level routers delegate management tasks to their subordinates on a lower level. Following this approach, each subordinate manages a part of the network autonomously and only contacts higher-level routers if it is not able to resolve conflicts or other problems itself.

The general idea is that the management hierarchy represents a spanning tree over all routers within the managed domain. This hierarchy tree is used to delegate management tasks from higher-level routers in the hierarchy to the lower-level ones. Each router is managed by its parent router within the hierarchy. Routers provide all their children with sufficient information to manage a subordinated part of the network themselves.

Subordinated routers try to manage their children with minimal interactions with their parents. Only when it is unable to manage the subordinated part itself, does it contact a parent for assistance. An example could be that the assigned address space gets exhausted due to a large number of children. In this case, a subordinated router might request a new or additional address space from its parent.

12.3.3.4 Routing Configuration

The routing configuration component is responsible to configure proper routing protocols when a new router node is added to an AN or when the environment requires reconfiguration of the current protocol.

The routing configuration consists of two parts: the PnP server and the PnP client. The basic idea is that a router – after its PnP process has finished – can act as PnP server for other routers. Each router – if turned on – starts a PnP client to request configuration data from already existing, neighbour routers within the AN. The architecture allows communication between the PnP servers for updates and distributing information related to routing configuration changes, etc.

Some routing protocols – especially the OSPF, which is the first target for this PnP solution – have multilevel hierarchy and other domain like features. For OSPF, it is an important configuration question that how the OSPF areas will be set. It has a two-level hierarchy – backbone area connects non-backbone areas – and non-backbone areas have many types (normal, stub, NSSA). When adding a new router into an existing AN, the area level set-up is not a big issue, the new router can join the area where the serving router belongs.

In the case of interconnecting to ANs, both with existing OSPF configuration, the configuration is nontrivial. Area configuration must be somehow decided and synchronization of the two OSPF set-ups must be done. This process must be governed by the AN network where the new network would like to join. For this process, information about current area structure and area level configuration should be known by the PnP system. Setting up new area configuration is based on this knowledge, on the new network structure and other policies, e.g. how many routers can be in one area. The routing configuration therefore implements a dedicated PnP server where this information is stored and the proper area configuration is calculated for the process. For robustness, a backup dedicated PnP server that replicates the dedicated server is also a part of the architecture.

For providing a flexible, fast and reliable means for AN routers to configure their routing protocol driver(s) automatically during an AN PnP process, the ANRC (Ambient Network routing configuration) framework has been designed.

The ANRC provides architecture and protocol(s) for AN routers. The architecture assures the automatic and reliable execution of routing protocol configuration in a composition scenario, independently of the number of participating routers. The architecture is also capable to handle many types of routing protocols. It was designed to be general and not tied to any routing protocol. It means that the new routers inserted into an AN domain will use the routing protocol used in the accepting AN domain and the ANRC mechanism will configure this routing protocol for the new routers.

The main software parts of the ANRC framework are the server, the client and the protocols between them. Both the server and the client should be implemented into an AN router. Both of them are defined by state machines, so their implementation is easy in any kind of programming language and on any kind of router platforms.

The whole mechanism is governed by a well-designed process that handles the possible situations during the ANRC process. During the process, two types of protocol are used: the client–server and the server–server protocol. The client–server protocol is used by the new router to initiate the process and obtain necessary information. The server–server protocol is used by the routers that are configured and runs ANRC server already. When router accepting requests from an ANRC client needs more information than it has already, then this protocol is used to obtain the required information from the domain-level ANRC server.

These protocols are also made to be general and flexible. The defined protocol itself is a framework protocol, providing messages and mechanisms to handle the communication between ANRC components, but does not define the concrete message formats for a concrete routing protocol. Because different routing protocols need different information for their configuration, this information is carried by the ANRC protocol in a flexible way. The ANRC framework gives a method how to define and understand a request/answer message for a concrete routing protocol.

12.3.3.5 Super-Peer Election

As already introduced in Section 12.3, each AN is represented by an elected super-peer for management purposes. The super-peer election component takes care of electing the most suitable candidate among all peers of an AN. When electing a super-peer, we always have to take into account the properties and attributes of the nodes participating in the AN. Such properties can be, for example, battery life, memory capacity and link capacity. These properties represent each node in an n-dimensional space; thus, it will be difficult to compare two nodes' goodness. For this, we created a function that maps these properties into a single one-dimensional goodness metric that enables us to compare our nodes. The actual election process has to be executed in certain situations, which also influence the mapping function. These are shown in the following subsections.

We can assume that the actual super-peer is always the one that is the most eligible from the nodes in the AN. When a new node joins the AN, it first has to negotiate with the existing super-peer. This negotiation is also used to check the newly arriving nodes *goodness* measures, and if it is a better one, it can take over the role of the super-peer. The other nodes in the AN are notified about the change by the old super-peer with an intelligent broadcast message.

It is possible that the properties of the peers change over time. Thus, periodic regathering of the measures and the topology is necessary to ensure that the super-peer is always the most eligible node. This aggregation can be carried out by the super-peer using pattern-based aggregation. After it has the complete information and finds a peer more fitting for the role, it initiates a role change and notifies the rest of the AN's nodes about it.

The loss of the super-peer is detected when one of the nodes tries to contact it, but the request times out. In this case, the node detecting the loss initiates a pattern-based aggregation as in the previous case. The nodes receiving such a message will recognize that it does not originate from the super-peer, so their attempt to also initiate an aggregation is cancelled. If several aggregations are started in parallel, the one with the lowest peer ID gets precedence, whereas the other ones are cancelled upon receipt of the other peer's aggregation request.

An AN can get partitioned into separate subnetworks, due to the loss of a gateway-like node, e.g. when two parts of the network depart too far away from each other. In this case, the partition containing the super-peer 'just' discovers the loss of some of the member peers. In the other subnetworks, however, the loss of the super-peer will be detected and the same procedure should be used for electing a new super-peer.

If the networks compose with absorption, the two current super-peers have to check their properties against each other's, and the better one will remain a super-peer in the composed network. Of course, this node might not be optimal because of topology reasons, so an aggregation should be initiated to see if there is an even better node in the now composed network.

12.3.4 Closed-Loop Traffic Control

The closed-loop traffic control approach explores the use of traffic engineering mechanisms as a way to optimize the network resource usage. The objective of traffic engineering is to avoid congestion in the network and to make better use of available network resources by adapting the routing to the current traffic situation. More efficient operation of a network means more traffic can be handled with the same resources, so enabling a more affordable service. For a network operator, it is important to tune the network in order to accommodate more traffic and meet service level agreements (SLAs) made with their customers. In addition, as new bandwidth demanding and also delay and loss sensitive services are introduced, it will be even more important for the operator to manage the traffic situation in the network.

In Ambient Networks, we can expect both conditions similar to current IP backbone networking and conditions where the topology changes are similar to ad hoc networks and traffic demands shift due to the mobility of networks and network compositions. Rapidly shifting traffic demands and more dynamic network topologies means that one cannot rely only on long-term network planning and dimensioning that are done when the network is first built. Traffic engineering mechanisms are needed to adapt to changes in topology and traffic demand and dynamically distribute traffic to benefit from available resources. It is from this perspective our research effort should be seen.

Traffic engineering is the process of performance evaluation and optimization of an operational network. The first step of the traffic engineering process is the collection of necessary information about network state, i.e. the current traffic situation and network topology. This information is then used for optimization of routing parameters and to adapt the packet forwarding based on the state of the network. Depending on the type of traffic engineering mechanism, different information is needed and different ways to manage this information are used. The amount of network state information used by the TE mechanism is just one of several design trade-offs. Others include centralized versus distributed solutions, online versus offline mechanisms and whether we try to optimize legacy routing protocols or develop novel routing mechanisms [181].

The traffic engineering work is here divided into three approaches. Two of them, the global and the local view, focus mainly on handling traffic dynamics, and the third approach focuses on handling dynamic topologies. In the global approach, it is generally assumed that the optimization is centralized and has a complete network-wide view of topology and an estimation of traffic demands. The focus here is on the robustness of weight-setting algorithms subject to changing traffic demands (or outdated measurements). With the local approach on the contrary, the optimization is decentralized. Multipath routing with dynamic variance (MRDV) together with the loop avoidance protocol (LAP) has been designed to make use of local information only. In the third approach, we study how policy constraints can be applied to address the dynamics of topology in Ambient Networks.

12.3.4.1 Traffic Engineering Using Global Information

An important goal of traffic engineering is to avoid congestion in the network and to make better use of available network resources by adapting the routing to the current traffic situation. Today, the main alternative for intradomain traffic engineering in IP networks is to use different methods for setting the weights (and so decide upon the shortest paths) in the routing protocols OSPF (Open Shortest Path First) and IS-IS (Intermediate System to Intermediate System). These are both link state protocols and the routing decisions are based on link costs and a shortest (least-cost) path calculation. With the equal-cost multipath (ECMP) extension to the routing protocols, the traffic can also be distributed over several paths that have the same cost. These routing protocols were designed to be simple and robust rather than to optimize the resource usage. They do not by themselves consider network utilization and do not always make good use of network resources. The traffic is routed on the shortest path through the network even if the shortest path is overloaded and there exist alternative paths. It is up to the operator to find a set of link costs (weights) that is best suited for the current traffic situation and avoids congestion in the network.

The first step in the traffic engineering process is to collect the necessary information about network topology and the current traffic situation. Global view traffic engineering methods generally need as input a traffic matrix describing the demand between each pair of nodes in the network. But today the support in routers for measuring the traffic matrix is limited. Instead, an often-suggested approach is to estimate the traffic matrix from link loads and routing information [175,176]. Link loads are readily obtained using the Simple Network Management Protocol (SNMP) and routing information is available from OSPF or IS-IS link state updates. The traffic matrix is then used as input to the routing optimization step, and the optimized parameters are finally used to update the current routing. In this study, this means that the traffic matrix is used together with heuristic search methods to find the best set of link weights.

From a network-wide perspective, the traffic engineering problem can best be modelled as a multicommodity flow optimization problem. This type of optimization technique takes as input global information about the network state (i.e. traffic matrix and link capacities) and is able to calculate the global optimal solution. Here, though, the main focus is on what can be achieved with the legacy routing protocols. The optimal solution is mainly used for comparison. When taking the restrictions of shortest paths or equal-cost multipaths in the OSPF and IS-IS protocols into consideration, the problem of finding the optimal routing becomes harder. The problem of finding weights that optimize the routing is NP hard [173,176]. This means that one usually has to rely on heuristic methods to find the set of weights.

An often-proposed method to determine the best set of link weights is to use local search heuristics [173,174,177]. Given network topology, link capacities and the demand matrix, the heuristics evaluate points in a search space, where a point is represented by a set of weights. A neighbour to a point is another set of weights produced by changing the value of one or more weights from the first point. In the heuristics, different neighbours are produced and the cost of each one is calculated using a cost function. From each heuristic, the neighbour with the best cost is the one that will be the output.

In this study, we have selected two heuristics, local search [173] and strictly descending search [177], and investigate how these perform when they are applied to estimated and shifting traffic demands.

12.3.4.2 Traffic Engineering using Local Information

In today's core IP networks, a wide range of traffic patterns have to be carried while complying with different performance requirements and highly dynamic environments due to the growth of the use of mobile technologies. Therefore, traffic engineering solutions should consider both the optimization of the network resources and the use of decentralized management approaches. In this scenario, new routing algorithms are required in order to achieve an efficient use of the network resources: traffic flows should be optimally directed to the available network resources in such a way that no link becomes overloaded and congestion is avoided.

MRDV was presented in as a technique to achieve these goals: a MRDV router splits the traffic by routing along secondary paths (that means, nonoptimal paths) in order to balance the load in the network according to a local criterion based on the link occupancy in its output interfaces. Therefore, MRDV does not require any information exchange about link load: each router running the MRDV algorithm weights the cost of each path towards the destination with a variance factor reflecting the load on the next hop.

In a dynamic environment such as Ambient Networks, with sudden changes in traffic demands, MRDV is an interesting approach in order to use only local information for traffic engineering. Instead of flooding the network with load information and waiting for a new routing to be calculated, a node can make a local decision and adapt to the situation. A node that experiences a sudden increase in traffic demand can directly shift load from heavily loaded links to under-utilized paths.

A possible drawback is that the local improvements might create loops or overload somewhere else in the network if care is not taken. So, a careful evaluation of this type of mechanism is needed. The loop avoidance protocol has been developed as a way to enhance the performance of routing schemes that are designed to balance the load in the network according to local information such as MRDV. In simulation studies, LAP has been shown to significantly improve the performance of MRDV [185].

12.3.4.3 Constraints on Ad Hoc Networks

In the global and local traffic engineering approaches presented above, it is assumed that the topology is much more stable than the traffic demands. In the third approach, the focus is on the dynamic topologies in ad hoc networks. We study how policy constraints can be applied to address the dynamics of topology in Ambient Networks.

Almost all research on ad hoc networks has assumed that there is one ad hoc network. If additional nodes or networks come within range of the existing ad hoc network and want to join this ad hoc network, they will fully join this network, i.e. there is only one network that has been extended. Furthermore, some research has been performed on how these ad hoc networks can set up connectivity with external networks and the Internet. However, these approaches are either based on Mobile IP with a home agent and a foreign agent with a modified protocol between the host and the foreign agent (to allow multi-hop routing between the foreign agent and the host) or using network address translators (NATs) in the gateways between the ad hoc network and the surrounding networks.

Within Ambient Networks composition is one of the key innovations, where networks join and share resources 'ad hoc', but resources and capabilities should not be seen as something that must be shared upon composition. Rather, resources and capabilities are to be seen as valued input to the composition negotiation. This is also valid for ad hoc networks that meet and compose within Ambient Networks, they do not have to perform an absorption model composition where they irreversibly form one new Ambient Network with one Ambient Control Space controlling and managing the new network. Instead, it is likely that most ad hoc network compositions will be partial compositions or gateway compositions. One network might not want to reveal its internal topology, including addresses of all internal infrastructures and where different functionality may reside, to another network that it will compose with for a longer or shorter period. As with other types of networks, where access to resources and routing through networks are negotiated, but the location of the resources or the topology of the network through which traffic is routed is not revealed, this will hold true for ad hoc networks as well. The amount of information regarding the network being shared with other networks should be restricted, the information that will be shared should be well controlled and the negotiation of sharing this information should follow well-defined rules or policies.

This research describes how different ad hoc networks can remain separate networks even when they come within radio range of each other. Furthermore, this work is investigating how traffic can be transited through other, intermediate ad hoc networks, when a source node and a destination node reside in two different ad hoc networks. Finally, we investigate how nodes within an ad hoc network will be able to communicate with nodes in the Internet, potentially via intermediate ad hoc networks.

12.4 Conclusions

The Ambient Networks concept envisages the composition and cooperation of heterogeneous networks, on demand and transparently, without the need for manual (pre or re)configuration or offline negotiations between network operators. To achieve these goals, Ambient Network management functionality must become dynamic, distributed, self-managing, self-policing and autonomically responsive to the network and its ambience.

The four complementary system approaches developed in this chapter (i.e. P2P, pattern-based, PnP and TE (also known as closed-loop-based approach)) are focused to solve different management problems. The P2P and pattern-based approaches target the management layer composition challenges, whereas the PnP and TE approaches target the self-management challenges. They represent the first step towards the design and realization of the AN autonomous management systems.

12.4.1 Integrating the Four Approaches

As we have described earlier, some of the characteristics of Ambient Networks are the high level of dynamicity in combination with the introduction of multitude at all levels: We will have multiple different link technologies to choose from, we will have multiple different network technologies, we will have multiple different networks that will compose and decompose, we will have multiple different operators and service providers that will cooperate and compete in a dynamic fashion and we will have a multitude of different applications that will run on a multitude of different terminals with different capabilities. All this provides the user with the best connectivity possible based on the users' requirements at that time and the current capabilities of the networks available.

In order to manage networks and network nodes in such an environment, radical changes have to be made to the management paradigms of today. This will not be feasible using a centralized, hierarchical solution with a high degree of hands-on control of networks and nodes. Instead, new network management systems are required. These must be distributed over the different networks, highly autonomous and self-managing, and require only interaction from operators and specialists if the system itself cannot resolve conflicts.

We have shown that enhancements made to traffic engineering will allow us to use the existing network resources in a more efficient way by distributing the traffic over all links and that this can be done automatically and that these algorithms will also work effectively in a dynamic environment. We have shown that implementing PnP functionality allows us to introduce new equipment as well as to adapt to changes in current configurations in a self-adapting way. We have shown that peer-to-peer technologies are well suited to allowing different elements and different parts of network management systems to communicate, negotiate and implement changes in the networks in an autonomic way, thereby enabling composition of different management systems with no human interaction. We have also shown that pattern-based management is an effective, decentralized tool to discover the current condition of different networks in near-real time.

A full integration of different management approaches is not foreseen and further work is required to achieve that goal. However, we have already analysed how the different approaches relate to each other at a high level and how they will interwork with each other to become a network management system, as well as how this network management system will interact with other parts of the ACS service management to provide support for service management and overlay set-up, when requested. Depending on the request, service management asks the PnP management and/or P2P management to adapt the current network configuration to meet the new requirements. Service management uses patterns to define and detect preferred routes depending on the requirements from overlays and services.

12.4.1.1 Extending the Functionality

There are still a number of open questions to be resolved. How we will be able to identify root causes for alarms being sent to and from different nodes and networks and domains? When we have a distributed system there will not be a single point, which will see all alarms. How we will be able to differentiate between what should be reported as alarms and what should be reported as regular reconfigurations? Which parts can be resolved by the network(s) themselves, and which parts must be resolved by an operator? Which operator shall be contacted if there are multiple operators available? How will an operator be able to get hands-on control over (parts of) the network if she finds it necessary?

We have not yet investigated the effects that different aspects of security will have on these management systems. With an increased autonomicity and internetwork/interdomain interaction, this will put completely new requirements on how nodes identify each other, how they can trust each other, how they will authorize actions initiated in perhaps other networks and how different instructions and requests can be saved in an unhampered way to be able to trace what actually happened, and in what order, at a later stage.

These and many other aspects must be identified, verified and incorporated into a management system in order to provide a management system that is complete and efficient. The AN management systems should be modular, but some parts must be present in all systems, and we need to show how different modules will work together in an efficient way.

12.4.1.2 Interacting with ACS Functions

Network management systems must interact with all other parts of the ACS if it is to fulfil its work in an efficient way. This means that to enable network management system to work, these interactions must be identified, the work division between the network management system and the other parts must be resolved and agreed upon, and the solutions must be tested and verified.

Some of the interactions we can see today are with the generic link layer and different link layer technologies. The following unanswered questions remain: What information will be sent to and from the NMS in order to initiate a bootstrap? How will the results be communicated back? How will the NMS be involved in selection of link layer technology? Where will the division between control and management spaces go with regards to composition? Who will do what, and how will they communicate with each other? What mobility actions will be cleared with the NMS before being initiated?

Context management and network management are and will be closely interleaved with each other. Context management provides the means for distributed NMSs to get access to policies and information, and context management will require interaction from NMSs to know how to update, synchronize and distribute their information within and between networks and domains.

12.4.2 Requirements Analysis for Future Autonomic Management Systems for Ambient Networks

One of the main drivers behind AN autonomous management systems is that the industry is finding that the cost of technology is decreasing, yet IT costs are not (i.e. IT gap). Also, as systems become more advanced, they tend to become more complex and increasingly difficult to maintain. To complicate matters further, there has been, and will be for the foreseeable future, a scarcity of IT professionals to install, configure, optimize and maintain these complex systems.

One other important driver behind AN autonomous management system is the increasing network operational costs and the cost of introducing new services (i.e. service gap). The rapid deployment of IT infrastructure technology and hardware, and its accelerating increase in performance (each generation improves by a factor of 1.5–2 per annum), gives rise to new challenging problems. On the one hand, the availability of computation performance and network bandwidth at a reasonable cost stimulates demand for products with more functionality and the demand for increasingly powerful services. On the other hand, developers cannot keep up with the demand for software, resulting in an ever-increasing service gap: the time lag and high cost of adapting/engineering end system software and the diversion of highly qualified staff to 'administrative' functions.

The main reason for this service gap, and the mismatch between technological potential and its realization as software and services, is to be found in the management separation between the computational and communication resources and the lack of focus on complexity. Complexity is currently the main factor preventing highly qualified workers from being productive, as many of them are occupied with system administration, configuration and maintenance, resulting in costs but no return of investment. At the same time, complexity limits developers' productivity as it increases project time scales and encourages specialization and inflexibility. It is widely agreed that the high costs of the initial project stages and its associated administrative overheads inhibit the future growth of IT and its advance into new areas. If software and services are to progress at a faster pace than computational performance and bandwidth, which is the only way to narrow the service gap, managing complexity becomes the key issue. In part because of the service gap these deficiencies remain to be addressed, and their solution using traditional technology is not getting nearer.

Therefore, the aim of AN autonomic management systems is to reduce the amount of maintenance and management needed to keep systems working as efficiently as possible, as much of the time as possible, i.e. it is about making systems self-managing for a broad range of activities.

The main aim of the AN management autonomous systems is that they exhibit self-awareness properties, in particular self-contextualization, programmability and self-management (i.e. self-optimization, self-organization, self-configuration, self-adaptation, self-healing, self-protection and policing) as depicted in Figure 12.7.

Self-contextualization – Context is any information that can be used to characterize the situation of an entity (a person or object) that is considered relevant to the interaction between a user and an application. A context-aware system is capable of using context information ensuring it successfully performs its expected role and also maximizes the perceived benefits of its use. Nevertheless, this is a user-centric view and reflects the fact that most research on context and context awareness up to now has been focused on 'user context'. In contrast, a new

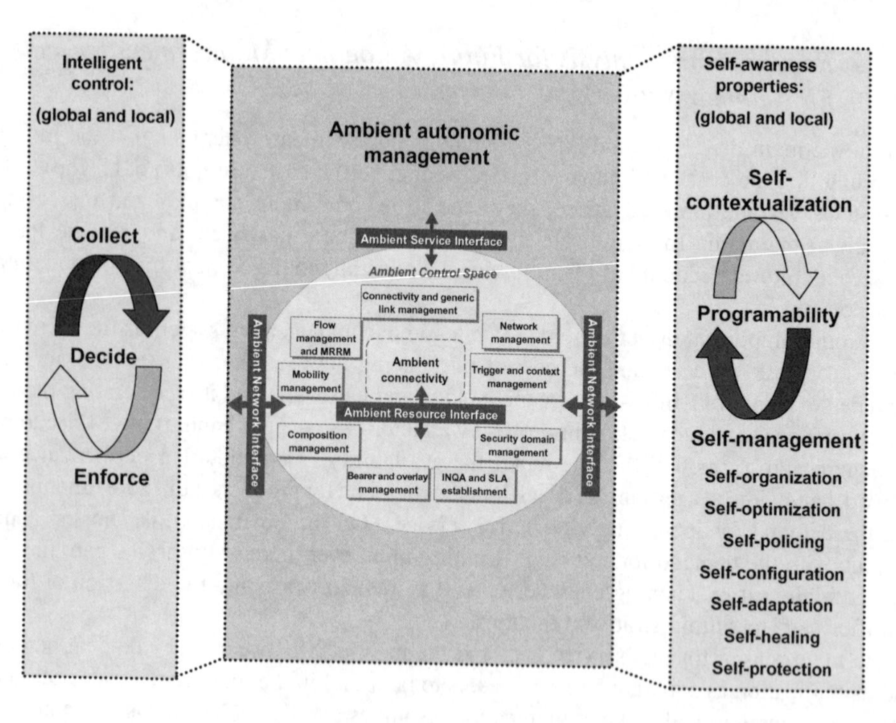

Figure 12.7 Ambient Networks autonomic management systems

generation network gives context a much broader scope and renders it universally accessible as a basic commodity provided and used by the network. In this way, context becomes a decisive factor in the success of future AN autonomous rule-based systems adaptive to changing conditions. As such contextualization is a service/software property. Self-contextualization means that a management service/software component autonomously becomes context aware. It represents the ability of a management system to describe, use and adapt its behaviours to its context. Network context for supporting service/software components should be made available, so that multiple service/software components may take advantage of the available network context. In order to do so in the complex environment of the large and heterogeneous Internet, the service/software component must be equipped with certain self-management capabilities. Once a management service/software component becomes context aware, it can make use of context information for other self-management tasks that depend on context information.

Programmability – Recent research on distributed systems and network technologies has focused on making service networks programmable. The objectives of programmable service networks are to take advantage of network processing resources and to promote new service models allowing new business models to be supported. The resulting service models do not, however, target development of services, but, rather, their deployment. Dynamic service programming applies to executable service code that is injected into the AN autonomic systems elements to create the new functionality at run-time. The basic idea is to enable third parties (users, operators and service providers) to inject application-specific services into the autonomic systems platform. Applications and services are thus able to utilize required network support in terms of optimized network resources and as such they can be said to be network aware, i.e. a service-driven network. This means that network programmability is following autonomous flows of control triggered and moderated by network events or changes in network context. The network is self-organized in the sense that it autonomically monitors available context in the network and provides the required context and any other necessary network service support to the requested services and self-adapts when context changes.

Self-management – Management is an essential topic when dealing with utilization of network context information for supporting services and contextualized services. Managing context from the perspectives of the context information provider means dealing with a number of processes related to manipulation of network context information, for instance, the creation, composition and inference of context of diverse quality; also quality of context (QoC)-based storage, distribution and caching are relevant. Clearly, a context source must be trustworthy and the information it provides must be sufficiently precise for the task in hand, and in this way, the new concept of QoC becomes important.

Among AN autonomous capabilities, self-optimization, self-organization, self-configuration and self-adaptation are highly relevant. Moreover, the management of network context information must be addressed in the framework of the autonomous computing paradigm. This means that a key element for contextualization is the addition of intelligence and self-management capabilities to facilitate network context self-management and thus eliminating unnecessary multilevel configurations as in conventional hierarchical management systems. With such embedded intelligence, it is necessary only to write or specify the high-level design goals and management constraints, so that the network and service overlay should make the low-level decisions on its own. The system should reconfigure itself according to changes in the high-level requirements (i.e. use of cognitive and knowledge networks principles). This requires the ability to express rules within each configuration level and also between levels.

Currently, network management faces many challenges: complexity, data volume, data comprehension, changing rules, reactive monitoring, resource availability and others. AN self-management aims to automatically meet these challenges, in the following ways.

Self-optimization – In the large and heterogeneous Internet, heterogeneous and distributed network context information and resources and their availability are rapidly changing. There is a need for an AN autonomous management tool for consistent monitoring and control of network context information and resources, so that service/software components may be executed or deployed in the most optimized fashion. AN autonomic systems must seek to improve their operation every time. They must identify opportunities to make themselves more efficient from the point of view of strategic policies (performance, cost, etc.).

Self-organization – Network elements and context information and resources are distributed across heterogeneous networks. In order for services to make use of these distributed information and resources, they must be structured or referenced in an easy-to-access-and-retrieve structure in an automatic fashion. All these network context information and resources must be autonomously organized and reserved through a service layer. The autonomous structuring of network context information and resources is the essential work of self-organization.

Policing – Because of the ability of Ambient Networks to adapt to evolving situations and conditions, new network resources can become active, policies can change and the business needs and models can vary accordingly. These changes are actuated through high-level policies that need to get appropriately translated into low-level localized actions allowing achievement of overall system goals.

Self-configuration/self-adaptation – Autonomous structuring of network context information and resources makes them available to services. User services and the underlying supporting services must be reconfigured in order to make use of new network context information and resources. The new network context information and resources also trigger changes such as reconfiguration in the network context-aware overlay. Self-configuration is therefore desirable. AN autonomic systems aim at developing and assessing a novel open programmable infrastructure for enabling self-configuration. AN autonomic systems must configure themselves in accordance with high-level policies representing service agreements or business objectives, rules and events. When a component or a service is introduced, the system will incorporate it seamlessly, and the rest of the system will adapt to its presence. In the case of components, they will register themselves and other components will be able of use it or modify their behaviour to fit the new situation.

Self-healing – AN autonomic systems will detect, diagnose and repair problems caused by network or system failures. Using knowledge about the system configuration, a problem-diagnosis embedded intelligence would analyse the monitored information. Then, the network would use its diagnosis to identify and enforce solutions or alert a human in the case of no solutions available.

Self-protection – There are two ways in which an AN autonomic system must self-protect. It must defend itself as a whole by reacting to, or anticipating, large-scale correlated problems arising from malicious attacks or cascading failures that remain uncorrected by self-healing measures.

The realization of self-awareness properties revolves around increasing the level of automation of the intelligent control loop described as *collect–decide–enforce*.

- *Collect* is about monitoring that allows construction of a picture about the surrounding environment in order to build self-awareness.
- *Decide* involves inference and planning. The former refers to a process whereby the problem is diagnosed based on the collected information, whereas the latter refers to the process whereby a solution is selected.
- *Enforce* comprises deployment, which adds functionality by means of new components, and configuration, which changes the existing functionality by means of programmability of networks.

The realization of this intelligent loop eventually leads to a distributed, adaptive, global evolvable system capable of fostering continuous changes as it depicted in Figure 12.7.

References

[1] The 3rd Generation Partnership Project, http://www.3gpp.org.
[2] IEEE 802 LAN/MAN Standards Committee, http://grouper.ieee.org/groups/802/index.html.
[3] Long term evolution (LTE)/system architecture evolution (SAE), 3rd Generation Partnership Project, http://www.3gpp.org/Highlights/LTE/LTE.htm.
[4] Embracing the future – BT's vision for a 21st century network, BT Plc., http://www.btplc.com/21cn.
[5] NewArch project: future-generation Internet architecture, http://www.isi.edu/newarch.
[6] D. Clark *et al.*, NewArch: future generation Internet architecture, Final Technical Report, Air Force Research Laboratory, 2003.
[7] The Ambient Networks Project, http://www.ambient-networks.org.
[8] R. Prasad, W. Mohr and W. Konhauser, *Third Generation Mobile Communication Systems*, Artech House, London, 2000.
[9] P. Díaz, The RAINBOW concept for the UMTS access network, *PIMRC*, 8–11 September 1998.
[10] The Wireless World Initiative, http://www.wireless-world-initiative.org.
[11] The Wireless World Research Forum, http://www.wireless-world-research.org.
[12] E. Gustafsson and A. Jonsson, Always best connected, *IEEE Wireless Communications*, **10**(1), 49–55, 2003.
[13] D. D. Clark, J. Wroclawski, K. Sollins and R. Braden, Tussle in cyberspace: defining tomorrow's Internet, *Proc. ACM SIGCOMM 2002*, Pittsburgh, PA, August 2002.
[14] I. Stoica, D. Adkins, S. Zhuang, S. Shenker and S. Surana, Internet indirection infrastructure, *Proc. ACM SIGCOMM 2002*, Pittsburgh, PA, August 2002.
[15] G. Camarillo and M. A. García-Martín, *The 3G IP Multimedia Subsystem (IMS)*, John Wiley & Sons, Ltd, Chichester, 2006.
[16] Digital Living Network Alliance, www.dlna.org.
[17] D. Moro (ed.), *Migration Strategies to Ambient Networks*, IST Ambient Networks deliverable 1-3, December 2004, IST-2002-507134-AN/WP1/D13.
[18] I. G. Gjerde, A framework for analysing end-to-end QoS in a multi-provider environment, PhD thesis, University of Zagreb, Croatia, June 2003.
[19] D. Clark *et al.*, Addressing reality: an architectural response to real-world demands on the evolving Internet, *Proc. ACM SIGCOMM FDNA 2003 Workshop*, Karlsruhe, Germany, August 2003.
[20] I. Stoica *et al.*, Internet indirection infrastructure, *Proc. ACM SIGCOMM 2002*, Pittsburgh, PA, August 2002.
[21] J. Crowcroft *et al.*, Plutarch: an argument for network pluralism, *Proc. ACM SIGCOMM FDNA 2003 Workshop*, Karlsruhe, Germany, August 2003.
[22] Future Internet Network Design (FIND), http://find.isi.edu.
[23] Wireless World Initiative New Radio (Winner), https://www.ist-winner.org.
[24] Mobilife, http://www.ist-mobilife.org.
[25] SPICE (Service Platform for Innovative Communication Environment), http://www.ist-spice.org.

Ambient Networks: Co-operative Mobile Networking for the Wireless World Norbert Niebert (Ericsson GmbH), Andreas Schieder (Ericsson GmbH), Jens Zander and Robert Hancock

[26] M. Kampmann, M. Vorwerk, M. Kleis, S. Schmid, S. Herborn, R. Aguero and J. Choque, Multimedia delivery in Ambient Networks, *ACM Multimedia*, 2005.
[27] IEEE P802.21/D00.01: Draft IEEE Standard for Local and Metropolitan Area Networks: Media Independent Handover Services, July 2005.
[28] ITU-T Rec. H.248.1: Gateway Control Protocol: Version 2, May 2002.
[29] C. Groves *et al.*, *Gateway Control Protocol Version 1*, RFC 3525, June 2003.
[30] ITU-T Rec. G.805: Generic Functional Architecture of Transport Networks, March 2000.
[31] ITU-T Rec. G.809: Functional Architecture of Connectionless Layer Networks, March 2003.
[32] M. Abadi and R. Needham, Prudent engineering practice for cryptographic protocols, *IEEE Transactions on Software Engineering*, **22**, 6–15, 1996.
[33] F. Kohlmayer (ed.), *Ambient Network Security Architecture*, Ambient Network WP7 deliverable 2, December 2005, IST-2002-507134-AN/WP7/D02.
[34] M. Prytz (ed.), *MRA Architecture*, Ambient Network WP2 deliverable 4, December 2005, IST-2002-507134-AN/WP2/D04.
[35] 3G security: security architecture, 3rd Generation Partnership Project, TS 33.102 (Release 6), 2004.
[36] Part 11: Wireless LAN Medium Access Control (MAC) and Physical Layer (PHY) Specifications, IEEE 802.11 Working Group, P802.11-REVma/D4.0, August 2005.
[37] P. Eronen and J. Arkko, Role of authorisation in wireless access network security, Extended abstract presented at the *DIMACS Workshop*, New Jersey, USA, November 2004.
[38] I. E. Svinnset and B. Viken (eds.), *D3.3 Connecting Ambient Networks – Final Architecture, Protocol Design and Evaluation*, Ambient Network WP3 deliverable 3, December 2005, IST-2002-507137-AN/WP3/D/3-3.
[39] P. Eronen (ed.), *Initial Security Requirements and Concepts for Secure Access and Mobility Procedures*, Ambient Networks WP7 Report 2 Version 1.2, November 2004, IST-2002-507134-AN/WP7/R002 (Annex 1 of [NewSelander]).
[40] I. Herwono (ed.), *D7.3 Security and Trust Cross Issue*, Ambient Network WP7 deliverable 3, December 2005, IST-2002-507134-AN/WP7/D03.
[41] Liberty Alliance, http://www.projectliberty.org.
[42] Trusted Computing Group, https://www.trustedcomputinggroup.org.
[43] R. Shirey, *Internet Security Glossary*, IETF, RFC 2828, May 2000.
[44] M. Berg (ed.), *Non-Conventional/Low-Cost Concepts*, Ambient Network WP2 deliverable 3, June 2004, IST-2002-507134-AN/WP2/D03.
[45] Y. Ismailov (ed.), *Mobility Architecture and Framework*, Ambient Network WP4 deliverable 2, April 2005, IST-2002-507134-AN/WP4/D4.2.
[46] T. Aura, M. Roe and J. Arkko, Security of Internet location management, *Proc. 18th Annual Computer Security Applications Conference (ACSAC)*, Las Vegas, NV, December 2002, pp. 76–8.
[47] C. Vogt and J. Arkko, *Taxonomy and Analysis of Enhancements to Mobile IPv6 Route Optimization*, IRTF Internet draft: draft-irtf-mobopts-ro-enhancements-03.txt, October 2005 (work in progress).
[48] A. Jonsson (ed.), *Ambient Networks ContextWare – First Paper on Context-Aware Networks*, Ambient Network WP6 deliverable 1, January 2005, IST-2002-507134-AN/WP6/D61.
[49] A. R. Prasad (ed.), *Security Input to Final Ambient Network Concepts*, Ambient Networks WP7 Report 4, May 2005, IST-2002-507134-AN/WP7/R04 (Annex 1 of [38]).
[50] C. Ellison, B. Frantz, B. Lampson, R. Rivest, B. Thomas and T. Ylonen, *SPKI Certificate Theory*, RFC 2693, September 1999.
[51] R. Moskowitz and P. Nikander, *Host Identity Protocol Architecture*, RFC 4423, May 2006.
[52] B. Carpenter and S. Brim, *Middleboxes: Taxonomy and Issues*, RFC 3234, IETF, February 2002.
[53] J. Loughney, M. Nakhjiri, C. Perkins and R. Koodli, *Context Transfer Protocol (CXTP)*, RFC 4067, IETF, July 2005.
[54] H. Tschofenig and P. Eronen, *Analysis of Options for Securing the Generic Internet Signaling Transport (GIST)*, IETF Internet draft: draft-tschofenig-nsis-gist-security-00.txt, October 2005 (work in progress).
[55] F. Alfano, P. McCann, H. Tschofenig and T. Tsenov, *Diameter Quality of Service Application*, IETF Internet draft: draft-alfano-aaa-qosprot-05.txt, October 2005 (work in progress).
[56] H. Tschofenig, A. Mankin, T. Tseno and A. Lior, *RADIUS Quality of Service Support*, IETF Internet draft: draft-tschofenig-radext-qos-02.txt, October 2005 (work in progress).

[57] B. Carpenter (ed.), *Architectural Principles of the Internet*, RFC 1958.
[58] Y. Rekhter, T. Li and S. Hares (eds.), *A Border Gateway Protocol 4 (BGP-4)*, RFC 4271.
[59] J. L. Sobrinho, Network routing with path vector protocols: theory and applications, *SIGCOMM 2003*.
[60] T. Griffin, What can we unlearn from BGP?, *Proc. 1st ACM Workshop on Dynamic Interconnection of Networks*, Cologne, 2005.
[61] 3GPP, Mobile application part (MAP) specification, TS 29.002.
[62] 3GPP, 3GPP system to wireless local area network (WLAN) interworking: functional and architectural definition, TS 23.934.
[63] R. Campos, C. Pinho, M. Ricardo, J. Ruela, P. Pöyhönen and C. Kappler, Dynamic and automatic interworking between personal area networks using composition, *Proc. 16th IEEE International Symposium on Personal Indoor and Mobile Radio Communications*, Berlin, Germany, September 2005.
[64] J. Rosenberg *et al.*, *SIP: Session Initiation Protocol*, RFC 3261, June 2002.
[65] P. Calhoun, J. Loughney, E. Guttman, G. Zorn and J. Arkko, *Diameter Base Protocol*, RFC 3588, September 2003.
[66] R. Hancock, G. Karagiannis, J. Loughney and S. Van den Bosch, *Next Steps in Signalling (NSIS): Framework*, RFC 4080, June 2005.
[67] M. Stiemerling, H. Tschofenig and C. Aoun, *NAT/Firewall NSIS Signalling Layer Protocol (NSLP)*, Internet draft: draft-ietf-nsis-nslp-natfw-13, October 2006 (work in progress).
[68] J. Manner, G. Karagiannis, A. McDonald and S. Van den Bosch, *NSLP for Quality-of-Service Signalling*, Internet draft: draft-ietf-nsis-qos-nslp-12, October 2006 (work in progress).
[69] Draft IEEE Standard for Local and Metropolitan Area Networks: Media Independent Handover Services, IEEE LAN/MAN draft IEEE P802.21/D00.01, July 2005.
[70] H. Schulzrinne and R. Hancock, *GIST: General Internet Signalling Transport*, Internet draft: draft-ietf-nsis-ntlp-11, August 2006 (work in progress).
[71] A. Fessi, C. Kappler, C. Fan and A. Klenk, Framework for Metering NSLP, Internet draft: draft-fessi-nsis-m-nslp-framework, November 2005 (work in progress).
[72] R. Campos, N. Akhtar, C. Kappler, P. Paakkonen, P. Poyhonen and D. Zhou, On the evaluation of the extended Generic Internet Signalling Transport Protocol, *15th IST Mobile Summit*, June 2006.
[73] P. Pääkkönen, N. Akhtar, R. Campos, C. Kappler, P. Pöyhönen and D. Zhou, Scalability of name resolution for Ambient Networks, *4th International Conference on Wired/Wireless Internet*, Bern, May 2006.
[74] R. Braden, L. Zhang, S. Berson, S. Herzog and S. Jamin, *Resource Reservation Protocol (RSVP) – Version 1: Functional Specification*, RFC2205, Internet Engineering Task Force, September 1997.
[75] P. Chimento and B. Teitelbaum, *Simple Interdomain Bandwidth Broker Signaling*, Internet2, January 2000 (work in progress).
[76] M. Boucadair, *QoS-Enhanced Border Gateway Protocol*, Internet draft, Internet Engineering Task Force, July 2005 (work in progress).
[77] Y. Rekhter and T. Li, *A Border Gateway Protocol 4 (BGP-4)*, RFC 1771, Internet Engineering Task Force, March 1995.
[78] N. Niebert, H. Flinck, R. Hancock, H. Karl and C. Prehofer, Ambient Networks – research for communication networks beyond 3G, *IST Mobile and Wireless Communications Summit 2004*.
[79] J. Sachs *et al.*, Future wireless communication based on multi-radio access, *Proc. WWRF11*, Oslo, Norway, 2004.
[80] J. Lundsjö *et al.*, Multi-radio access architecture for Ambient Networking, *IST Mobile and Wireless Communications Summit 2005*.
[81] IETF Working Groups Mobility for IPv4 (mip4) and IPv6 Working Groups (mip6), home pages http://www.ietf.org/html.charters/mip4-charter.html, http://www.ietf.org/html.charters/mip6-charter.html.
[82] IRTF Research Group IP Mobility Optimizations (Mob Opts), home page http://www.irtf.org/charters/mobopts.html.
[83] IETF Working Group Context Transfer, Handoff Candidate Discovery, and Dormant Mode Host Alerting (seamoby), http://www.ietf.org/html.charters/OLD/seamoby-charter.html.
[84] F. Berggren, I. Karla, R. Litjens, P. Magnusson, F. Meago, R. Veronesi and H. Tang, Multi-radio resource management for communication networks beyond 3G, *VTC Fall 2005*.
[85] G. P. Koudouridis, R. Agüero, E. Alexandri, J. Choque, K. Dimou, H. R. Karimi, H. Lederer, J. Sachs and R. Sigle, Generic link layer functionality for multi-radio access networks, *IST Mobile and Wireless Communications Summit 2005*.

[86] K. Dimou *et al.*, Generic link layer: a solution for multi-radio transmission diversity in communication networks beyond 3G, *VTC Fall 2005.*

[87] A. Furuskär and J. Zander, Multiservice allocation for multiaccess wireless systems, *IEEE Transactions on Wireless Communications*, **4**(1), 174–84, 2005.

[88] A. Tölli, P. Hakalin and H. Holma, Performance of common radio resource management (CRRM), *IEEE ICC 2002.*

[89] F. Malvasi *et al.*, Traffic control algorithms for a multiaccess network scenario comprising GPRS and UMTS, *VTC Spring 2003.*

[90] M. Siebert and B. Walke, Design of generic and adaptive protocol software (DGAPS), *Third Generation Wireless and Beyond 2001 (3Gwireless '01).*

[91] Ambient Networks, *Draft Multi-Radio Access Architecture*, Project deliverable D2-2, January 2005.

[92] F. Berggren and R. Litjens, Performance analysis of access selection and transmit diversity in multi-access networks, submitted for publication.

[93] G. P. Koudouridis, H. R. Karimi and K. Dimou, Switched multi-radio transmission diversity in future access networks, *Proc. Vehicular Technology Conference, Fall 2005*, Dallas, TX, 25–28 September.

[94] R. Karimi, G. P. Koudouridis and K. Dimou, On the spectral efficiency gains of switched multi-radio transmission diversity, *Proc. WPMC '05*, Aalborg, Denmark.

[95] E. Alexandri and M. Bortnik, Performance of multi-radio transmission diversity on IP packets in UMTS and WLAN, in press.

[96] P. Magnusson *et al.*, Multi-radio resource management for communication networks beyond 3G, *IEEE Vehicular Technology Conference (VTC 2005), Fall*, Dallas, TX, 25–28 September.

[97] A. Baraev, L. Jorguseski and R. Litjens, Performance evaluation of radio access selection procedures in multi-radio access systems, *Proc. WPMC '05*, Aalborg, Denmark, 2005.

[98] J. Hultell and M. Berg, Generalized roaming and access selection in multi-operator environments, *Proc. RadioVetenskap och Kommunikation*, Linköping, Sweden, 14–16 June 2005.

[99] O. Rietkerk, G. Huitema and J. Markendahl, Business roles enabling access for anyone to any network and service with Ambient Networking, *Helsinki Mobility Roundtable*, Helsinki, 1–3 June 2006.

[100] M. Berg and J. Markendahl, A concept for public access to privately operated cooperating local access points, *Proc. IEEE 61st Vehicular Technology Conference, VTC'05, Spring*, Stockholm, 2005.

[101] J. M. Pereira, Fourth generation: now it is personal!, *Personal, Indoor and Mobile Radio Communications*, September 2000.

[102] K. Johansson *et al.*, Integrating user deployed local access points in a mobile operator's network, *Proc. WWRF12*, Toronto, Canada, 4–5 November 2004.

[103] D. C. Schultz, B. Walke, R. Pabst and T. Irnich, Fixed and planned relay based radio network deployment concepts, *Proc. 10th Wireless World Research Forum*, New York, USA, October 2003.

[104] IEEE 802.21: Draft IEEE Standard: Media Independent Handover Services.

[105] Y. D. Lin and Y. C. Hsu, Multi-hop cellular: a new architecture for wireless communications, *Proc. IEEE INFOCOM 2000*, March 2000, pp. 1273–82.

[106] E. Asmare, S. Schmid and M. Brunner, Setup and maintenance of overlay networks for multimedia services in mobile environments, *Proc. 8th International Conference on Management of Multimedia Networks and Services (MMNS 2005)*, Barcelona, Spain, 24–26 October 2005, pp. 82–95.

[107] A. Yegin, "Link-layer Triggers and Hints for Detecting Network Attachments", draft-yegin-dna-l2-hints-01.txt, February 2004 (work in progress), Internet Draft

[108] Yegin A., Njedjou E., Veerepalli S., Montavont N., and Noel T, "Link-layer Hints for Detecting Network Attachments", draft-yegin-dna-l2-hints-00.txt, October 2003 (work in progress), Internet Draft

[109] Specifications of the 3GPP (http://www.3gpp.org/specs/specs.htm): GSM/GPRS specifications: TS 04.18 (Radio Resource control Specifications) TS 08.08 (Mobile-services Switching Centre - Base Station system (MSC-BSS) Interface Layer 3 Specification); UMTS Specifications: 23.009 (Handover procedures), 23.060 (General Packet Radio Service (GPRS), 24.008 (Mobile radio interface Layer 3 specification; Core network protocols), 25.304 (User Equipment (UE) procedures in idle mode and procedures for cell reselection in connected mode), 25.331 (RRC protocol), 25.413 (RANAP), 25.423 (RNSAP), 25.433 (NBAP), 25.931 (UTRAN functions, examples on signalling procedures)

[110] Specifications of the 3GPP2 (http://www.3gpp2.org/Public_html/specs/): "Introduction to cdma2000 Spread Spectrum Systems," C.S0001-D v1.0; "3GPP2 Access Network Interfaces Interoperability Specification

Revision A (3G-IOS v4.1.1) Revision A", A.S0001-A v2.0; "IP Based Location Services Stage 1 Requirements (384KB)", S.R0066-0 v1.0; "Cellular Radiotelecommunications Intersystem Operations", N.S0005-0; "Upper Layer (Layer 3) Signaling Standard for cdma2000 Spread Spectrum Systems", C.S0005-D v1.0

[111] V. Gupta & D. Johnston, "A Generalised Model for Link Layer Triggers", March 2004, [http://www.ieee802.org/handoff/march04_meeting_docs/Generalized_triggers-02.pdf].

[112] Bluetooth Specification (Core and Profiles) v1.1, Bluetooth SIG, February 2001.

[113] IEEE Specifications: Std 802.15.1-2002, "Wireless Medium Access Control (MAC) and Physical Layer (PHY) Specifications for Wireless Personal Area Networks (WPANs)", ISBN 0-7381-3068-0, IEEE, New York, June 2002; Std 802.15.3-2003, "Wireless Medium Access Control (MAC) and Physical Layer (PHY) Specifications for High Rate Wireless Personal Area Networks (WPANs)", ISBN 0 7381 3704 9, IEEE, New York, September 2003; Std 802.15.4-2003: "Wireless Medium Access Control (MAC) and Physical Layer (PHY) specifications for Low Rate Wireless Personal Area Networks (LR-WPANS)", ISBN 0-7381-3687-5 IEEE, New York, 2003

[114] "High Rate Ultra Wideband PHY and MAC Standard", ECMA Standard 368, http://www.ecma-international.org/, December 2005

[115] "Mobile Wimax standard" IEEE C802.16e-03/20r1, IEEE draft.

[116] "IEEE 802 Handoff ECSG", WWW page, [http://www.ieee802.org/handoff/].

[117] C. Perkins (ed.), "IP Mobility Support for IPv4", IETF, RFC 3344, August 2002

[118] D. Johnson, C. Perkins, J. Arkko, "Mobility Support in IPv6", RFC3775, June 2004

[119] MIPSHOP working group charter, http://www.ietf.org/html.charters/mipshop-charter.html

[120] H. Soliman, C. Castelluccia, K. El Malki, L. Bellier, "Hierarchical Mobile IPv6 Mobility Management", RFC4140, August 2005

[121] R. Koodli (ed.), "Fast Handovers for Mobile IPv6", RFC4068, July 2005

[122] C. Williams, "Goals for Localised Mobility Management", draft-ietf-mipshop-lmm-requirements-04.txt, July 2004 (work in progress), Internet Draft

[123] P. McCann, "Mobile IPv6 Fast Handover for 802.11 networks", RFC4260, November 2005

[124] T. Ernst et al., "Network Mobility Support goals and requirements", draft-ietf-nemo-requirements-06.txt, November 2006 (work in progress), Internet Draft

[125] V. Devarapalli, R. Wakikawa, A. Petrescu, P. Thubert, "Network Mobility Protocol", RFC3963, January 2005

[126] S. D. Park, E. Njedjou and N. Montavont, "L2 Triggers Optimised Mobile IPv6 Vertical Handover: The 802.11/GPRS Example", draft-daniel-mip6-optimized-vertical-handover-00.txt, January 2004 (work in progress), Internet Draft

[127] P. Bertin, T. Noël, and N. Montavont, "Parameters for Link Hints", draft-bertin-hints-params-00.txt, August 2003 (work in progress), Internet Draft

[128] F. André, JM. Bonnin, B. Deniaud, K. Guillouard, N. Montavont, T. Noël, and L. Suciu, "Optimised Support of Multiple Wireless Interfaces within an IPv6 End-Terminal", CYBERTE project

[129] J. Choi, G. Daley, "Detecting Network Attachment in IPv6: Goals", draft-jinchoi-dna-goals-00.txt, February 2004 (work in progress), Internet Draft

[130] J. Loughney, M. Nakhjiri, C. Perkins, R. Koodli, "Context Transfer Protocol", RFC 4067, July 2005

[131] M. Liebsh, A. Singh, H. Chaskar, D. Funato, and E. Shim, "Candidate Access Router Discovery", RFC4066, July 2005

[132] Mobil Ad-hoc Networks (manet) working group charter, http://www.ietf.org/html.charters/manet-charter.html.

[133] C. Perkins, E. Belding-Royer, and S. Das, "Ad hoc On-Demand Distance Vector (AODV) Routing", RFC 3561, July 2003.

[134] AN-D6.1: Ambient Networks ContexWare: First Paper on Context-Aware Networks, Ambient Networks Project, January 2005, Document number IST-2002-507134-AN/WP6/D61, www.ambient-networks.org. "Broadband Radio Access for IP-Based Networks", http://www.ist-brain.org/

[135] AN-D6.2: Proof of Concept Demos – Role and Opportunities for Context Management in Ambient Networks, Ambient Networks Project, July 2005, www.ambient-networks.org. "Mobile Internet Network Development", http://www.ist-mind.org/

[136] C. Reichert, M. Kleis and R. Giaffreda, Towards distributed context management for Ambient Networks, *IST Mobile and Wireless Communications Summit*, June 2005.

[137] D. Balakrishnan, M. El Barachi, A. Karmouch and R. Glitho, Challenges in modelling and disseminating context information in Ambient Networks, *CANET05 – MATA 2005*.

[138] J. Coutaz, J. Crowley, S. Dobson and D. Garlan, Context is key, *Communications of the ACM* 48(3), 49–53, 2005.

[139] N. Niebert *et al.*, Ambient Networks: an architecture for communication networks beyond 3G, *IEEE Wireless Communication Magazine*, 11(2), 14–22, 2004.

[140] P. Eardley *et al.*, Ambient internetworking: an architecture for extending 3rd generation mobile networks, *3G2004 – 5th International Conference on 3G Mobile Communication Technologies*, London, 18–20 October 2004, http://conferences.iee.org/3G2004.

[141] D1.2: Ambient Networks Scenarios, Requirements and Draft Concepts, March 2004, www.ambient-networks.org.

[142] D. Zhou, P. Pöyhönen, C. Pinho and N. Akhtar, Ambient Network interfaces and network composition, *Global Mobile Congress*, October 2005.

[143] A. K. Dey, Understanding and using context, *Journal of Personal and Ubiquitous Computing* 5(1), 4–7, 2001.

[144] T. G. Kanter, Hottown, enabling context-aware and extensible mobile interactive spaces, *IEEE Wireless Communications and IEEE Pervasive Computing*, special issue on context-aware pervasive computing, 2002, pp. 18–27.

[145] N. Samann and A. Karmouch, An evidence-based mobility prediction agent architecture, *Proc. 5th International Workshop on Mobile Agents for Telecommunication Applications (MATA 2003)*, Marrakesch, October 2003, ISBN 3-540-20298-6 (*Lecture Notes in Computer Science*, Springer, Berlin).

[146] T. G. Kantar, G. Q. Maguire Jr. and M. T. Smith, Rethinking wireless Internet with smart media, http://psi.verkstad.net/Papers/conferences/nrs01/nrs01-theo.PDF.

[147] J. M. Serrano, J. Justo, R. Marín, J. Serrat, N. Vardalachos, K. Jean and A. Galis, Framework for managing context-aware multimedia services in pervasive environments, *International Journal of Internet Protocol Technology (IJIPT)*, special issue on context in autonomic communication and computing, **2**(3), SSN (online): 1743–8217, ISSN (print): 1743–8209, 2006, http://www.inderscience.com/browse/index.php, http://flora.sourceforge.net.

[148] R. Ocampo, L. Cheng, and A. Galis, ContextWare support for network and service composition and self-adaptation, *IEEE MATA 2005 – Mobility Aware Technologies and Applications, Service Delivery Platforms for Next Generation Networks*, Montreal, Canada, 17–19 October 2005, ISBN-2 553-01401-5 (Springer, Berlin), www.congresbcu.com/mata2005, http://kaon2.semanticweb.org.

[149] M. Khedr and A. Karmouch, Negotiating context information in context-aware systems, *IEEE Intelligent Systems Magazine*, **19**(6), 21–9, 2004, http://www.w3.org/TR/swbp-xsch-datatypes/#sec-user-defined-problem.

[150] D. Raz, A. Juhola, J. Serrat and A. Galis, *Fast and Efficient Context-Aware Services*, John Wiley & Sons, Ltd, Chichester, 2006, ISBN 0-470-01668-X, 350 pp., http://eu.wiley.com/WileyCDA/WileyTitle/productCd-047001668X.html.

[151] R. Ocampo, A. Galis, H. De Meer and C. Todd, Context-aware networks using flow context tags, *IFIP TC6 Conference – NetCon'05 – Network Control and Engineering for QoS, Security and Mobility Conference*, Lannion, France, 14–18 November 2005, www.netcon05.org/comm.php.

[152] AN-D8.1: Ambient Network Management – Technologies and Strategies, December 2004, IST-2002-507134-AN/D8-1, www.ambient-networks.org.

[153] B. Hubert, T. Graf, G. Maxwell, R. Mook, M. Oosterhout, P. Schroeder, J. Spaans and P. Larroy, Linux advanced routing and traffic control, http://www.lartc.org.

[154] T. Bray, J. Paoli, C. M. Sperberg-McQueen, E. Maler and F. Yergeau (eds.), *Extensible Markup Language 1.1*, W3C Recommendation, 4 February 2004, http://www.w3.org/TR/xml11.

[155] M. Gudgin, M. Hadley, N. Mendelsohn, J. Moreau and H. Nielsen (eds.), *Simple Object Access Protocol (SOAP) 1.2*, W3C Recommendation, 24 June 2003, http://www.w3.org/TR/soap12.

[156] D. L. Tennenhouse and D. J. Wetherall, Towards an active network architecture, *Computer Communication Review*, 26(2), 5–17, 1996.

[157] A. T. Campbell *et al.*, A survey of programmable networks, *ACM SIGCOMM Computer Communication Review* 29(2), 7–23, 1999.

[158] D. Raz and Y. Shavitt, An active network approach for efficient network management, *International Working Conference on Active Networks 1999 (IWAN99)*, Berlin, Germany, July 1999.

[159] A. Galis, S. Denazis, C. Brou and C. Klein (eds.), *Programmable Networks for IP Service Deployment*, Artech House Books, London, 2004, ISBN 1-58053-745-6, 450 pp., www.artechhouse.com.

[160] T. Berners-Lee, R. Fielding and L. Masinter, *Uniform Resource Identifier (URI): Generic Syntax*, IETF RFC 3986, January 2005.

[161] P. Mockapetris, *Domain Names – Concepts and Facilities*, IETF RFC 1034, November 1987.

[162] P. Faltstrom, *Design Choices When Expanding DNS*, IAB Internet draft: draft-iab-dns-choices-02.txt, June 2005 (work in progress).

[163] D. Harrington *et al.*, *An Architecture for Describing Simple Network Management Protocol (SNMP) Management Framework*, RFC 3411, IETF, December 2002.

[164] I. Stoica, Chord: a scalable peer-to-peer lookup service for Internet applications, *ACM SIGCOMM*, San Diego, CA, August 2001, pp. 149–60.

[165] S. Ratnasamy *et al.*, A scalable content-addressable network, *ACM SIGCOMM*, San Diego, CA, August 2001, pp. 161–72.

[166] B. Y. Zhao *et al.*, Tapestry: an infrastructure for fault-tolerant wide area location and routing, Technical Report UCB/CSD-01-1141, UC Berkeley, April 2001.

[167] Geographic Location/Privacy (geopriv) Working Group Charter, available at http://www.ietf.org/html.charters/geopriv-charter.html (April 2005).

[168] J. Cuellar, J. Morris, D. Mulligan, J. Peterson and J. Polk, *Geopriv Requirements*, RFC 3693, February 2004.

[169] G. Chen and D. Kotz, Solar: a pervasive-computing infrastructure for context-aware mobile applications, Technical Report TR2002-421, Department of Computer Science, Dartmouth College, 2002.

[170] G. Chen and D. Kotz, Context aggregation and dissemination in ubiquitous computing systems, *Proc. 4th IEEE Workshop on Mobile Computing Systems and Applications*, IEEE Computer Society Press, Los Alamitos, CA, 2002.

[171] A. C. Snoeren, K. Conley and D. K. Gifford, Mesh-based content routing using XML, *Proc. 18th ACM Symposium on Operating Systems Principles*, Banff, Canada, 2001.

[172] AN-D6-3: Ambient Networks ContextWare: Second Paper on Context-Aware Networks, Ambient Networks Project, December 2005, www.ambient-networks.org.

[173] B. Fortz and M. Thorup, *Internet traffic engineering by optimizing OSPF weights*, *Proc. IEEE INFOCOM*, 2000.

[174] B. Fortz and M. Thorup, Optimizing OSPF/IS-IS weights in a changing world, *IEEE Journal on Selected Areas in Communications*, **20**(4), 756–67, 2002.

[175] A. Gunnar, M. Johansson and T. Telkamp, Traffic matrix estimation for a global IP backbone – a comparison on real data, *Proc. IMC 2004*, Taormina, Italy, October 2004.

[176] A. Medina, N. Taft, K. Salamatian, S. Bhattacharyya and C. Diot, Traffic matrix estimation: existing techniques and new directions, *ACM SIGCOMM 02*, August 2002.

[177] K. G. Ramakrishnan and M. A. Rodrigues, Optimal routing in shortest path data networks, *Lucent Bell Labs Technical Journal*, **6**(1), 117–38, 2001.

[178] R. Droms, *Dynamic Host Configuration Protocol*, RFC 2131, March 1997, http://www.ietf.org/rfc/rfc2131.txt.

[179] B. Hubert, T. Graf, G. Maxwell, R. Van Mook, M. Van Oosterhout, P. Schroeder, J. Spaans and P. Larroy (eds.), Linux advanced routing and traffic control, http://www.lartc.org.

[180] S. Schmid, F. Hartung, M. Kampmann and S. Herborn, SMART: intelligent multimedia routing and adaptation based on service specific overlay networks, *Proc. Eurescom Summit 2005*, Heidelberg, Germany, 27–29 April 2005, in press.

[181] H. Abrahamsson and A. Gunnar, Traffic engineering in Ambient Networks: challenges and approaches, *Second Swedish National Computer Networking Workshop (SNCNW 2004)*.

[182] K. Zimmermann, S. Felis, S. Schmid, L. Eggert and M. Brunner, Autonomic wireless network management, *2nd IFIP TC6 International Workshop on Autonomic Communication (WAC 2005)*, Vouliagmeni, Athens, Greece, 3–5 October 2005.

[183] E. Asmare, S. Schmid and M. Brunner, Pattern-based setup and maintenance of overlay networks, *8th IEEE International Conference on Management of Multi-Media Networks and Services (MMNS)*, Barcelona, Spain, October 2005.

[184] M. Erdei, P. Kersch, Z. L. Kis, B. Kovacs and R. Szabó, Self-organizing Ambient Control Space – an Ambient Network architecture for dynamic network interconnection, *ACM Mobicom Workshop on Dynamic Interconnection of Networks (DIN 2005)*, Cologne, Germany, 5 September 2005.

[185] N. A. C. Rodríguez, J. Colás, G. G. B. Blas, F. J. R. Salguero and J. E. Gabeiras, A decentralized traffic management approach for Ambient Networks environments, *16th IFIP/IEEE International Workshop on Distributed Systems: Operations and Management (DSOM 2005)*, Barcelona, Spain, 24–26 October 2005.

[186] L. Cheng, R. Ocampo, A. Galis, R. Szabó, C. Simon and P. Kersch, Self-management in Ambient Networks for service composition, *The 2005 IFIP International Conference on Intelligence in Communication Systems (INTELLCOMM 2005)*, Montreal, Canada, 17–19 October 2005.

[197] A. P. Gonzalez and R. Stadler, Distributed real-time monitoring with accuracy objectives, *IFIP Networking 2006*, Coimbra, Portugal, May 2006.

[188] A. Galis, S. Denazis, C. Brou and C. Klein (eds.) *Programmable Networks for IP Service Deployment*, Artech House Books, London, 2004, ISBN 1-58053-745-6, www.artechhouse.com.

[189] J. Markendahl, J. Werding and Ö. Mäkitalo, Company asset analysis of candidates for novel access provisioning, *Proc. of RVK 05*.

[190] J. Eisl (ed.), Ambient Network Mobility Scenarios & Requirements, Ambient Network WP4 deliverable 1, July 2004, IST-2002-507134-AN/WP4/D4.1.

[191] T. Jokikyyny and J. Eisl (eds.). Mobility Framework and Mechanisms, Ambient Networks WP4 deliverable 3, December 2005, IST-2002-507134-AN/WP4/D4.3.

[192] M. Georgiades, N. Akhtar, C. Politis, R. Tafazolli, "Enhancing mobility management protocols to minimize AAA impact on handoff performance", Elsevier International Journal for the Computer and Telecommunications Industry, Computer Communications, Volume 30, Issue 3, 608–618, Special Issue: Emerging Middleware for Next Generation Networks, February 2007.

[193] C. Politis, T. Diagiuklas, N. Akhtar, K. Chew, M. Georgiades, R. Tafazolli, "Hybrid multilayer mobility management with AAA context transfer capabilities for All-IP networks", IEEE Wireless Communications Magazine, 76–88, August 2004.

[194] G. Selander (ed.), "Ambient Network Intermediate Security Architecture Ambient Network WP7 deliverable 1, January 2005, ISTÑ2002-507134-AN/ WP7/D01."

[195] B. Aboba, D. Thaler, L. Esibov, *Link-Local Multicast Name Resolution (LLMNR)*, RFC 4795, January 2007.

[196] R. Hancock, C. Kappler, J. Quittek, M. Stiemerling, A Problem Statement for Partly-Decoupled Signalling in NSIS, Internet Draft: draft-hancock-nsis-pds-problem-04, October 2006 (work in progress).

Abbreviations

AAA	Authentication, authorization and accounting
ACS	Ambient Control Space
AD	Administrative domain
AMS	Ambient Management System
AN	Ambient Network
ANI	Ambient Network Interface
ANSD	Ambient Network security system
AP	Access point
API	Application Programming Interface
ARI	Ambient Resource Interface
ASI	Ambient Service Interface
ASF	Ambient support function
AVP	Ambient virtual pipe
BFS	Breadth-first search
CA	Certificate authority or composition agreement
CBR	Constant bit rate
CC	Context client
CIB	Context information base
CLA	Context level agreement
CM	Context manager
CMIP	Common Management Interface Protocol
ConCord	Context coordination
ConCoord	Context coordinator
CPU	Common processing unit
CS	Context source
CW	ContextWare
DAD	Duplicate address detection
DB	Database
DHCP	Dynamic Host Configuration Protocol
DHT	Distributed Hash Table
DIM	DIM spanning tree algorithm

Ambient Networks: Co-operative Mobile Networking for the Wireless World Norbert Niebert (Ericsson GmbH), Andreas Schieder (Ericsson GmbH), Jens Zander and Robert Hancock

DINA	DINA programmable network platform
DNS	Domain Name System
DoS	Denial of service
DTD	Document type definition
ECMP	Equal-cost multiple path
ESSID	Extended service set identifier
FE	Functional entity
GAP	Generic Aggregation Protocol
GEOPRIV	Geographic location/privacy
HI	Host identity
HIT	Host identity tag
IETF	Internet Engineering Task Force
IP	Internet Protocol
IPTV	Internet Protocol Television
ID	Identifier
IS-IS	Intermediate System to Intermediate System
ISP	Internet service provider
L1, L2, L3	OSI layers 1, 2, 3
LAN	Local area network
LG	Location generator
LI	Location information
LS	Location server
LR	Location recipient
MAC	Medium access control
MC	Media client
MCF	Multicommodity flow
MIB	Management information base
MP	Media port
MPLS	Multiprotocol label switching
MRDV	Multipath routing with dynamic variance
MS	Media server
NACK	Negative acknowledgment
NAT	Network address translator
NS-2	Network simulator
NSSA	Not-so-stubby area
OCS	Overlay control space
ODP	Open distributed processing
OMP	Optimized multipath
ONodes	Overlay nodes
OSL	Overlay support layer
OSPF	Open Shortest Path First
OSI	Open Systems Interconnection
P2P	Peer-to-peer
PAN	Personal area network
p-AMO	Peer-to-peer Ambient Management Overlay
PAP	Peer-to-peer Ambient Control Space Prototype

PBX	Private branch exchange
PDA	Personal digital assistant
PDF	Probability density function
PnP	Plug-and-play
QoC	Quality of context
PTT	Push-to-talk
QoS	Quality of service
RM	Rulemaker
RM-ODP	Reference model of open distributed processing
SDM	Security domain manager
SIMPSON	Simple pattern simulator for large networks
SLA	Service level agreement
SMART	Smart Multimedia Routing and Transport
SNMP	Simple Network Management Protocol
SoA	State of the art
SSL	Secure socket layer
SSON	Service-specific overlay network
TCP	Transport Control Protocol
TE	Traffic engineering
TLS	Transport layer security
TMN	Telecommunications management network
TTL	Time-to-live
TTP	Trusted third party
UDP	User Datagram Protocol
UCI	Universal context identifier
UMTS	Universal mobile telecommunications system
(Un)PnP	(Un)plug-and-play
URI	Uniform resource identifiers
VPAN	Virtual private Ambient Network
VPLS	Virtual private LAN services
VPN	Virtual private network
WEP	Wired equivalent privacy
WiFi	Wireless fidelity
WLAN	Wireless local area network
XML	Extensible Markup Language
xDSL	Digital subscriber loop (e.g. ADSL)

Index

Ambient Networks: Co-operative Mobile Networking for the Wireless World Norbert Niebert (Ericsson GmbH), Andreas Schieder (Ericsson GmbH), Jens Zander and Robert Hancock